Ecosystems and Living Resources of the Baltic Sea

Evald Ojaveer

Ecosystems and Living Resources of the Baltic Sea

Their assessment and management

Springer

Evald Ojaveer
Estonian Marine Institute
University of Tartu
Tallinn, Estonia

ISBN 978-3-319-53009-3 ISBN 978-3-319-53010-9 (eBook)
DOI 10.1007/978-3-319-53010-9

Library of Congress Control Number: 2017933477

Printed on acid-free paper

This Springer imprint is published by Springer Nature
The registered company is Springer International Publishing AG
The registered company address is: Gewerbestrasse 11, 6330 Cham, Switzerland

Preface

Over many millennia, the systems of the Baltic Sea have developed under very variable climate conditions in fresh, marine, and brackish waters into present-day ecosystems and stocks. Today, the basic environmental conditions vary dramatically throughout the Baltic Sea. The most substantial ecophysiological parameter for aquatic organisms – salinity – ranges from 35 psu in the Kattegat to 1–2 psu in the northernmost Bothnian Bay. Therefore, compared to a number of other ("normal") seas with a constant salinity of 33–37 psu, in the Baltic Sea, ecosystems and environment are very different.

In recent decades, the state of the biota in the Baltic Sea has modified. Since the last deglaciation, the organisms which have found their home in the brackish Baltic Sea and adapted to its environmental conditions have become increasingly affected by humans – by their intemperate exploitation, pollution of the marine environment, and facilitation of invasion of alien species. The management usages have frequently not taken into account the differences between the natural areas of the Baltic Sea. As a result, a number of important protection measures have become useless. Therefore, the ecosystems and resources of the Baltic Sea have deviated from their ordinary state, and their services and goods are no longer available in the quality and quantity they were before.

Differentiation and separate assessment of Baltic Sea natural systems are necessary for correct management of the ecosystems and living resources of this sea. It is essential that a clear understanding of the structure of the Baltic Sea ecosystems and the diversity of the brackish Baltic Sea from the eumarine areas of the World Ocean is fully accounted for. This would allow ecosystem- and population-based consideration of living resources. The ecosystem-based treatment assumes that we have to deal with living alliances of vegetation, microorganisms, animals, and their abiotic environment behaving together as a functional unit. Such an approach to the environmental units related to the natural regions and population-based treatment of exploited living resources as presented below is the realistic possibility that can lead to the responsible management of ecosystems and resources in the Baltic Sea.

This monograph consists of six chapters. An overview of the evolution of the Baltic Sea and its living organisms after the last Ice Age is given in Chap. 1. Abiotic characters of the contemporary Baltic Sea, the natural regional system based on currents, and other hydrological and environmental features are discussed in Chap. 2. Vegetation and primary production, heterotrophic bacteria, zooplankton, zoobenthos, fish (the most important populations of the marine pelagic, marine demersal, diadromous, freshwater, and relict fishes), marine birds, and marine mammals are considered in Chap. 3.

In addition to the fish species of marine origin (cod, herring, and sprat) which presently provide over 90% of the exploitable fish resources of the Baltic Sea, migratory, freshwater, and relict fish species constituting the remaining exploitable resources in the Baltic ecosystems should be studied to a certain level to develop the possibility of their assessment and management in the future. Recent dynamics of ecosystems and biological resources are analyzed in Chap. 4, with the salinity- and climate-related driving forces and anthropogenic influences differentiated. In Chap. 5, the basic principles of routine assessment and management of existing fish resources are addressed. Also, the possibility of composition of the long-term (10–20 plus years in advance) qualitative forecasts on ecosystems and fish resources is discussed. An overview of the international collaboration in the assessment and management of ecosystems and living resources in the Baltic and corresponding organizations is given in Chap. 6.

The bulk of the information presented in this monograph is taken from a book written in Estonian (*Läänemeri* by Evald Ojaveer, Tallinn: Teaduste Akadeemia Kirjastus, 2014). The material was thoroughly reviewed and condensed. The consideration of the management of the living resources in the Baltic Sea was notably extended.

The overview is based on the information published in various languages since the nineteenth century. This information was substantially complemented with new data collected during the long-term series of detailed hydrographic and biological complex investigations carried out in the Baltic Proper and the gulfs of Finland and Riga from 1957 to 1997 by the Tallinn and Riga marine laboratories. Thanks to the participation of Dr. Margers Kalejs (†) in this work, a very strong effort was devoted and rich material collected during 420 specific scientific cruises (on monthly to quarterly bases, the total count of stations visited numbering 10, 300, Fig. 1). Therefore, the most complicated marine ecological problems involving multiple sophisticated interrelations are treated below on the basis of the results of these original complex studies avoiding simplifications involved in model calculations.

Assessment and management of ecosystems and natural resources are closely related to the application of the results of scientific studies in practice. Applicability of the means and solutions discussed below has been the touchstone for deciding upon their value, as the author has devoted over half century to studies of the ecosystems and fish resources of the Baltic Sea. Also, one of the most outstanding experts in the assessment and management of the Baltic Sea ecosystems and fish stocks, Dr. Hans Lassen, strongly supported the monograph in the form of many and lengthy discussions of the material and the fundamental theses presented in this book. He read the manuscript at different stages and made critical comments with

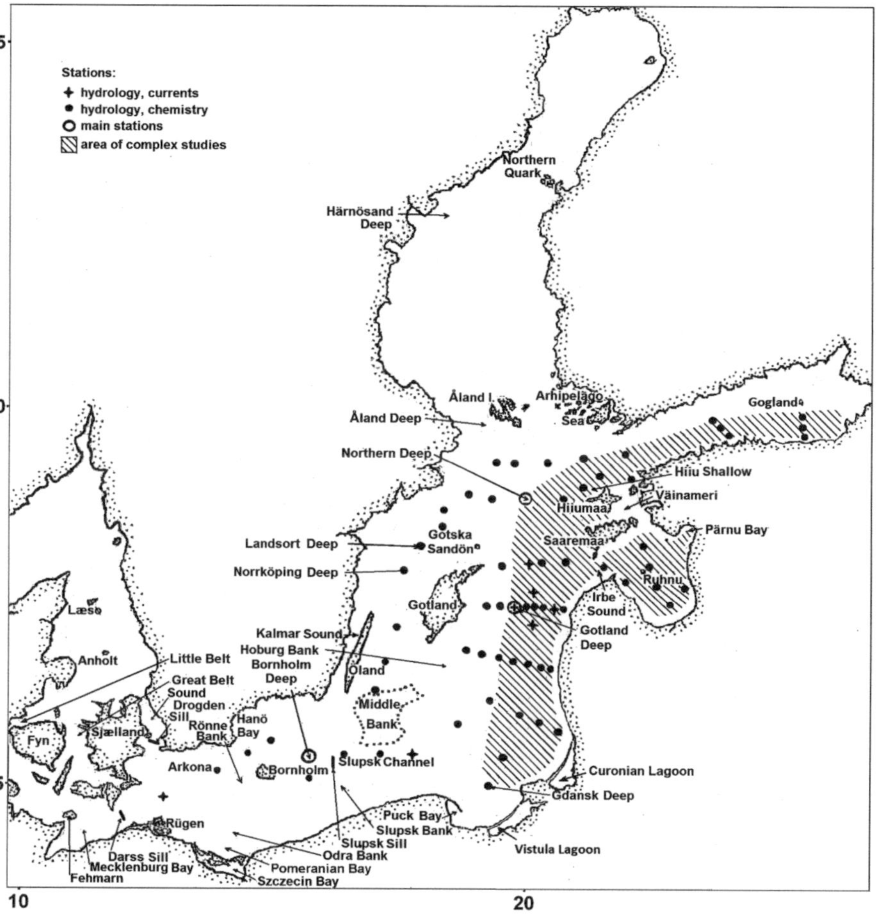

Fig. 1 The area of observations and collection of materials by the Tallinn and Riga marine laboratories in 1957–1997 (author's data)

constructive proposals on how to improve the text, especially that of Chaps. 5 and 6. I am very grateful to him.

The book is intended for graduate students, researchers, and managers involved with the Baltic Sea ecosystems and living resources.

Tallinn, Estonia
January 2016

Evald Ojaveer

Contents

Chapter 1
Evolution of the Baltic Sea

Abstract Extensive warming and cooling periods in the Pleistocene that resulted in alteration of glacial and interglacial epochs, preceded formation of the contemporary Baltic Sea. The first stage in the development of the Baltic Sea was formation of a large glacial lake about 12,000 years ago. Organisms having survived in severe conditions in various refugia colonized the newly formed basin. This stage was followed by the Yoldia Sea stage between the years 11,700 and 10,700 BP. After this the freshwater Ancylus stage with boreal climate lasted up to the year 9500 BP. The warmest brackish-water Litorina stage followed. It transferred to the colder and more fresh present-day Limnea Sea stage about 4500 years ago. The marine fauna survived the last Ice Age in oceanic waters and the species extruded by the severe conditions repopulated the North European seas after the change of the conditions. The postglacial freshwater fish fauna of the North Europa consists basically of the same species as in the preglacial time. As the time in refugia during the Ice Age and also after it has been too short for significant genetic differentiation, the present North European fish fauna is poor in true species but rich in intraspecific forms.

1.1 Development of the Baltic Sea After the Last Ice Age

Formation of the Baltic Sea region of our planet can be seen over a period of at least three billion years, from the Precambrian to the late Paleozoic, when a continent named Baltica might have existed. The present-day Baltic Sea is located in the depression where prior to its establishment, during a series of glacial and interglacial epochs in the Pleistocene, sat the Eemian Sea, a water body much larger in area than the present Baltic Sea and spaciously connected to the oceanic waters from two sides.

Extensive warming and cooling periods in the Pleistocene caused alternation of glacial and interglacial epochs with basic changes both in environmental conditions and biota in the present-day Baltic Sea area. The most important variations in the living conditions for organisms were induced from the glaciers, which periodically expanded and retreated. During the last, Weichselian Ice Age (120,000–15,000 years ago), a glacier covered the area of the Eemian Sea, which existed during the Eemian interglacial, and a large part of the Northern European continent. In the central area of the glacier, a mass of ice with an estimated thickness of over 3 km pressed the

E. Ojaveer, *Ecosystems and Living Resources of the Baltic Sea*,
DOI 10.1007/978-3-319-53010-9_1

crust of the Earth beneath it down. Ice-dammed lakes were formed on the southern and eastern margins of the advancing glacier and were gradually pushed southwards. Periglacial lakes discharged westwards through a large system of rivers.

Data on the duration of glaciations and interglaciations, as well as on the boundaries of the forming sea and the evolution of the environmental conditions in the sea, are incomplete, particularly concerning earlier periods. Conclusions have been derived on the basis of geological, geomorphological, paleontological, sedimentological and dating materials. The reconstructions are mainly based on the data of the development of the coastline and the sediments collected from coastal areas, and much less on the sediment samples taken from the deep-sea areas. With this in mind, it is not surprising that our understandings are continuously being updated based on new material that has become available.

It is probable that within each glacial and interglacial period, there were climate variations – warmer and colder times. Obviously the climate changes in the Baltic Sea basin, which were the basic cause of the melting of the ice cap during the deglaciations in the Holocene, were due to various external and internal factors. Borzenkova et al. (2015) are of the opinion that the important alterations in solar radiation took place due to slow changes in the Earth's orbit, variations in the concentration of stratospheric aerosol caused by volcanic activity, in the content of greenhouse gases (carbon dioxide, methane and nitrous oxide), in the atmosphere due to natural factors, in surface albedo of the water body itself, in the surrounding land vegetation and in the intensity and type of circulation due to variations in basin salinity.

The decline of the Fennoscandian ice sheet (which had previously accumulated as part of the 100-thousand-year-long glacial cycle) started around 18,000 years ago, but due to the long timescales involved in land–ice dynamics and the slow crustal readjustment to the disappearance of the weight of an ice sheet 3 km deep, its effects have been felt through most of the Holocene. The deglaciation of the Baltic Sea basin started about 15,000 years ago when the border of the glacier was south of the area of the contemporary sea. The evolution of the new Baltic Sea area has undergone major changes due to two interrelated factors: (1) the most important effect of the melting of the ice sheets is an interplay between global sea-level rise due to the increase in the ocean volume and a regional isostatic uplift of the Earth's crust; (2) changes in the orbital configuration of the Earth, which are thought to be the major trigger of the glacial to interglacial transition, but which also modulated solar insolation to the high northern latitudes during the Holocene and thus strongly influenced the energy balance of the Baltic Sea area (Borzenkova et al. 2015).

With climate warming and the melting of the glaciers, the lakes that had formed around the frozen area merged into one system. As a result, a large glacial lake was formed approximately 12,000 years ago; it represented the first stage in the development of the Baltic Sea (Fig. 1.1a). Organisms that had survived in severe environmental conditions of the ice age in various refugia colonized the newly formed basin (Paaver and Lõugas 2003). At that time, diatoms and invertebrates of marine background as well as Arctic cod (*Boreogadus saida*) occurred in the Baltic basin.

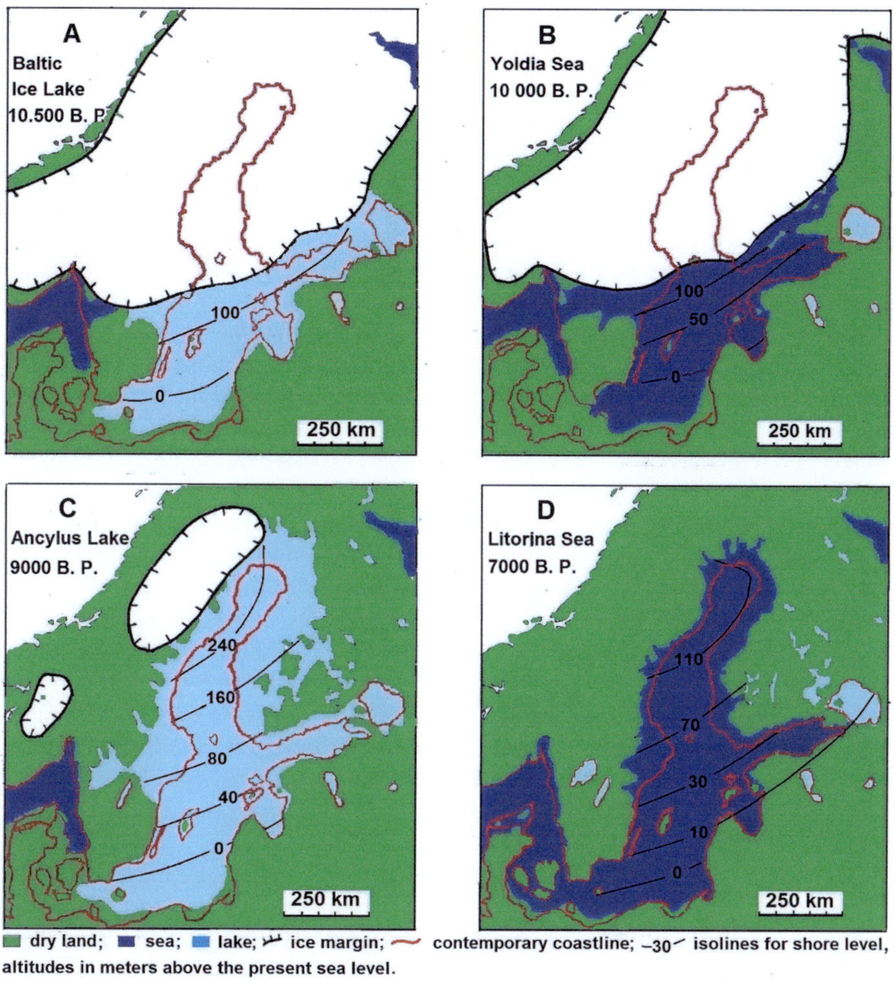

Fig. 1.1 Stages of early development of the Baltic Sea (Ojaveer 2014; Raukas and Hyvärinen 1992)

Some scientists (e.g., Sauramo 1958) have expressed the view that at that time, there existed a period in the development of the Baltic Sea when the basin of the sea was connected to marine waters via the cold White Sea basin in the east and, most likely, through an additional connection with the Atlantic Ocean in the west. This view was opposed by another Finnish geologist, Ignatius, who was of the opinion that the evidences for the differentiation of such a saltwater period in the history of the Baltic Sea are insufficient. He was in favour of acknowledging the Baltic connection with the saltwater sea in the west, in case the connection could be proven to exist (cited in Voipio 1981).

A subarctic climate reigned. The landscape was influenced by the glacial activity and the isostatic land uplift. Continuously additional water could leave the Baltic, probably through the connection with the oceanic waters in the area of the contemporary Danish Sounds. After the melting of ice and the retreat of the glacier to the north of Billingen Mountain in central Sweden, the water accumulated as a result of the melting of the glacier and broke through to the ocean via the newly formed Närke Sound situated at Billingen in central Sweden. According to the data of the Swedish sediment chronology, the sound opened 8213[1] years ago. This event denoted the end of the Ice Lake stage of the Baltic Sea (Fig.1.1b). The Baltic Sea gradually got rid of the glacial ice, beginning mainly from its southern edge. According to the estimate by Borzenkova et al. (2015), the stage of the Baltic Ice Lake lasted 11,550 calendar years BP.

The new stage of the sea was named the Yoldia Sea after the mollusc *Yoldia arctica*. This stage lasted from the year 11,700 to 10,700 BP (Borzenkova et al. 2015). During some of this time, the basin had the character of an internal sea. The salinity was low, as the inflow of large masses of fresh melting waters continued and the connection with the ocean was not deep. At this stage, the near-bottom layers of salt water intruding into the Baltic Sea carried with them marine diatoms, ostracods, crustaceans (including *Yoldia arctica*) and other organisms. The surface current going in the opposite direction carried cold fresh water into the ocean. At this stage, the whale *Balaena mystacetis*, seal *Phoca groenlandica* and varieties of whiting (*Merlangius merlangus*), haddock *(Melanogrammus aeglefinus)* and herring *(Clupea harengus)* immigrated into the Baltic Sea. The skeleton of a 21-cm-long herring was found at the location of the former Närke Sound at 8 m depth in glacial clay at the height of 88 m above sea level (Munthe 1956). In the northern part of the Yoldia Sea, arctic conditions reigned during those times and the biota was very poor. In the southern part of the sea, the production was somewhat higher. The climate of Eastern Europe was continental, with relatively warm summers and cold winters.

The continuing warming and melting of the ice cap and the elevation of the Earth's crust reduced the depth of the Närke Sound step by step and limited the saltwater inflow. Therefore, the salinity in the Baltic Sea dropped and the marine stage was replaced with a freshwater stage. The interval between the marine and freshwater stages of the sea has been named the Echeneis Sea stage (after the diatom *Campylodiscus echeneis* found in the littoral of the southern part of the sea), but most often the Echeneis Sea is not treated as a separate developmental stage of the Baltic Sea. The connection between the Baltic Sea and the ocean discontinued in the Yoldia stage, creating Ancylus Lake (Fig. 1.1c; the name is derived from the scientific name of the freshwater worm *Ancylus fluviatilis*, which populated the basin at that time). The Ancylus Lake stage of the Baltic Sea started from the year 10,700 and lasted until 9500 calendar years BP.

During the relatively short freshwater Ancylus Lake stage, the climate was boreal and the temperature increase continued. Ancylus Lake was oligotrophic, with cold water in its deep parts. In summer the coastal areas were warm, with rich vegetation characteristic of large lakes. As the rise in the Earth's crust, which had gotten rid of

[1] ^{14}C years

the weight of the ice cap, was larger in the northern areas than in southern the water level in Ancylus Lake increased southwards, resulting in a new connection between the Baltic Sea and the ocean in the area of the present-day Danish Sounds.

About 9500 years ago, a new stage started in the development of the Baltic Sea, the brackish-water Litorina Sea stage. The Litorina Sea stage (Fig.1.1d) got its name from the mollusc *Littorina* spp. The beginning of the stage coincided with the start of the humid and relatively warm Atlantic climate period. The period between 7500 and 5500 years ago was the warmest in the Baltic Sea basin area as a whole, although the timing of maximum temperatures was not synchronous in different parts of the region (Borzenkova et al. 2015). This stage facilitated a marked widening of the area of marine organisms requiring higher salinity and milder temperature conditions. In response to the increased salinity, the relict species of freshwater background that had populated Ancylus Lake retreated from the southern and central parts of the sea. They found more acceptable living conditions in the gulfs of Riga, Finland and Bothnia, where the salinity remained lower because of freshwater discharge by rivers. Simultaneously, the relict populations that remained in these large gulfs occurred in the conditions relatively isolated from others. Such status has lasted up to the present time.

Some authors differentiate a specific transition stage between the Ancylus and Litorina stages, with increasing salinity and corresponding changes in the composition of biota naming it the Mastogloia Sea (after the brackish-water diatom *Mastogloia* spp.). Others do not see clear proof for the differentiation of a specific stage and prolong the Ancylus stage, estimating that it terminated about 9500 years ago. The stratigraphic boundary between Ancylus Lake and the Litorina Sea is clear. After the opening of the Danish Sounds, the salinity increased continuously in the whole sea. The salinity of the Litorina Sea was 5–6 psu higher than in the present-day Baltic Sea. It rose to 8 psu in Bothnian Bay, to 10 psu in the Bothnian Sea and the Gulf of Riga, to 5–10 psu in the Gulf of Finland, to about 15 psu in the central part of the sea, and over 20 psu in the SW Baltic. With the beginning of the Litorina stage, the glacial clays in the sediments were substituted with postglacial muds with an ample content of organic matter (Raukas and Hyvärinen 1992).

The reaction of the Earth's crust to the disappearance of the ice cap and its weight continued. Due to its rise, the depth of the Danish Sounds decreased and the inflows of saline water dropped. Therefore, about 4000 years ago, the salinity of the Baltic Sea began to decrease. The organisms needing higher salinity retreated towards the SW and the area of brackish-water and freshwater organisms widened. Thereafter, a trend towards the cooling of the northern hemisphere and increased climatic instability began, typical of the Late Holocene interval. Temperatures in the Baltic Sea region started to drop around 5000–4500 calendar years BP, coincident with decreased summer solar insolation due to the quasi-cyclical changes in the Earth's orbit. The modern configuration of the Baltic Sea as a brackish Mediterranean or inland sea of lower salinity and colder climate was established around 4500 years ago and is named the Limnea Sea (after the gastropods *Lymnea* spp.).The Earth's crust has continued rising, remaining one of the key processes in the evolution of the basin.

1.2 Formation of Biota in the Baltic Sea

In the periods of glaciation, a portion of the freshwater fish fauna of Northern Europe was ruined. The remaining part retreated to the south and survived in refugia, where their genetic differentiation began to take place. The freshwater fish fauna of the present Baltic Sea basin originates from the boreal Neogene fauna of the Palearctic region. It was impoverished during the Pleistocene when the whole preglacial fauna of Northern Europe was exterminated or forced to retreat southwards several times. In warm interglacial periods, the species returned to ice-free areas. The process of retreat and recolonization was repeated several times. The postglacial freshwater fish fauna in Northern Europe basically consists of the same fish species as in the preglacial time. However, during the ice age, the fish became isolated in different refugia, where they were subject to genetic differentiation, although the time of isolation was too short for speciation to be completed. Also, the postglacial period has been too short for significant genetic differentiation. As a result, the present Northern European fish fauna is poor in true species but rich in intraspecific forms, i.e., the composition and interrelations of fish populations have gotten entangled (Paaver and Lõugas 2003).

In early times of the development of the Baltic Sea after the long-lasting ice age, the most important problems were related to the mechanical influence of ice and the severe thermal conditions. However, the fauna of the contemporary Baltic Sea consists of marine, brackish-water and freshwater species. Since the beginning of the development of life on our planet, in the aquatic systems, water salinity (mediated by the osmotic pressure) has been the basic environmental factor determining the composition of ecosystems. Two main types of aquatic biota exist, constituting either marine (usual salinity 33–37 psu) or freshwater (< 0.5 psu) ecosystems. The young, species-poor brackish Baltic Sea differs from both of these types in physicochemical and ecological conditions, as well as in biological characteristics of organisms. The main reason for much smaller numbers of species in brackish-water systems is supposedly that brackish basins are short-lived in geological terms, and also that in a number of exploited organisms, speciation has only reached the stage of differentiation of populations with diverse abundance dynamics. Considering this and the possible changes in immigration routes to the Baltic Sea during the postglacial history make understanding the origin and qualities of new communities in the Baltic Sea complicated.

The marine fauna survived the last ice age in oceanic waters and the species extruded by the severe conditions repopulated the North European seas after the change in the conditions. Marine biota began to settle in the brackish Baltic basin only after its salinity had risen to such a level that adaptation of populations of marine origin was possible. The higher salinity is required for the facilitation of adaptation with the osmotic problems for the species of marine origin. Naturally, when a connection had formed between the Baltic basin and the ocean, the euryhaline populations at the west coast of Scandinavia had the best chances to spread into the Baltic area. After the ice retreated from the southern and central parts of the sea and the cold marine water flowed through the Närke Sound into the Baltic, fish species of marine origin began adaptating to the conditions in the new region. Arctic and arctic–boreal cold-water species, such as the common sea snail (*Liparis liparis*), the snake blenny (*Lumpenus lampretaeformis*) and the eelpout (*Zoarces viviparus*) are thought to have migrated into the Baltic Sea in its early saltwater stage. It is probable that bullrout (*Myoxocephalus scorpius*), haddock (*Melanogrammus aeglefinus*), whiting (*Merlangius merlangus*) and herring (*Clupea harengus*; most likely spring spawning herring) also immigrated into the Yoldia Sea. The earliest remains of herring in the Baltic were found at the former Närke Sound (Munthe 1956).

The first fish species that colonized the Baltic Sea after the ice age were arctic species: smelt (*Osmerus eperlanus*), whitefish (*Coregonus lavaretus*), fourhorned sculpin (*Triglopsis quadricornis*) and vendace (*Coregonus albula*). These and the species related to them were probably common in Northern Europe before the glaciation. Presently, they populate the Baltic Sea basin and regions north-east of it. These species probably retreated with the advancing glacier and survived in glacial lakes. Occurrence of whitefish in glacial lakes is proved with archaeological findings (Paaver and Lõugas 2003). Smelt and vendace are currently found in lakes of the region of the former glaciers in Russia, northern Poland and Germany. The species of this group of marine background made use of different methods of immigration. Fourhorned sculpin used of the periglacial lakes where they were trapped during the period of the advancing glacier and established freshwater populations in these lakes. However, the species could also immigrate through the marine connection between the Baltic and Atlantic Ocean. The species also occurred in the Baltic during the Ancylus Lake stage (Lepiksaar 1938, 1984; Munthe 1956).

Freshwater species that are cold-adapted or moderately cold-tolerant are widely distributed in Europe, among them grayling (*Thymallus thymallus*), pike (*Esox lucius*), roach (*Rutilus rutilus*), perch (*Perca fluviatilis*), ruffe (*Gymnocephalus cernuus*) and burbot (*Lota lota).* These species were widely distributed in the Baltic area before the ice age. During the glaciation, they lived at rather low temperatures. Most of the species of this group are adapted to life conditions in relatively shallow waters with soft bottoms and rich vegetation. Remains of pike, ide, perch, ruffe and burbot have been found from late glacial sediments in southern Sweden and Denmark. These species seem to have spread into the Baltic Sea from different refugia and during different periods of deglaciation.

The majority of freshwater fish species (bream (*Abramis brama*), minnow (*Phoxinus phoxinus*), bleak (*Alburnus alburnus*), roach (*Rutilus rutilus*), ruffe (*Gymnocephalus cernuus*), dace (*Leuciscus leuciscus*), ide (*Leuciscus idus*), nine-spined stickleback (*Pungitius pungitius*), brook lamprey (*Lampetra planeri*)) and other moderately cold-adapted species evidently survived during the glaciation in ice-free areas and returned via the water system connected to the ice lake. The most rapid spread of freshwater fish may have occurred during the Ancylus period.

Later immigrants into the Baltic Sea were the species adapted to the temperate conditions, having populated areas with a milder climate during glaciation. These include tench (*Tinca tinca*), chub (*Leuciscus cephalus*), rudd (*Scardinius erythrophthalmus*), gudgeon (*Gobio gobio*), rifle minnow (*Alburnoides bipunctatus*), white bream (*Blicca bjoerkna*), bream (*Abramis brama*), spined loach (*Cobitis taenia*), mud loach (*Misgurnus fossilis*), stone loach (*Barbatula barbatula*), and bullhead (*Cottus gobio*). They immigrated from different directions, especially during the periods of lower salinity, and chiefly populated lakes, rivers and the coastal zone of the Baltic (Lepiksaar 1938, 1984; Ojaveer 2014).

Thermophilic freshwater fish immigrated into the Baltic area from the refugia at the upper reaches of the rivers of the Black Sea basin. These include asp (*Aspius aspius*), sunbleak (*Leucaspius delineatus*), razorfish (*Pelecus cultratus*), vimba bream (*Vimba vimba*), sheatfish (*Silurus glanis*), pikeperch (*Sander lucioperca*), nase (*Chondrostoma nasus*), and barbel (*Barbus barbus*), and are considered new-comers from the Caspian and Black Sea basins. It is probable that these species immigrated into the North European fauna before the last glaciation, survived the ice age by retreating to the south, and later spread again northwards. Archaelogical data show that pikeperch and sheatfish occurred in the Baltic area as early as the Yoldia Sea stage around 10,000 years ago (Paaver and Lõugas 2003). They increased their area of occurrence in the warmer freshwater Ancylus Lake stage.

Salmon (*Salmo salar*), sea trout (*Salmo trutta*) and Atlantic sturgeon (*Acipenser sturio*) are species with an Atlantic background. It is probable that the river lamprey (*Lampetra fluviatilis*) also belongs to this group. Colonization of the Baltic by this group probably started from freshwater refugia.

It is not certain if the marine species survived in the Baltic when the salinity in the Ancylus stage decreased substantially, but the deep water layers in Ancylus Lake may have retained a higher salinity. However, whiting (*Merlangius merlangus*) and haddock (*Melanogrammus aeglefinus*) disappeared from the Baltic Sea during the Ancylus period.

During the Litorina stage, with its higher salinity and temperature, colonization of the sea by boreal species of marine origin obviously quickened. Apparently at this stage, there was immigration by four-bearded rockling (*Rhinonemus cimbrius*), lesser sandeel (*Ammodytes tobianus*), greater sandeel (*Hyperoplus lanceolatus*), lumpsucker (*Cyclopterus lumpus*), flounder (*Platichthys flesus*), turbot (*Scophthalmus maximus*), common goby (*Pomatoschistus microps*), sprat (*Sprattus sprattus*) and cod (*Gadus morhua*). It is probable that the Litorina stage was the first period in the development of the Baltic Sea when a number of populations of marine origin were able to start reproduction there.

Marty (1966) is of the opinion that spring and autumn spawning herring were disunited in the Atlantic Ocean a long time before their immigration into the Baltic Sea. The cold-adapted spring spawning herring differentiated during the last ice age in a closed basin in the North Atlantic. It can be assumed that amongst these populations, some were more euryhaline and eurytherm than the oceanic herring, and they settled Norwegian fjords. Jorstad et al. (1994) showed that the Balsfjord herring population is genetically closer to the Pacific herring than to the oceanic herring, which spawns at a depth of 40–60 m in the North Atlantic. Therefore, it is probable that the Norwegian fjord herring is the ancestor of the White Sea coastal herring, the Pacific herring and the Baltic spring spawning herring. All these stocks have a relatively small number of vertebrae and they spawn on vegetation in rather shallow water.

It is possible that herring immigrated into the Baltic Sea in several waves. In the herring populating the Baltic Sea today, three larger groups can be distinguished: spring spawning sea herring in the open part of the sea, gulf herring which lives in the gulfs of Bothnia, Finland and Riga and autumn spawning herring. Gulf herring forms the herring group best adapted to the Baltic conditions while autumn spawning herring is the least well-adapted. Therefore, it can be assumed that the earliest immigrants into the Baltic were the ancestors of spring spawning gulf herring and that they probably occurred in the Baltic Sea during the Yoldia stage. If in the Ancylus stage, in some part of the sea, some salinity persisted, then they could have survived in the Baltic basin adapting to a very low salinity at reproduction (this is peculiar to the current spring spawning gulf herrings). In a case in which herring was exterminated from Ancylus Lake, the herring groups having adapted to the Baltic conditions to some extent might have persisted outside the Baltic Sea in shallow coastal areas of the ocean and re-immigrated into the Baltic after the opening of a new opportunity. Spring spawning sea herring is associated with a higher salinity found in the open part of the Baltic; therefore, sea herring populations might be later immigrants. They might have come to the Baltic Sea at its highest salinity level in the Litorina stage. Autumn spawning herring differs from spring spawners on the species level. It is possible that autumn spawning herring, as the herring group needing higher salinity and temperature than its sibling species spring herring, immigrated into the Baltic Sea at its highest salinity and temperature level in the Litorina stage some three thousand years ago (Munthe 1956).

References

Borzenkova I, Zorita E, Borisova O, Kalnina L, Kiseliene D, Koff T, Kuznetsov D, Lemdahl G, Sapelko T, Stancikaite M, Subetto D (2015) Climate change during the Holocene (past 12,000 years). In: The BACC Author Team (ed) Second assessment of climate change for the Baltic Sea basin. Springer, Cham/Heidelberg/New York/Dordrecht/London, pp 25–49

Jorstad KE, Dahle G, Paulsen O (1994) Genetic comparison between Pacific herring (*Clupea pallasi*) and a Norwegian fjord stock of Atlantic herring (*Clupea harengus*). Can J Fish Aquat Sci 51(Suppl. 1):233–239

Lepiksaar J (1938) Eesti kalastiku kujunemise loost jääajast tänapäevani. Eesti Kalandus 11:285–290; 12:302–305

Lepiksaar J (1984) Eesti kalastiku kujunemise loost jääajast tänapäevani. Eesti kalanduse minevikust 1:327–338

Marty YY (1966) Vzglyadõ na formirovaniye morfobiologitšeskikh osobennostei morskikh seldei Atlantitšeskogo i Tikhogo okeanov (Views on the formation of morpho-biological features of marine herrings of the Atlantic and Pacific oceans). Trudy PINRO 17:303–316

Munthe H (1956) On the development of the Baltic herring in the light of the late quaternary history of the Baltic. Arkiv för Zoology 9:333–341

Ojaveer E (2014) Läänemeri. TA Kirjastus, Tallinn

Paaver T, Lõugas L (2003) Origin and history of the fish fauna in Estonia. In: Ojaveer E, Pihu E, Saat T (eds) Fishes of Estonia. Estonian Academy Publishers, Tallinn

Raukas A, Hyvärinen H (eds) (1992) Geologija Finskogo zaliva (Geology of the Gulf of Finland). Estonian Academy of Sciences, Academy of Finland, Tallinn

Sauramo M (1958) Die Geschichte der Ostsee. Suomalaisen Tiedeakatemian Toimituksia, Sarja A, III. Geologica–Geographica 51. Suomalainen Tiedeakatemia, Helsinki

Voipio A (ed.) (1981) The Baltic Sea. Elsevier, Amsterdam/Oxford/New York

Chapter 2
Abiotic Conditions in the Contemporary Baltic Sea

Abstract Abiotic conditions in the Baltic Sea including the bottom relief and the properties of the water in various parts of the sea are characterized. The most important natural environmental conditions for the organisms are water salinity and the halocline, temperature and the thermocline, oxygen and light conditions. The causes of different natural conditions in sea areas and the importance of water movements in creation of these differences are considered. Based on the phenomenon of existence of areas continuously differing in environmental conditions and biota, the natural regional system consisting of macro-regions (Transition Area, Open Part of the Sea and Large Gulfs) and ten regions has been described. Due to significant variations in the bottom relief, salinity, current systems, location of nutrient depots etc. the regions have clearly differing ecophysiological and other environmental conditions. Therefore populations of species in these units differ and their condition cannot be assessed together by species to grant their correct management. The populations significantly differing in stock dynamics should be considered separately based on the natural boundaries of the stock units.

The development of the Baltic Sea, its environment and biota from the original freshwater and cold Baltic Ice Lake via the Yoldia, Ancylus and Litorina stages to the contemporary sea has left its footprints both on its abiotic nature and the life forms that have persisted up to the present. The past manifests itself in the present-day ecosystems of the Baltic Sea. The most important global engines of this development – geological and climatic processes – simultaneously constitute the most important background of the contemporary environmental conditions. Besides the less noticeable but continuously operating geological processes, climate variations are the primary factor that has brought about deviations in the everyday life, causing larger or smaller, clearly periodical or seemingly irregular variations in environmental conditions.

The bottom relief of the Baltic Sea – a row of deeps and thresholds between them, proceeding to the east and north from the Danish Sounds that connect the sea and the ocean - formed beginning in the Cambrian. This has determined the features of the ecosystems located in the sphere of influence of both marine and continental climates.

E. Ojaveer, *Ecosystems and Living Resources of the Baltic Sea*,
DOI 10.1007/978-3-319-53010-9_2

The brackish-water Baltic Sea is separated from its neighbour, the North Sea, which is of normal oceanic salinity, by the Transition Area in which the marine environment and biota transfer to the brackish-water surroundings in a comparatively limited area. Moving from the North Sea through the Skagerrak, Kattegat and Danish Sounds eastwards, the water salinity and the importance of marine organisms considerably decrease (Figs. 2.1 and 2.2). The depth of the Drogden sill in the sound separating the Baltic Sea from the areas with higher salinity is only 8 meters. The Darss sill represents a comparatively clear border between the distribution of marine and brackish-water organisms. Naturally, the importance of organisms of marine background is higher in the southwestern and southern parts of the sea (Fig. 2.2). Among the bottom animals that have penetrated into the Baltic, species characteristic for the littoral of the North Sea, Scandinavia, Murman and White Sea coasts (*Pygospio elegans*, *Nereis diversicolor*, *Halicryptus spinulosus*, *Macoma baltica*, *Cardium edule*, *Mytilus edulis*, *Gammarus locusta*, *Jaera albifrons*, *Balanus improvisus*, *Asterias rubens* etc.) clearly dominate.

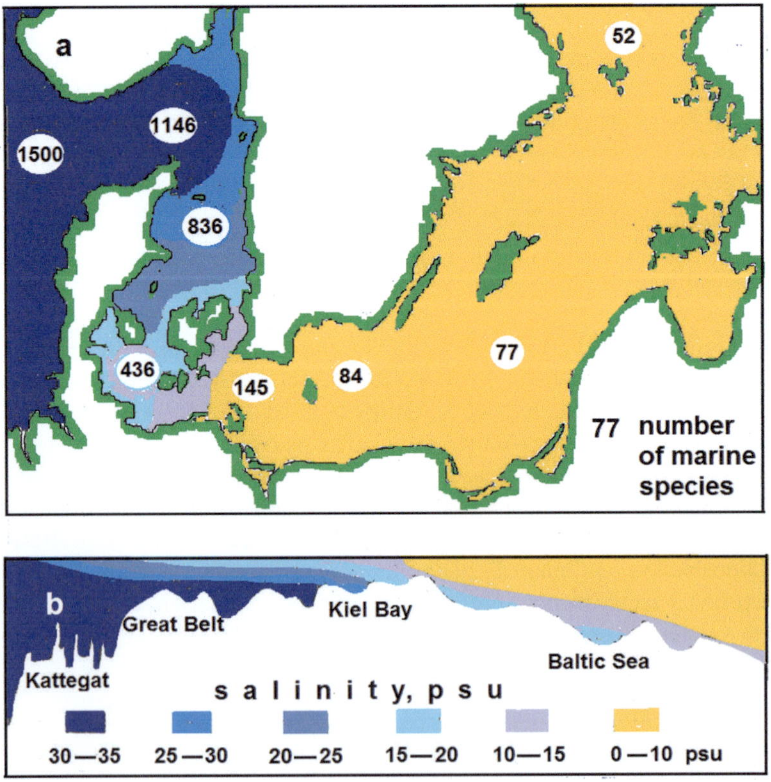

Fig. 2.1 Changes in salinity and the number of species of marine origin between the North and Baltic Seas after Zenkevich (1963)

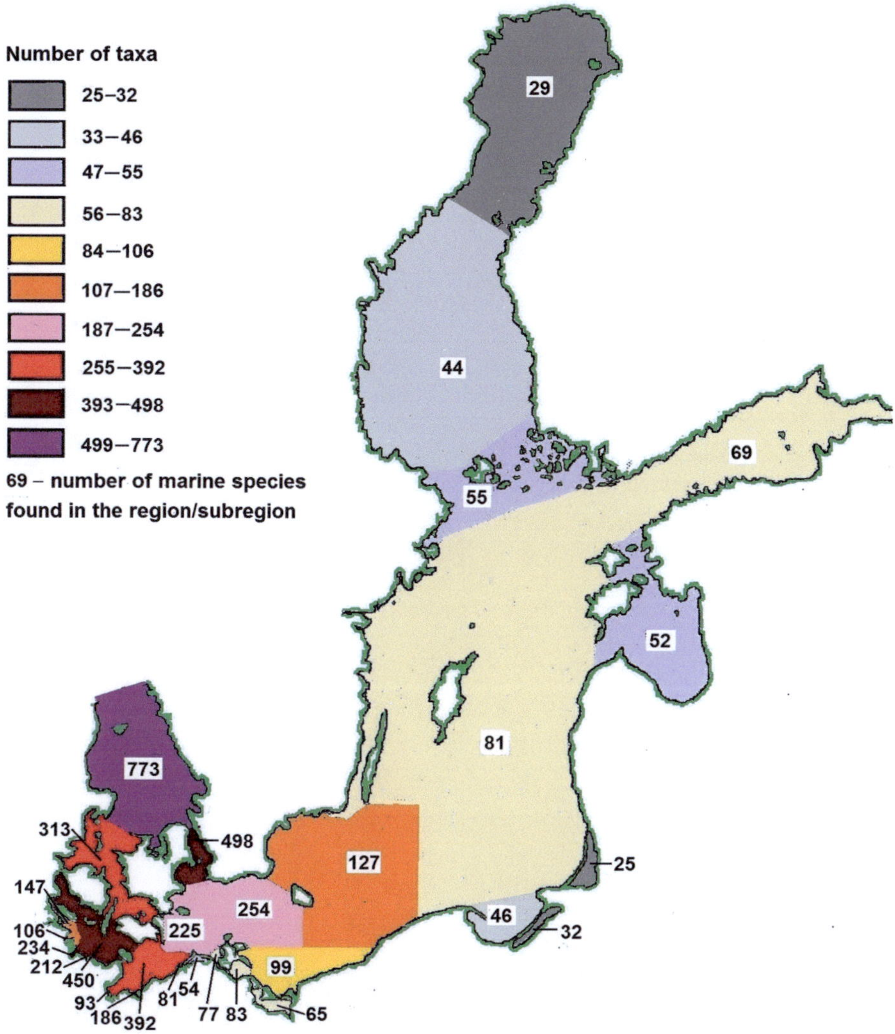

Fig. 2.2 Distribution of marine taxa of macrozoobenthos in the Baltic Sea (Ojaveer et al. 2010)

During the colonization of the Baltic Sea, the bottom fauna descended from the littoral to the sublittoral, probably seeking the highest possible salinity. This has given Zenkevich (1963) the basis to state that rivalrybetween species for populating the sublittoral of the Baltic Sea was very weak or actually absent. In the Litorina period, the *Macoma baltica* biocoenosis, eurybiontic regarding salinity, temperature and also oxygen conditions, penetrated into the sublittoral of the Baltic Sea. It travelled far into the freshened water and occupied the upper zone of the sublittoral. In the lower horizon of the sea, the relict arctic cold-water complex dominated. This complex has also descended deeper than its earlier habitat.

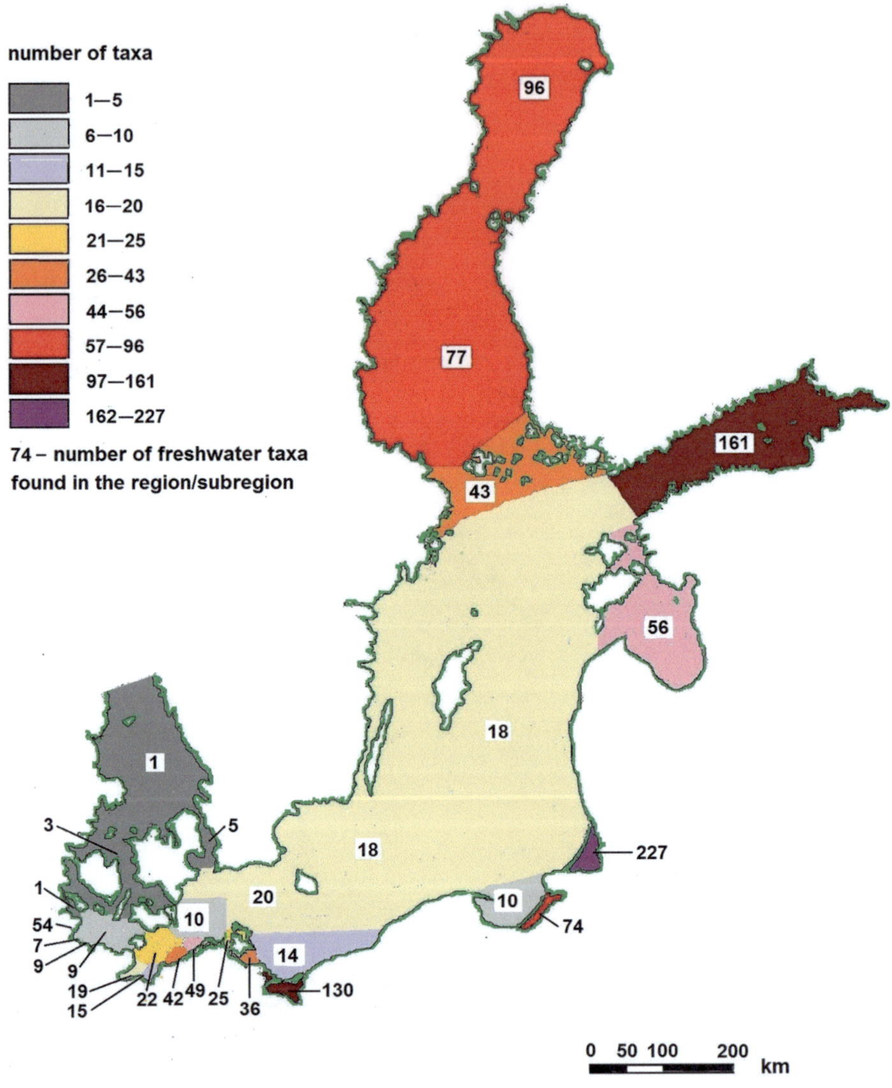

Fig. 2.3 Distribution of freshwater taxa of macrozoobenthos in the Baltic Sea (Ojaveer et al. 2010)

The northern areas of the sea and large gulfs that have been much less approachable for marine organisms than the open Baltic are presently populated with species (so-called glacial relicts) which immigrated into the basin of the sea during its development and formation. The same probably applies to the organisms of freshwater background (Fig. 2.3). This state of affairs represents a very important difference between the living organisms of the southwestern and northern parts of the Baltic Sea. If the planktonic species populating the Belts or Sound differ relatively slightly from the corresponding ones of the North Sea, the plankton in the gulfs of

Finland and Bothnia, composed in significant part of freshwater species, is completely different from the plankton of seas with normal salinity. The same can be said about the bottom animals, fish fauna and other components of the ecosystems.

The Baltic Sea stretches from the south to the north for about 1500 km. The mean depth of the sea, with an area of 415,266 km^2 and a volume of 21,721 km^3, is 52.3 m. The sea with its main basins – the Kattegat with a maximum depth of 109 m, the Sound and Belts (38 m), Arkona Basin (55 m), Bornholm Deep (105 m), Gotland Deep (249 m), Gdansk Deep (113 m), Northern Deep (180 m), Landsort Deep north of Gotland Island (459 m), the narrow depression connecting the Åland Sea with the northern part of the open Baltic (301 m), Harnösand Deep in the Bothnian Sea (230 m), Bothnian Bay (146 m), the Gulf of Finland (123 m), the Gulf of Riga (62 m), the bights, lagoons, boddens, etc. – presents widely varying combinations of salinity and other environmental characteristics for the organisms (Voipio 1981; HELCOM 1986, 1990, 1993, 2002, etc.).

The state of the environment in the Baltic Sea is determined by the main climatic conditions, the wealth and composition of water, and human activity. All organisms in this sea – having lived there since its formation or immigrated into it from other marine or freshwater bodies – can at present time live there only under current conditions. As the brackish-water conditions vary both temporally and spatially, the distribution and living conditions of organisms are limited chiefly by salinity (Figs. 2.1, 2.2, and 2.3), which determines, through osmotic conditions, whether and in which area/ecosystem a certain organism can live or reproduce. In addition to salinity, the survival of immigrants also considerably depends on temperature, oxygen conditions/presence of hydrogen sulphide, biological productivity, water acidity/alkalinity (pH) and a number of other environmental parameters.

2.1 Water Balance

The water environment in the Baltic Sea has formed as a result of a mixing of the fresh water mainly discharged by rivers and the water of marine salinity inflowing from the North Sea through the Danish Sounds. The amount of added fresh water depends on the climatic periods. From various parts of the large catchment area of 1721,233 km^2, rivers discharge into the Baltic Sea fresh water of very different quality. On average, the rivers bring into the Baltic Sea (including the Kattegat) 472 km^3 of fresh water annually. After Voipio (1981), about 257 km^3 should be added as a result of precipitation, whereas the Baltic Marine Environment Protection Commission (1986) found a summary precipitation for the Baltic Sea of 204.71 km^3 for the year 1975 and 223.83 km^3 for 1976.

It is essential to determine the water balance of the sea to be able to understand the long-term water renewal and the various ways in which it influences the sea's environment. The analysis includes a separate consideration of the interconnected subbasins of the Baltic Sea. In general, the Baltic Sea is characterized by a positive freshwater balance. The external forcing functions governing the water exchange are (HELCOM 1986):

- freshwater supply from the land drained;
- outside sea level and salinity (North Sea);
- meteorological forcings.

The renewal of the water mass is described by the water balance. This concept includes the quantitative relationships and the interconnections between the elements responsible for the renewal of the water by the external processes: precipitation (P), evaporation (E), input by land runoff (L), salt water inflow through the straits (M), water storage or retention (ΔV) and water outflow (H). The water balance is expressed by the following equation: $(P - E + L) + (H - M) = \Delta V$.

The net fresh water inflow $(P - E + L)$ is influenced by climatic conditions, and the hydrography of the drainage basin, the net freshwater outflow $(H - M)$ and the water storage (ΔV) could be seen as secondary elements constituting the effects of the primary elements with the sea water inflow (M) from the North Sea. The residence time (years) of marine water in bottom layers in various areas of the Baltic Sea have been estimated as follows (HELCOM 2009): Gulf of Bothnia 34–42, Gulf of Finland 24–32, Gulf of Riga 30–32, Baltic Proper 10–28, Danish Straits 8–10.

In addition to the water quantity, the recipient ecosystems are also influenced by the amount of chemicals and other compounds in the water. Evaporation is most intense in the second half of the year and in the southern part of the sea. It removes on average 208 km^3 water a year from the Baltic (HELCOM 1986). There are regional differences in the composition of both the natural water and the anthropogenic additions. Some preliminary data presented by the Helsinki Commission (HELCOM 1986) indicate that the total input of suspended matters to the Baltic Sea can be estimated as being about 7.5 mill. tonnes a year. The input per km^2 was estimated as follows: from Denmark, 8.4 t/km^2, from Poland, 6.7 t/km^2, from Sweden, 4.5 t/km^2, from Finland, 4.0 t/km^2, and from the USSR, 3.6 t/km^2 a year. The river discharge is generally largest in April–May. However, the time of the largest inflow varies by areas. In the southwestern part the river, discharge is the highest in March, in the central part of the open Baltic, from March to May, in the gulfs of Riga and Finland, in April–May, and in the Gulf of Bothnia, in May–June. In the summer months, the river discharge lowers, only to increase again in the southern rivers in November–December (Fig. 2.4). The largest amount of water is discharged into the Baltic Sea by the Neva River – on average, about 281 cubic kilometres annually. The Neva is followed by the Vistula with 194 km^3 and the Daugava with 88 km^3 (Voipio 1981).

2.2 Water Salinity

Saline water comes from the North Sea, either as a result of the regular continuous moderate water exchange between these seas or by intrusions of large saltwater masses through the Danish Sounds into the Baltic Sea related to meteorological processes. The addition of smaller water quantities (less than 10–20 km^3), even of

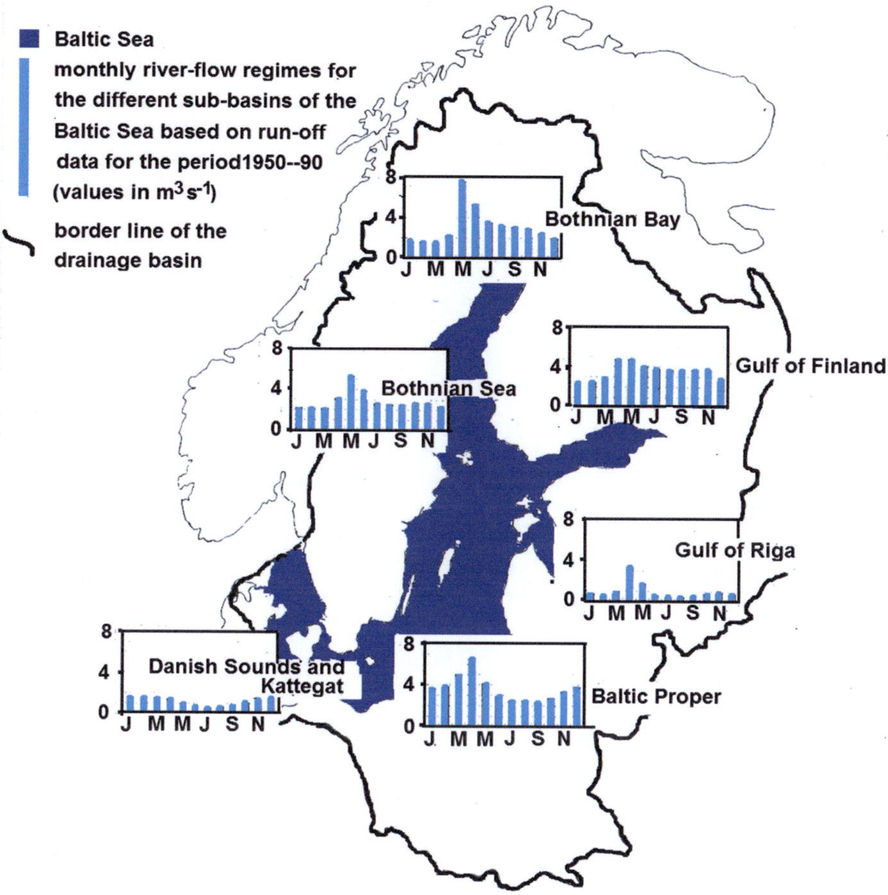

Fig. 2.4 Average annual freshwater inflow into the Baltic Sea from different parts of the catchment area by months in 1950–1990 (HELCOM 1996)

high salinity, lacks a notable influence on the oceanographic status of the Baltic Sea. However, the inflows with salinity as high as 17–25 psu exceeding 100–250 km^3 in volume can essentially change the conditions in the Baltic Sea. Intrusions of such large water masses are facilitated by fresh western gales when the sea level in the Baltic is low, especially after the periods of limited river discharge. Intrusions of North Sea saline water have been very important for the hydrography and ecosystems of the Baltic Sea, therefore they have been carefully followed since the 1880s, excluding the years of the World Wars (Fig. 2.5). Noteworthy intrusions of saline water have been numerous, but they have been distributed rather irregularly. However, extraordinarily large saltwater intrusions into the Baltic Sea supported by storms have not been very numerous. The most important of them took place at the beginning of the 1920s and the 1950s. Larger inflows have occurred in groups, during successive years, e.g., in 1948–1952 (Matthäus 1993). Noteworthy is the small

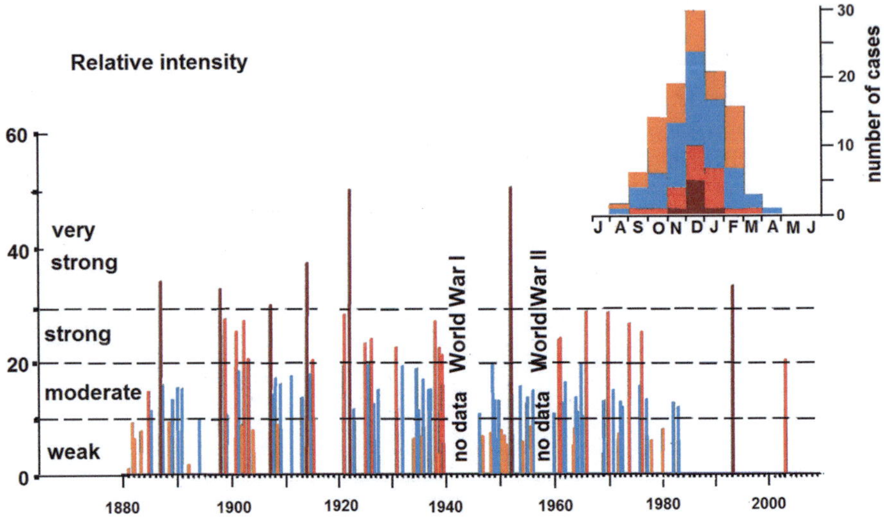

Fig. 2.5 Larger saline water inflows from the North Sea to the Baltic in the period 1880–2005. Above, the number of inflows from August to April by months (HELCOM 2007)

number of larger inflows since 1977. All large inflows have taken place in the period from late August to late April, most frequently from October to February (Fig. 2.5). Depending on the strength of the inflow and the salinity of the intruding water, the North Sea saline water needs 4–9 months to move from the Arkona Basin to the Gotland Deep. After a larger inflow and in the absence of new intrusions, the saline water remains in the deeps and its density decreases due to gradual mixing with less saline water. In such a case, the water movement is limited, and therefore such periods are called stagnation periods.

The straits connecting the Baltic Sea with the North Sea are of relatively limited depth and narrow; their bottom relief is very rough. The bottom of the Baltic Sea, already partly formed by the pre-glacial periods, is of variable depth (Fig. 2.6). As the connection between the North and Baltic seas through the Danish Sounds is narrow and shallow, in this area, the water flowing into the Baltic Sea is subjected to considerable changes. The surface water masses of low salinity leave the Baltic Sea mostly moving southwards in the western part of the open sea and later on, through the Transition Area. Saline water of higher density flows into the Baltic Sea mainly near the bottom. The volume of the salt water flowing into the Baltic Sea depends chiefly on the horizontal salinity gradient between the Baltic and North Seas as well as on the strength of the westerly winds. In the Transition Area, the inflowing saline water mixes with the more brackish outflow. This takes place mainly in the 10–15 m deep surface layer. These mixing and transportation processes are shown in Fig. 2.7, in a scheme composed by Steemann-Nielsen (1940). This scheme clarifies why the saline water flowing into the Baltic Sea is usually of much lower salinity than the North Sea water. However, if large North Sea water masses are rapidly pressed

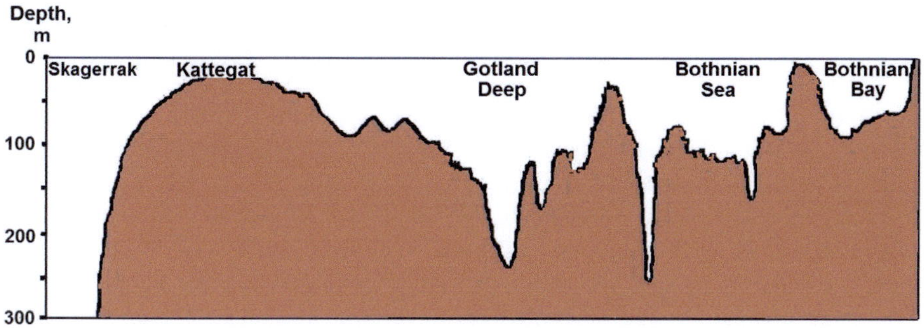

Fig. 2.6 Crosscut of the Baltic Sea depression (Ojaveer 2014)

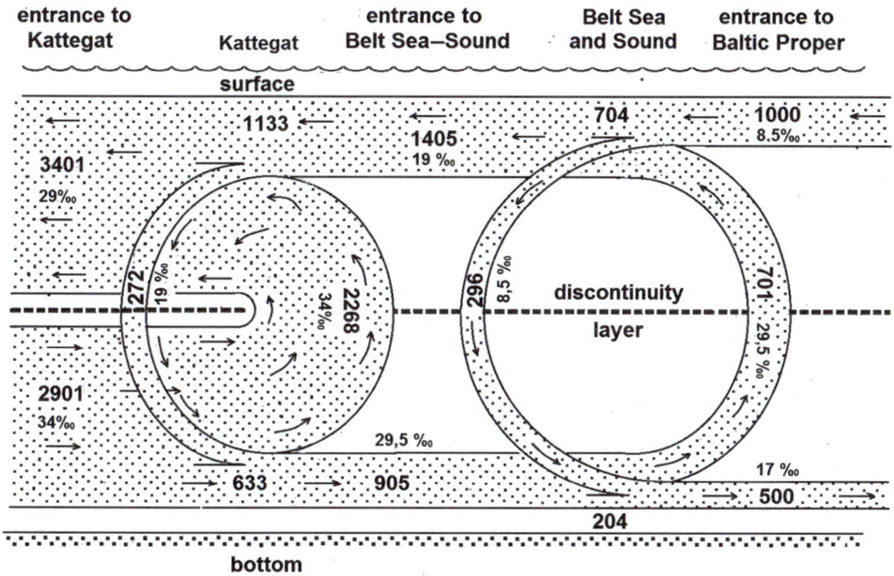

Fig. 2.7 Water exchange through Danish Sounds (From Ojaveer 2014, after Steemann-Nielsen 1940)

through the whole crosscut of the Transition Area by strong western winds, the salinity of the inflow can be higher than usual.

To the east and north of the Danish Belts, the bottom relief divides the sea into regions consisting of deeper basins separated by thresholds/sills. Every basin has its own size, bottom relief and salinity, which determine environmental conditions in the basins. Our data show that in the conditions of intensification of the flow of saline water from the Danish Belts to the Gotland Deep, in the eastern part of this deep, the upper limit of saline water can be up to 15 m higher, the salinity up to 1 psu and the oxygen concentration up to 3 cm dm^{-3} higher than above the bottom of the same depth in the western part of the deep. This shows that saline water first arrives in the deep areas situated at the eastern coast of the sea.

The water of higher salinity has higher density than fresher water and therefore moves to the north and east in the near-bottom layers of the sea. The fresher surface water of the Baltic Sea remains on top of the salty water. The water of intermediate salinity/density, which results from the mixing of saline bottom water and fresher surface water, is situated between the surface and bottom waters simultaneously separating them. In this intermediate layer of limited depth – the primary halocline – salinity increases relatively rapidly with depth. The primary halocline effectively separates saline deep water from the lower-density Baltic surface water in the area of its existence – in the Transition Area and the open part of the Baltic Sea. The depth of the location of the halocline in the Baltic Sea increases steadily with increasing distance from the Danish Sounds, reflecting the higher volume of deeper layers in the southwestern and southern parts of the sea and ever increasing domination of low-salinity water in the northern and eastern areas of the Baltic Sea. Consequently, the depth of the primary halocline is not uniform throughout the Baltic Sea. In the Bornholm Basin, it is found at the 40–50 m depth, but its depth increases northwards. In the Gotland and Northern deeps, the halocline occurs at 60–120 m (Fig. 2.8). However, the depth of the halocline and its vertical diameter can also have temporal variations reflecting changes in the river discharge and the volumes of saltwater influxes from the North Sea (Fig. 2.9). Usually, the depth of the primary halocline is least in the central areas of deeps and increases towards coasts. In addition to the primary halocline, a secondary halocline can also be present in the Baltic. It can originate between the inflowing saline water and the older, stagnant, high-salinity water, if the salinity of inflowing water is less than that of the older saltwater in deeps. It may develop at depths of 110–130 m.

In addition to the Transition Area and the Baltic Proper, the primary halocline is also present in the western Gulf of Finland. As there is no threshold between the open part of the Baltic and the Gulf of Finland, the bottom water of high salinity can penetrate into the gulf and the salinity in the demersal layers of the western part of this gulf can rise as high as 11 psu. The relatively unstable halocline in the western Gulf of Finland can, from time to time, become strongly eroded and the saline water can be mixed with the low salinity surface water. In the shallower eastern part of this gulf, no permanent halocline exists. As the impact of the fresh water discharged by the Neva increases, in the shallow eastern part of the gulf, the salinity continuously drops from area to area.

Between the Baltic Proper and the Bothnian Sea, a threshold exists. The water in the bottom layers of the Bothnian Sea originates mainly from the layers of 50–70 m depth in the Northern Baltic Proper. It flows across the threshold between the open part of the sea and the Bothnian Sea and forms the bottom water of the Bothnian Sea with a salinity of 6.5–7 psu, about 1–3 units higher than in the surface layers at the same place. In this gulf, stratification is weaker than in the open Baltic, and the halocline can be absent in certain periods (e.g., in winter). Water exchange between the Gulf of Bothnia and the open Baltic is comparatively intense. The northernmost part of the Gulf of Bothnia – Bothnian Bay – receives its deep water from surface

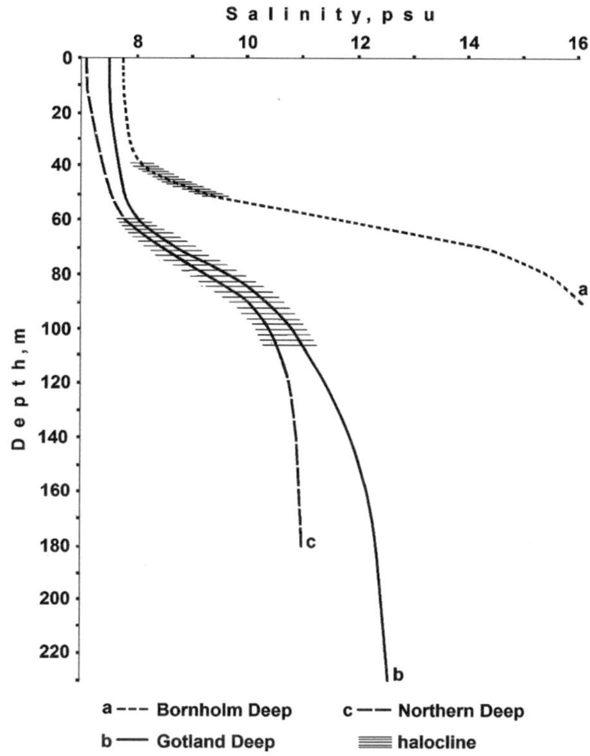

Fig. 2.8 Salinity of water layers and the average location of halocline in the Bornholm, Gotland and Northern Deeps in 1964–1990 (Ojaveer 2014)

layers of the Bothnian Sea via the 25-m deep Norra Kvarken. Therefore, the salinity in the bottom layers of Bothnian Bay is still lower (about 4.3 psu) than that in the Bothnian Sea, and the surface salinity slightly less than 3.5 psu (HELCOM 1996).

The Gulf of Riga receives its bottom water from the open sea through the Irbe Sound with a depth of 25 m and the even shallower Väinameri. After the influx of open sea water, the salinity of the bottom layers of this gulf (6–7 psu) can for some time rise 0.7–1.0 psu higher than the salinity of the surface layers. The water exchange between the open Baltic and the large gulfs is mainly based on the pressure of fresh water discharged by rivers and the impact of meteorological factors. In the near-coast peripheral areas of the gulfs of Bothnia, Finland and Riga, the water salinity can be as low as 2–3 psu or even less.

In terms of water stratification, considerable variety exists in the Baltic Sea. In some areas of this sea, stratification is rather stable and the environment is relatively uniform. However, such areas are usually separated by zones of both temporal and topical diversity of the environment. The comparatively rigid stratification of water in the open Baltic means a lower salinity in the surface water (6–8 psu) and higher salinity (mostly 10–14 psu or more) in the deeper water layers. The occurrence of the halocline of limited permeability has a very great impact on the life of organisms in the Baltic. The pelagic species are influenced mainly by environmental condi-

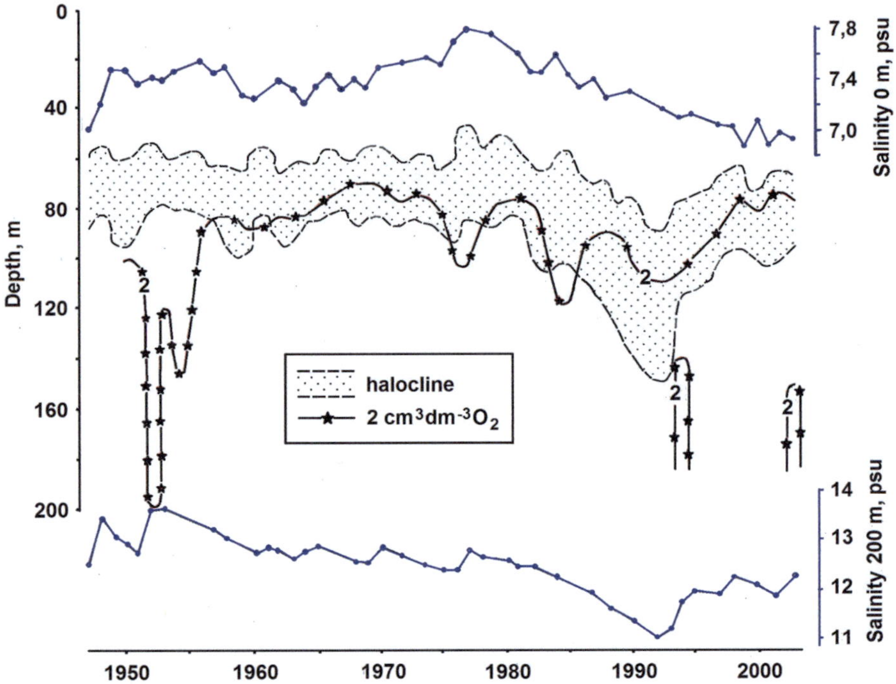

Fig. 2.9 Variation of depth of the upper and lower limits (the salinity 8.0 and 10.5 psu, correspondingly) of halocline and of the depth of the isoline of oxygen concentration 2 cm^3 dm^{-3} in the Gotland Deep during 1947–2002 (Ojaveer 2014)

tions in the upper water layers (above the primary halocline) where their main activity occurs. The properties of halocline and the life conditions in deep water layers below it have crucial importance for the demersal organisms and also for pelagic species in some seasons/periods of life. Especially substantial is the effect of the halocline on the formation of the thermic and oxygen conditions of water layers.

Due to the strong hindering influence of the halocline, the deeper water layers of higher salinity do not receive oxygen by way of mixing from the surface layers. The oxygen supply of the North Sea water at the inflow to the Baltic can only decrease – it goes for the satisfaction of the oxygen requirements of organisms, decomposition (mineralization) of organic matter, etc. Therefore, the water that has spent a certain amount of time in stagnant demersal layers loses oxygen, and in the case of an absence of new saltwater inflows, after depletion of oxygen reserves, hydrogen sulphide (H_2S) is formed. Oxygen deficiency does not allow for respiration of higher organisms; therefore, the situation means closure of the deep water for fish, zooplankton, zoobenthos and other organisms which need oxygen for life. The oxygen content at the halocline and in the layers below it is especially important for the reproduction of the organisms of marine background (which need higher salinity for multiplication than is in the water layers above the halocline). To enable the life of the organisms, the lower limit of the oxygen content in these water layers of higher salinity should be at least 1.5–2 $cm^3\ dm^{-3}$ or higher.

Also, generally, the increase in temperature of the surface layers during the warm period of the year, as well as its decrease in autumn, mainly concerns the layers above the halocline. The higher salinity and higher temperature (compared to the upper layers in winter) of the water layers at the halocline and immediately deeper are very important for the wintering, especially of the organisms of marine background, and most especially in severe winters in the northern areas. In this case, it is highly important whether the oxygen content renders organisms the use of the warmth at the halocline and in the upper layers of deep water below the halocline, for wintering.

Moreover, drastically complicating vertical exchange, the halocline very substantially aggravates the transport of the most important nutrient salts from deeper layers into the euphotic zone where they can be used for primary production. The nutrients can be linked to the food chain as far as the thermohaline stratification enables vertical circulation. Biological productivity is higher in such areas where the availability of nutrients from under the halocline is larger.

Consequently, the Baltic Sea environment with a number of water layers of different salinity, temperature and other conditions seriously complicates the distribution of life of organisms and their adaptation to the concrete habitats. Comparing the extent and the character of the temporal and spatial variations of the basic environmental parameters of the Baltic Sea with those in the neighbouring North Sea of normal persistent salinity (33–37 psu), one could see the following picture. In the North Sea, which was divided into eight different water types (Laevastu 1963), the overall spatial difference in salinity reaches 2.25 psu, and in the average temperature, 16 °C. In the Baltic Sea, the corresponding differences are much bigger, amounting to 33–34 psu and 22–23 °C, respectively.

The causes and consequences of temporal and spatial differences in salinity in the Baltic Sea areas and their relation to the climatic periods with different amounts of precipitation are illustrated in Fig. 2.10. For comparison of the salinity variations in the Bornholm, Gotland and Northern deeps during the years 1929–2002, the depth of the water layer of the salinity of 10 psu was taken as the indicator (the water layer with the salinity of 10 psu belongs to the halocline). During this period, the water layer with this salinity was situated in the Bornholm Deep at an average depth of 53.8 m, in the Gotland Deep, at 90.7 m, and in the Northern Deep, at 104.4 m. These differences indicate substantial discrepancy in hydrological and ecological conditions between these areas. In the Bornholm Deep, the temporal variability of the depth of the water layer of the salinity of 10 psu was comparatively small (Fig. 2.10). However, in the Gotland Deep and especially in the Northern Deep, the periods of relatively smaller variation of this depth alternated with sudden beginning and ending periods (the end of the 1920s and the early 1930s, from the early1980s to the late 1990s), when the water layer with a salinity of 10 psu sank to a depth of 120–130 m in the Gotland Deep and to 150–180 m in the Northern Deep. Consequently, in some periods of such salinity decrease, the salty water under the halocline disappeared and the processes requiring higher salinity could not occur in

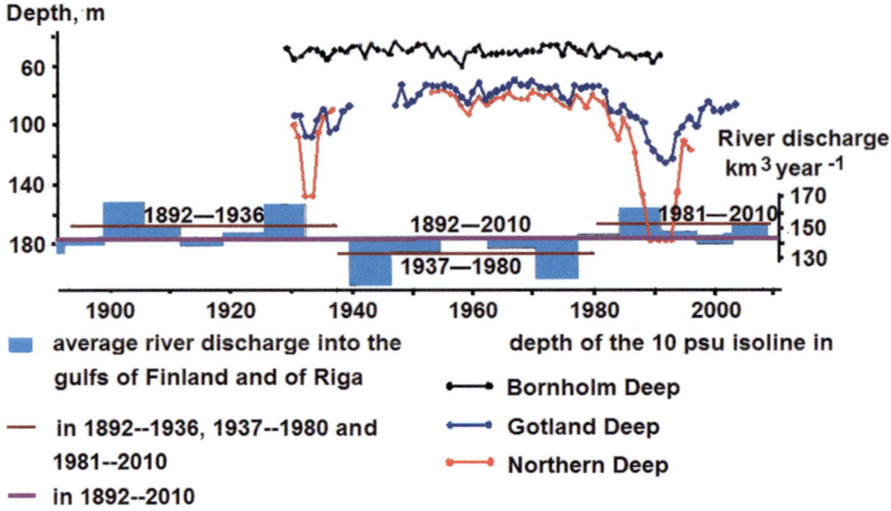

Fig. 2.10 Dependence of location of the 10 psu isohaline in the Bornholm, Gotland and Northern Deeps (Bulletin Hydrographique, 1921–1959) on the river discharge into the Gulfs of Finland and Riga during the period 1892–2000 (Ojaveer 2014)

the Northern Deep. Such periodical disappearance of saline bottom water has not been observed in the Bornholm or Gotland deeps. Therefore, concerning the living conditions of organisms, the area of the Northern Deep is principally different from the Bornholm and Gotland deeps because the presence of higher-salinity water below the halocline is not a continuous phenomenon there. However, in the Gotland area, the deep layer of higher salinity also thinned remarkably in the early 1930s and the late 1980s, limiting the area of possible distribution (and notably the possibilities for reproduction) of the organisms requiring higher salinity.

Changes in the freshwater discharge into the Northeast Baltic affect the thickness of the low-salinity water layer measured at the depth of the 10 psu salinity isoline (Fig. 2.10). For example, the increasing depth of the 10 psu isoline in the 1920s–1930s was preceded by an increase in the river discharge into the gulfs of Finland and Riga in the mid-1920s and early 1930s. Also, the increase in the freshwater inflow in the early 1980s caused a deepening of the 10 psu isoline. The water that pours into the Baltic Sea remains there until it either evaporates or is flushed out of the sea through the Danish Belts. The exchange of salty water through the Belts, and consequently the changes in the volume of the sea, depend very much on the atmospheric processes.

Among the processes forming the environment of organisms in the Baltic, the mixing processes of water of different salinity or different origin have an outstanding position. Moving constantly northwards and eastwards, the water masses of higher salinity (especially their coastward edge) are mixed with the water of lower salinity in surface layers by the internal waves, upwellings and other processes. In the coastal zone, mixing is most intense in the higher water layers. Mixing of the saline bottom water and the fresher surface water is most intense in autumn and at the beginning of winter under the conditions of homothermia in the upper water layers when the absence of the thermocline in the upper homohaline layer creates favourable conditions for mixing. In this process, the most important role belongs to currents and the vertical mixing (deepwater ascent or upwelling).

2.2.1 The Role of Currents

The high-salinity water flowing into the Baltic Sea along the bottom layers of the eastern part of the deep area does not run to the north/northeast directly. The flow is mainly determined by the combined effect of the seabed topography, local differences in density between the inflowing saline and the less saline Baltic waters, the Coriolis force, etc. When meeting a deep, the high-salinity water changes its original direction, participates in the cyclonic rotation and, after filling the deep, flows over the threshold into the next deep.

The cyclonic direction of currents occurring below the primary halocline also persists above the deeps in higher water layers. The direction of deep-water currents does not depend on the wind direction (Fig. 2.11). The strength of the cyclonic deep-water currents is variable and probably depends on the volume of saltwater inflow into the Baltic, as well as on the location of the deep in the sea. In the areas of deeps, the cyclonic current vortices create relatively homogeneous environmental conditions, resulting in large regions of conventionally homogeneous environments forming natural regional ecological systems. Between the cyclonic density-dependent currents (also involving surface layers above deeps) independent of winds and the wind-driven currents varying in their direction and velocity (Fig. 2.11) on the nearby shallow areas, zones of divergence above slopes of deeps exist (Fig. 2.12). Continuous renewal of these border zones between water masses of different properties is a basic precondition for the development and persistence of the limits of large, almost homogeneous habitats (regions) and ecological systems. As in every sequent deep from the southwest to the northeast, salinity is lower than that of the previous one, in the area of every deep in which a conventionally homogeneous water environment of different salinity exists. It should be mentioned that the above system of deep-water currents composed by direct observations is basically similar to the system of currents derived through model calculations (Elken and Matthäus 2008).

Water movements regulate the formation of sediments. In the system of circular currents, sediments accumulate in the deeper parts of the sea separated by thresholds. In addition to such sediment accumulations formed in deeps, phosphates exit from the sediments into water during the stagnation periods when the pH decreases approaching 7.0 and other necessary conditions have evolved. Certain parts of the biogenes gathered in the composition of sediments in deeps originate from the compounds of human activity. Consequently, in the area of every density-dependent current, conditions exist for formation of a local stock of nutrients (biogenes) and

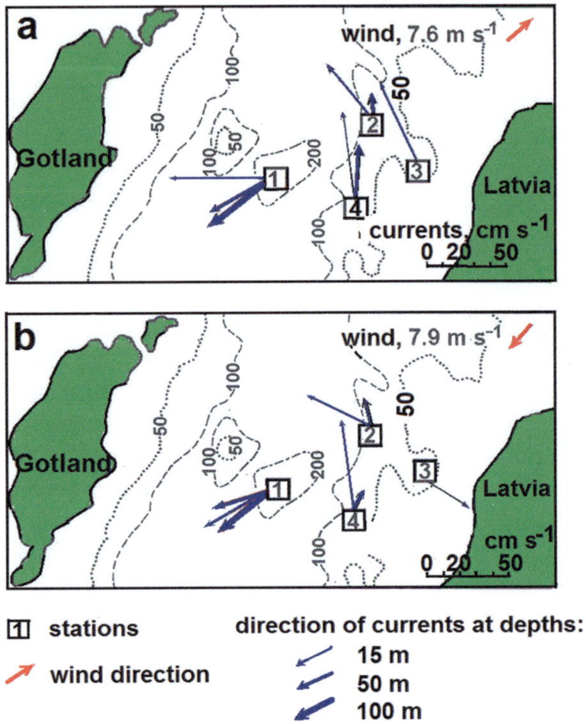

Fig. 2.11 Synchronous cyclonic currents in various depths at different wind directions at 11.00 on May 15, 1968 (**a**) and at 20.00 on May 17, 1968 (**b**), (Ojaveer 2014)

their transport to the organisms of that region (Fig. 2.13). This fact has been very important in the formation and stabilization of the Baltic ecosystems.

In addition to deep-water currents, wind-driven currents are of substantial importance in mixing and transporting surface layers. Both in the Baltic Proper and large gulfs, surface currents are generally also cyclonic (Fig. 2.14).

2.2.2 *Vertical Mixing of Water Layers*

Vertical mixing of water layers plays a very important role in the formation of salinity, temperature and oxygen content of water layers, and also in the creation of prerequisites for the development of the magnitude, content and other parameters of biological productivity. Vertical mixing moves nutrient-rich bottom water into the photic zone where nutrients are used for primary production. In the Baltic Sea, vertical mixing is directly boosted by wind energy and hindered by the stratification of water layers. Stratification of the sea, i.e., steady occurrence of the halocline and a seasonal thermocline, hinders vertical mixing of water layers and enrichment of the euphotic layer with biogenes. Vertical stratification is most stable in the warm period with both a thermocline and a halocline present. From autumn to spring, the

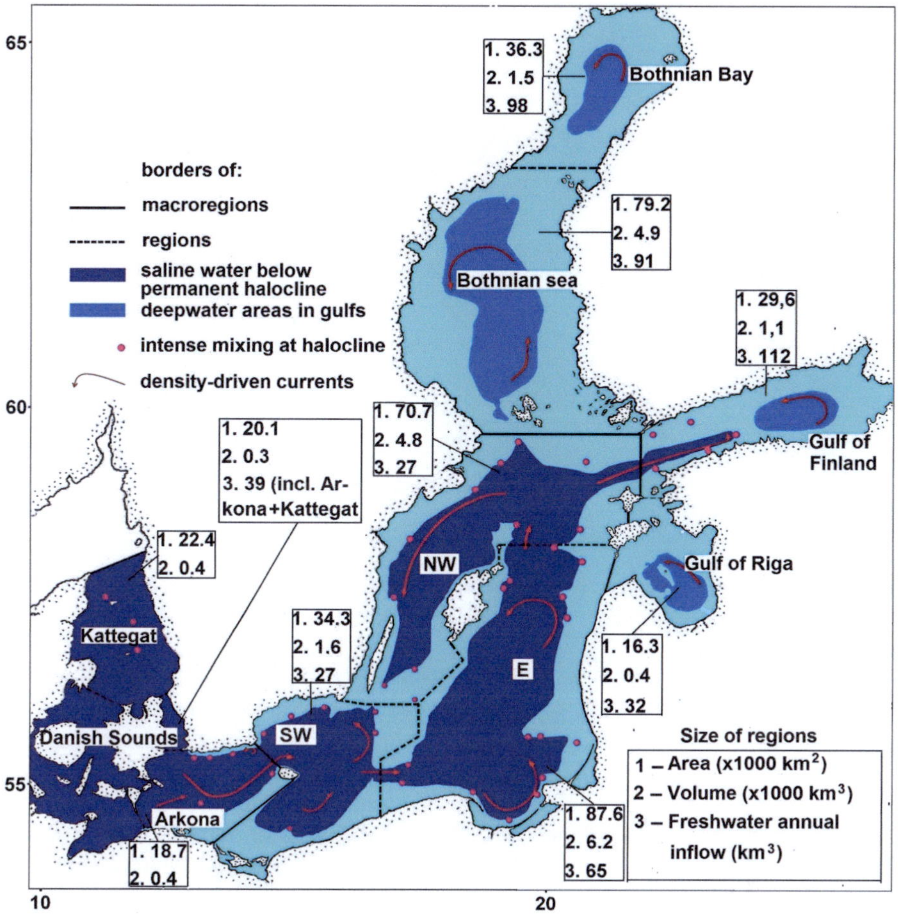

Fig. 2.12 System of density-dependent currents in the Baltic Sea (Ojaveer 2014)

thermocline is absent. Therefore, in the main area of the Baltic Sea, a large-scale enrichment of the euphotic layer with nutrients from the nutrient deposits under the halocline takes place mainly with intense convective vertical mixing during the autumn–winter homothermium.

However, all-year-round zones with intense vertical mixing can develop in the coastal zone, in particular, around banks in deep areas and at the edges of deeps on the coastal slope where the halocline and the thermocline, the latter in the warm period, are present. Such zones of deep-water ascents (upwellings) were studied in the Baltic Sea (in the Gotland Basin) in 1968 by Ojaveer and Kalejs (1974) and in Swedish coastal waters (Svansson 1973). Later on, they were very extensively followed, as they play a very important role in the mixing of water layers in the coastal zone and in the transportation of nutrients from the deep-water deposits to the neighbouring water layers (including to the euphotic zone) year round (Myrberg

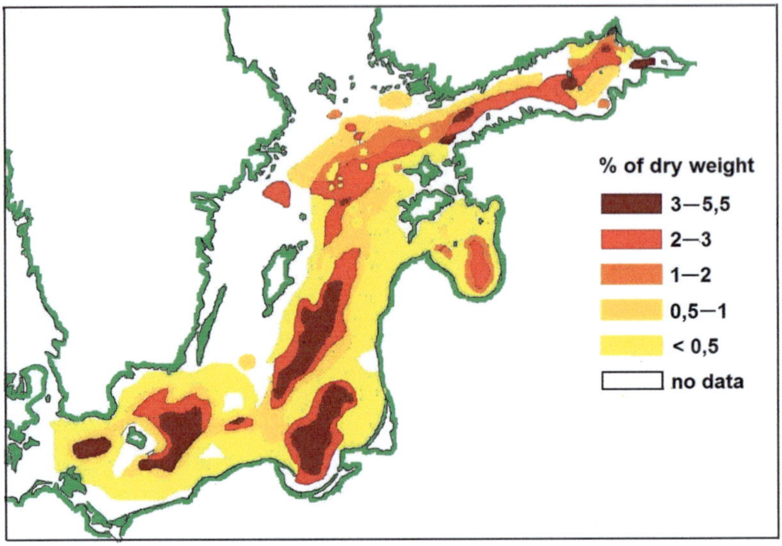

Fig. 2.13 Content of organic carbon in the upper layer of bottom sediments in the Baltic Sea (Järvekülg 1979)

and Andrejev 2003; Kowalewski and Ostrowski 2005; etc.). The intensity of the deep-water upwellings and downwellings under the influence of wind-induced processes in the coastal boundary zone, chiefly in the zone of divergence around the intersection of the halocline and thermocline with the bottom profile, is very variable. In such localities, deep-water layers mix with higher layers (Fig. 2.15) and, as a result, a certain amount of nutrients is washed from their deposits away into the nearby water layers. Therefore, in such zones (which could be called high-energy zones), the biological productivity is clearly higher and plankton and benthos biomasses are substantially larger than in the surrounding areas. This attracts feeding fish shoals and creates a basis for rich fish assemblages in the vicinity of coastal slopes of deeps. Such zones play a very important role in the integration of the regions.

The mixing zones inducing higher productivity occur at the halocline and thermocline and also in the comparatively narrow coastal zone. Such mixing areas of water layers involving considerable changes in salinity can also be found as hydrological fronts facing river estuaries, sounds and other near-coast turbulent areas (Fig. 2.16).

The above mixing processes are very important in the formation of the productivity of sea areas. They also contribute to the integration of regional systems. The size of energetic bases deposited as sediments in every region in accordance with the character of currents is an important indicator of the isolation and stability of regional ecosystems.

Fig. 2.14 Currents in the surface layer of the Baltic Sea (Ojaveer 2014)

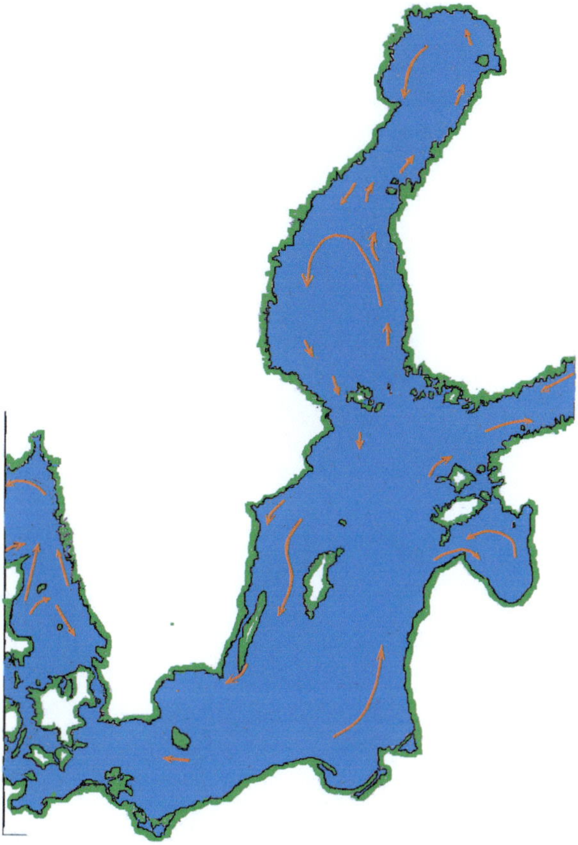

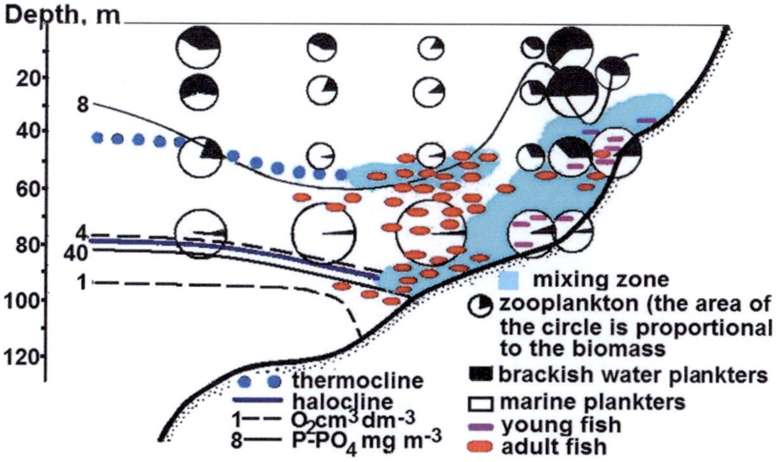

Fig. 2.15 Hydrographic and chemical parameters and distribution of zooplankton and fishes on the eastern slope of the Gotland Basin in autumn 1968 (Ojaveer 2014)

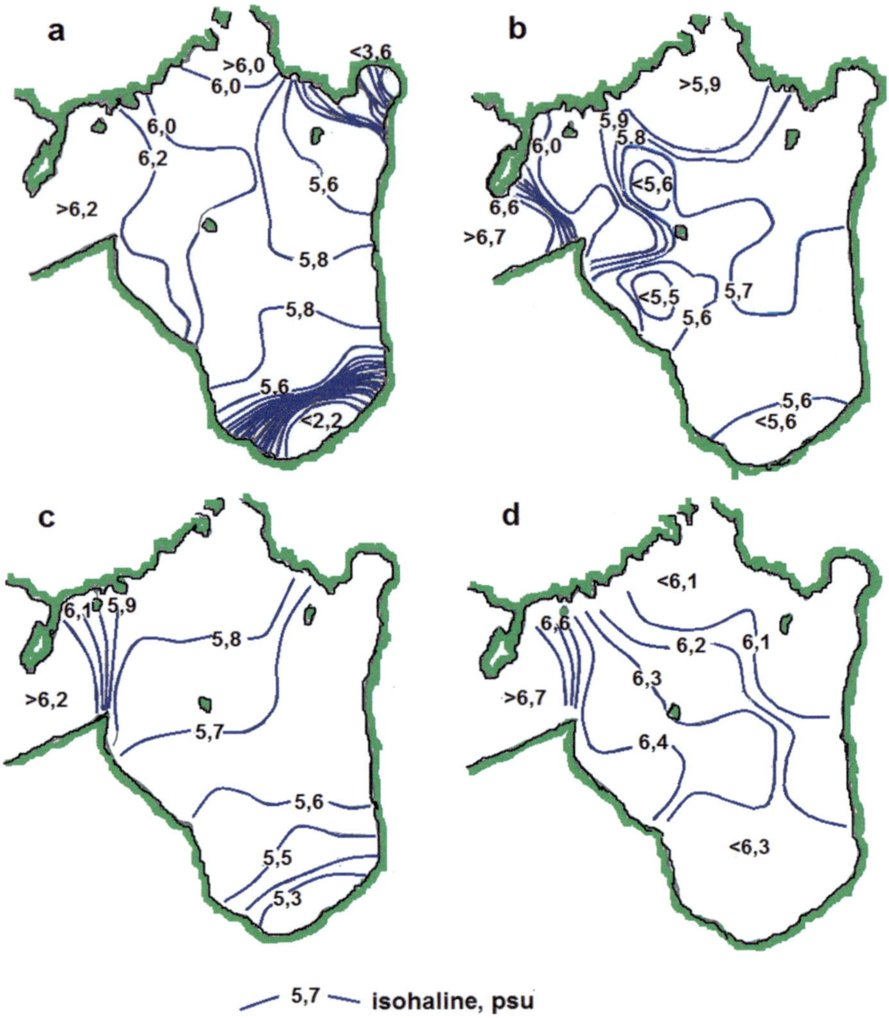

Fig. 2.16 Distribution of salinity in the surface layers of the Gulf of Riga (**a**) – 5–11. May 1976, hydrological fronts at the approaches of estuaries of the Daugava and Pärnu Rivers; (**b**) – 15–17. August 1979, hydrological front against the Irbe Sound; (**c**) – average salinity in surface and (**d**) in *bottom* layers in July–August in the period 1971–1993 (Berzinsh 1995)

2.3 Water Temperature

The Baltic Sea stretches in the north–south direction for more than 15 longitudinal degrees, climatic conditions varying throughout the sea. In summer, the highest surface temperatures range from 13 to 23 °C in the Kattegat, and from 13 to 22 °C in the Arkona region and in the SW part of the open sea, but only from 9 to 18 °C in Bothnian Bay. The most noticeable differences between the living conditions of

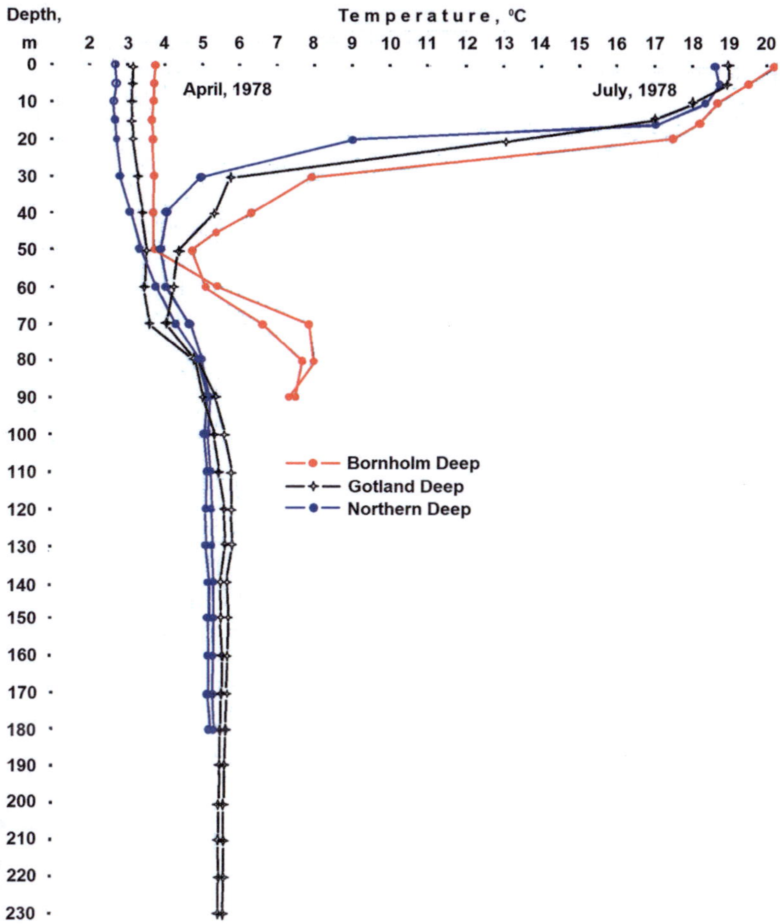

Fig. 2.17 Distribution of water temperature in the Bornholm, Gotland and Northern Deeps in spring (April) and summer (July) 1978 (Ojaveer 2014)

organisms in various parts of the sea can be stated in winter. Then, the circumstances in the northern gulfs situated partially in the realm of arctic climate or in its neighbourhood are much more severe than those in the southern and central parts of the sea belonging to the temperate zone and enjoying the proximity of large mild marine areas. Usually, some part of the sea is covered by ice in winter. This substantially diversifies environmental conditions in this sea. Primary production and the length of the growing period decrease from south to north, and regional variations are considerable.

The annual course of the average temperature of the water column in the Bornholm, Gotland and Northern deeps in July and April (the months with the highest and lowest mean temperatures, respectively) is shown in Fig. 2.17. It can be seen that the temperature of deeper water layers (below the halocline) does not undergo large seasonal fluctuations characteristic of the layers above the halocline. Therefore, their temperature in summer is usually lower, but in winter, it is higher than the temperature of the layers above the halocline. From summer months up to March–April, the water layers above the halocline grow cooler. Due to the falling air temperature, wind activity, mixing of water layers and the vanishing of the thermocline, in autumn and at the beginning of winter, the water from the surface to the halocline acquires a more or less homogeneous temperature. Further cooling in winter may cause the appearance of the so-called winter thermocline. On the surface, the temperature can be at the freezing point, below 0 °C. Then, the formation of ice begins. The water between the winter thermocline and halocline is more or less homothermic.

The warming of surface layers of the sea under the impact of solar energy leads to the origination of a thermocline, usually in April. The thermocline is formed in the open part of the sea at a depth of 10–20 m, and nearer to the surface in gulfs. The water temperature above the thermocline is nearly homogeneous. Cold winter water remains between the thermocline and the halocline; during the summer, it gradually gets somewhat warmer due to the mixing caused by macro- and microprocesses. Above deep areas of the open sea, the thermocline is generally strong and almost excludes vertical mixing. In the period of the existence of the halocline and thermocline, the mixing of surface waters with more saline deep layers, the rise of nutrients into the photic zone of the water column and other mixing processes are almost generally excluded. From April to August, the quasihomogeneous upper water layer heats up. This is followed by a gradual cooling of the same water up to the next winter. The main water masses of the open Baltic (the water from the surface to the primary halocline) show seasonal temperature fluctuations – they cool from autumn to March. Therefore, the water masses below the halocline, with their relative warmth, are of particular importance for the existence of the Baltic biota, especially for the thermophilic organisms of marine origin. In the period of the cooling of water in autumn, which starts from the coastal zone, organisms from shallower areas leave for larger deeps and warmer waters. In severe winters, they travel as close to the halocline as the oxygen content of these water layers allows.

Temperature conditions in the Bornholm Deep are different from those in other deeps in the open Baltic, because of their proximity to the warm waters of the Transition Area (the Kattegat and the Danish Belts) (Fig. 2.17). Therefore, the Bornholm Basin hosts a number of marine species that are not able to dwell permanently in the areas north of this basin. Here and in the western Baltic, the primary spring production starts in March–April, moving northward step by step thereafter. Because of the higher salinity and the warmer environment in this area, organisms grow faster than elsewhere in the Baltic.

In large gulfs, the temperature of surface water depends mainly on the air temperature (Figs. 2.18 and 2.19). In gulfs and marine areas without a halocline, and thus in the absence of saline and relatively warm water under the halocline (e.g., in the Gulf of Riga), the seasonal temperature pattern differs considerably from that in deep areas of the open Baltic. In the Gulf of Riga, the upper water layers start warming in March–April, after the disappearance of ice. The thermocline forms rather

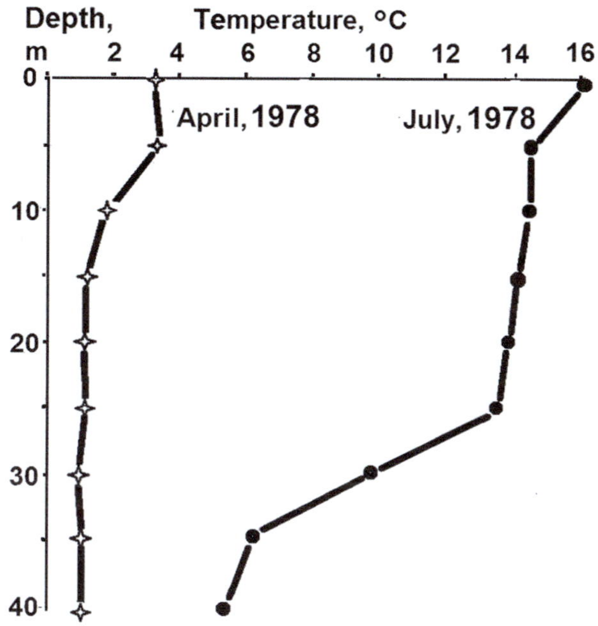

Fig. 2.18 Water temperature in the central Gulf of Riga in spring (April) and summer (July) 1978 (Ojaveer 2014)

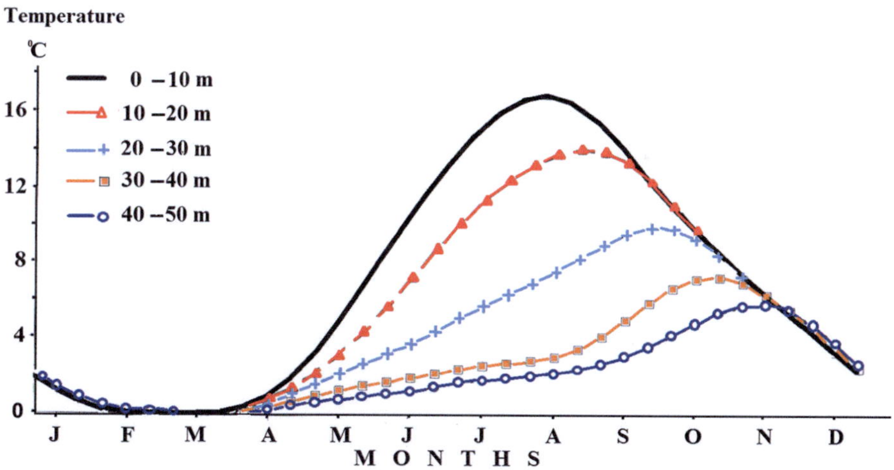

Fig. 2.19 Average annual course of water temperature in the Gulf of Riga during the years 1963–1985 (Berzinsh 1995)

close to the surface. The depth of the thermocline and its sharpness depend on weather conditions. Winds and micro-processes favour mixing of water layers, the increase in water temperature and deepening of the thermocline. In the absence of strong winds, currents and other movements that can mix near-bottom waters, the temperature in the bottom layers remains low until the autumn gales.

The character of winter in the Baltic area depends on the intensity of the inflow of warm air masses from the Atlantic Ocean (the situation inducing mild or temperate winters) or the intrusion of cold arctic air from the NE (leading to cold or severe winters). Depending on the climatic situation, the whole Baltic Sea, or only some part of it, can be covered with ice. In very mild winters, ice occurs only in the northern part of Bothnian Bay, in the eastern Gulf of Finland and in some straits of the NE Baltic. In moderate winters, ice usually covers the gulfs of Bothnia, Finland and Riga, the near-coast areas of the northern Baltic Proper and waters of archipelagos, but the southwestern, southern, central and open parts of the northern Baltic are nearly ice-free (Fig. 2.20). In very severe winters, the entire Baltic becomes covered with ice, and in extremely severe winters, part of the Skagerrak is similarly covered. Of the most recent 300 winters, 40 could be classified as very severe and 16 as extremely severe. In the large gulfs, the thickness of ice has been up to 1.5 m. In some winters, ice (pack ice) heaps/accumulations at the NE coasts of the sea have been as high as 20 m. The duration of the ice periods and the thickness of ice differ considerably between the regions of the sea. This is caused both by differences in the air temperature in the northern and southern areas and the hydrological conditions: the upper layers in the northern Baltic are influenced by the fresh and cold water flowing in from the gulfs of Bothnia, Finland and Riga, while those in the southern areas are under the impact of warmer deep waters.

Our data on winter severity in the northern part of the Baltic Sea and the gulfs of Finland and Riga reach back to the severe winter of 1363. Our estimates on winter conditions in earlier centuries are mainly based on data of ice formation and ice melting in the Stockholm, Riga and Tallinn ports. Since the middle of the eighteenth century, regular daily air temperature measurements, and since the second half of the nineteenth century, regular surveys on ice conditions have been conducted on the Baltic coasts. Figure 2.21, based on the estimates on the range of ice cover (averaged over decades), shows that several long periods of winter severity have occurred in this area since the fourteenth century. It is probable that the data refer to the externally forced climate variability in the Baltic Sea area, which is most likely due to changes in the impact of solar irradiance and other extraterrestrial factors.

It is obvious that in the period from the late sixteenth to the late nineteenth centuries, generally severe winters dominated, except for a short milder period in the first half of the seventeenth century. Beginning with the late nineteenth century, winters in the Northern Baltic became milder. However, extraordinarily cold winters have occurred not only during the small ice age when the number of very severe winters was high (in 1684, 1740, 1789, 1809, 1830, etc.), but also more recently, when the climate became milder (e.g., in 1893, 1940, 1942, 1947, 1987, etc.).

Large changes in winter severity have occurred in the past and will probably occur again in the future, presumably as a result of variation in the solar energy influx. It has been noticed that since the last decade of the nineteenth century, 14–16-year periods of on average mild winters have regularly been followed by a colder period spanning 6–8 winters (Ojaveer 2014).

The Baltic Sea is populated mainly by such species as are adapted to rather low temperature during their entire span of life. However, some species still have

Fig. 2.20 The average extent of ice cover in mild, medium and severe winter (Ojaveer 2014)

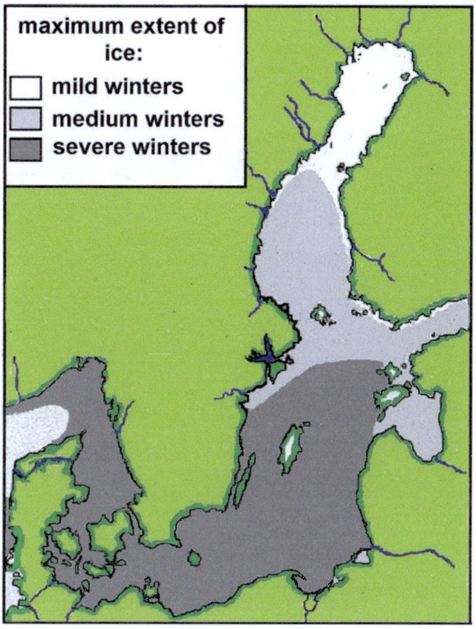

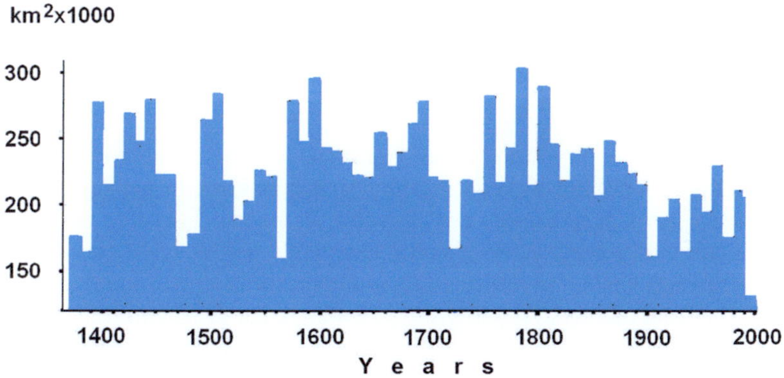

Fig. 2.21 The area of average ice cover in the Northern Baltic and the gulfs of Riga and Finland by decades beginning with the year 1363 (Ojaveer 2014)

problems with feeding or reproduction at low temperatures. Such species move into warmer water for the winter (to the central part of deeps or into deeper layers, closer to the halocline). As in deeper layers at the halocline, the oxygen content may be too low to enable normal respiration, and thus selection of a wintering area in severe winters requires thermophilic organisms to make a compromise between respiration conditions and the temperature of the environment (Fig. 2.22).

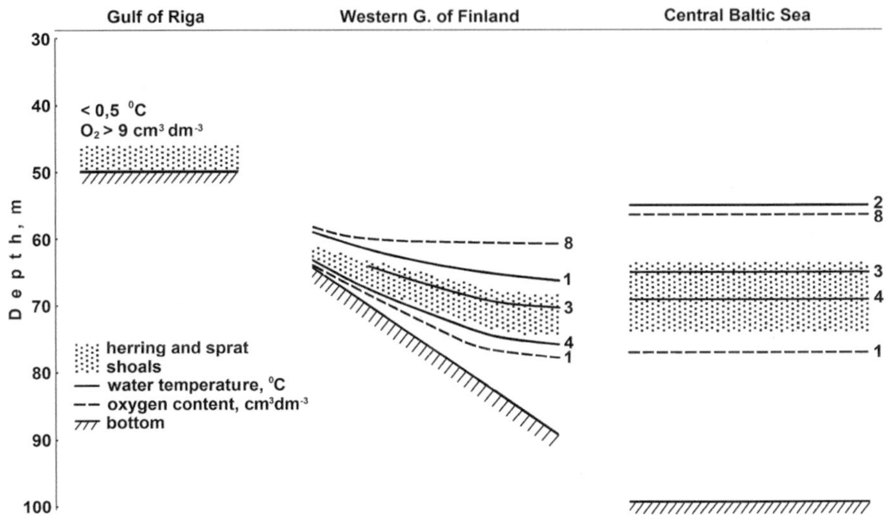

Fig. 2.22 Location of pelagic fish shoals in the Gulf of Riga, the western Gulf of Finland and the central Baltic Sea in the wintering period in relation to the water temperature and oxygen content (Ojaveer 2014)

2.4 Oxygen Conditions

Higher organisms require that the oxygen content in water be above a certain threshold value. Gametes, embryos and adult specimens as well cannot normally develop under conditions of sharp oxygen deficiency. Adaptations of herring, sprat, cod, flounder and other fish species do not allow for their normal development and existence in water with oxygen concentration constantly below 1.5–2 cm^3 dm^{-3}. For a number of invertebrates, the limiting oxygen concentration is somewhat lower (up to 1 cm^3 dm^{-3}, Järvekülg 1979). Figure 2.22 shows the distribution of pelagic fish shoals in the open part of the Baltic Sea, the Gulf of Finland and the Gulf of Riga in the wintering period and oxygen concentrations in the water layers at corresponding stations. The wintering shoals have moved as close to the warmer bottom water as the oxygen concentration allows. In the Gulf of Riga, with its comparatively high oxygen concentration from the surface to the bottom, the shoals are closely pressed to the bottom.

In the Baltic Sea, the occasions of oxygen deficiency in large water masses have been found mainly in deep layers. As oxygen-rich saline water flows into the Baltic through the Danish Sounds, and owing to the fact that in the Baltic Sea, oxygen cannot penetrate through the halocline into the deeper water layers, the deeper layers are oxygen-rich mainly in the Southwest and South Baltic. The saline water moving through the deeps in the Baltic loses its oxygen generally before reaching the Northern Deep. Consequently, the oxygen concentration in the saline water below the halocline depends on the location of the area in the Baltic. The average depth of the layer with an oxygen content of 2 cm^3 dm^{-3} varies in the Bornholm Deep

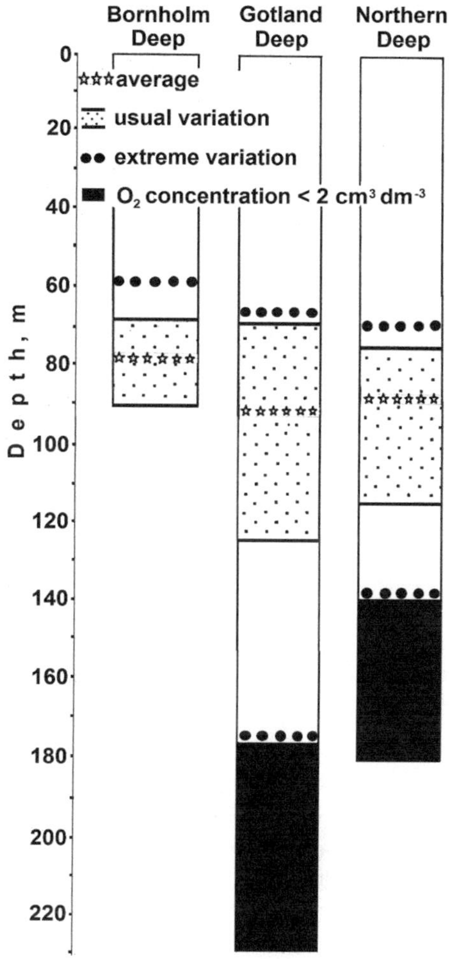

Fig. 2.23 Dynamics of location of the water layer with the oxygen content 2 cm^3 dm^{-3} in the Bornholm, Gotland and Northern Deeps during the period 1952–1996 (Ojaveer 2014)

between 68 and 90 m, but the oxygen concentration in the deeper water masses of the Gotland and Northern deeps is much lower (Fig. 2.23). The bottom of these deeps is covered with a 40–50-m thick water layer where the oxygen concentration is low or oxygen is absent.

2.5 Light Conditions

In all basins, the underwater light field as an influential component of the natural environment exerts a deciding influence on the rate of photosynthesis, distribution, behaviour and living conditions of organisms. It is known that the life in the sea (also in the brackish Baltic Sea) depends on the regular and irregular fluctuations of irradiation from various sources (seasonal and diurnal migrations of both invertebrates and vertebrates, adaptations to certain patterns of irradiation, etc.).

The intensity of solar radiation at the surface of the sea varies with latitude, season of the year, time of the day, cloudiness, winds (condition of the sea surface), etc. The surface irradiance of a basin under a cloudless sky is usually the maximum possible irradiance of such a surface at a given place and time. It can happen that under a partially cloudy sky, during short time intervals, irradiances from a few to even over 30 per cent higher are recorded, owing to the reflection of solar rays from clouds. With an overcast or partly overcast sky, the radiation is reduced compared with that on days with a clear sky. Due to the impact of the wind-driven clouds (of various structure and optical properties) and other weather conditions, the irradiation strongly fluctuates (Dera 1995). The determination of the irradiance transmittance through a cloudy atmosphere is complicated. Transmittance of solar irradiance into the water body depends on the absorbtion properties of water in natural basins. Near the very surface, we have a wavelength distribution such as that found on land. Below the surface, and especially in the lower part of the photic zone, considerably varied conditions concerning wavelength distribution occur. Both the depth and the optical type of the water are important in attenuation of light of various wavelengths. Red light penetrates rather badly into the water. In the Eastern Mediterranian, only 10% of red light (wavelength 675 nm) of the surface radiation was found at a depth of 5 m, for green light (550 nm), 10% of surface radiation was measured at 35 m, and for blue light (475 nm), at 82 m (Steemann Nielsen 1975). Absorption by the water molecules (chiefly concerning red light) and scattering of the short wavelength part due to particles (mud particles, dead and living organisms, humic matter, etc.) are the main acting factors (Trei 1991). Therefore, the spectra of the daylight irradiance in the clear oligotrophic Sargasso Sea are markedly different from those in the eutrophic waters of the Baltic (Fig. 2.24). Light curves of photosynthesis and the assimilation numbers noteworthily differ by season (Dera 1995).

The rates of photosynthesis sufficient to compensate for the rates of respiration both during the day and night in the tropics and subtropics and also during summer at high latitudes take place down to a depth where approximately 1% of the surface light is found. The lower limit of the photic zone is also called the compensation depth. Naturally, photosynthesis also takes place below the lower limit of the photic zone (Steemann Nielsen 1975).

It has been found that the lower boundary of the euphotic layer is situated at a depth 2.5 times the Secchi disc transparency. The Secchi disc readings are, naturally, very approximate, but they are comparable under similar conditions (in waters of more or less the same type).

2.6 Natural Regional System of the Baltic Sea

In recent millennia, the Baltic Sea has passed several developmental stages during which the conditions in the sea have substantially changed. The present-day Baltic Sea is not at all a homogeneous ecosystem. It is composed of a number of natural

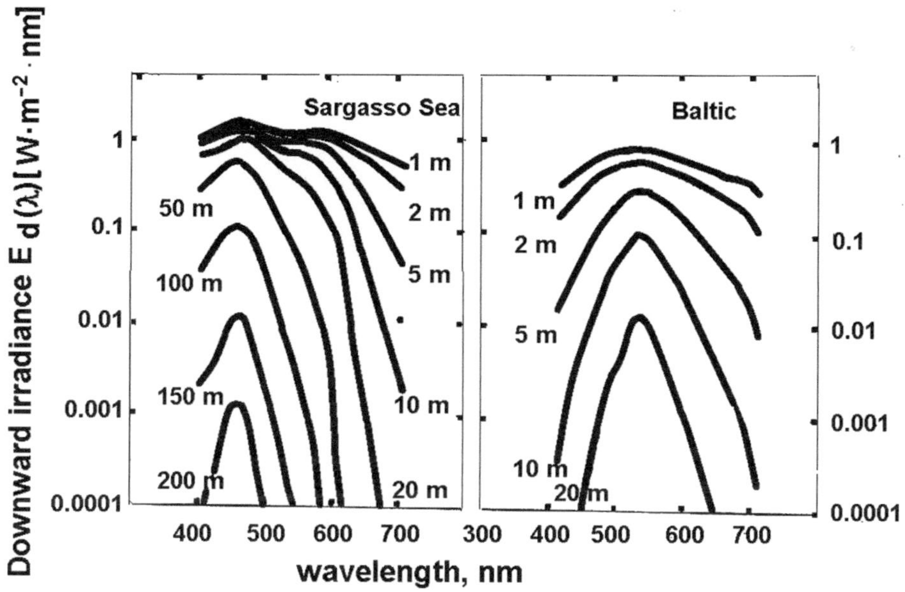

Fig. 2.24 Spectra of the daylight irradiation in the oligotrophic Sargasso and eutrophic Baltic Seas (Dera 1995)

regions continuously and substantially differing in environmental conditions. It is evident that this development in the ecosystems of the sea has not concluded, and changes in the subsystems will likely also be seen in the future. Differences in ecosystems and organisms between certain parts of the Baltic Sea (Transition Area, the Gulfs of Riga, Finland and Bothnia, etc.) have been generally known for rather long time, and every now and then, they have been treated separately (ICES 1932; HELCOM 1986, 1990; etc.). Separate treatment of clearly differing regions acquired special practical importance with the beginning of the international collaboration in the assessment of fish stocks in the Baltic Sea for improving their management.

As the estimation of parameters for the assessment of commercial fish stocks was begun by the Working Groups of the International Council for the Exploration of the Sea (ICES), it was taken as natural to apply the regions and Subdivisions created by this organisation in the Northern Atlantic and the Baltic Sea as units. However, the units created by the ICES in the 1930s were chiefly meant for the collection of fish catch statistics, and for this purpose, the system was suitable (Fig. 2.25). Nowadays, creation of units for the development of the population-based assessment and management of fish resources requires consideration of the natural boundaries of the exploited units. Therefore, macroregions and regions of the sea meant for the assessment and management of the ecosystems and living resources should be differentiated based on the natural boundaries of the stock units.

In the Baltic Sea, the number of groups of organisms at the typological level is comparatively small (i.e., at the level of typological species, the diversity is rather low), but their number at the applied level (the number of groups at the realistically important population level) is considerably higher. Separate populations of a

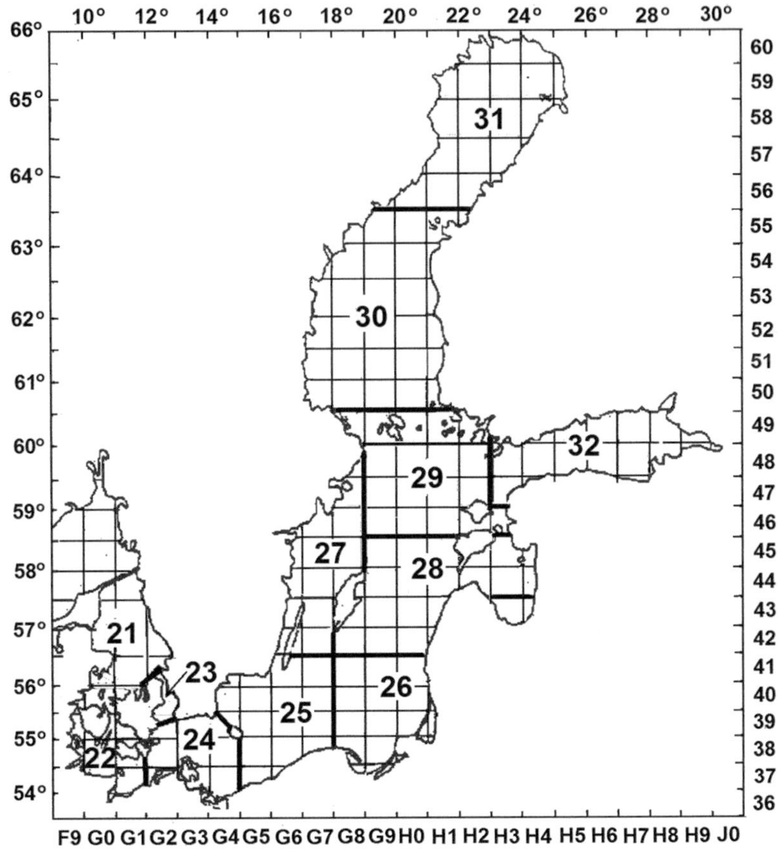

Fig. 2.25 The map of ICES Subdivisions and statistical squares (ICES 2006)

species, including exploited fish species, may develop generations differing significantly in their relative abundance in various environments/areas in the same year. Stock assessment methods presently in use do not allow for correct treatment of the data consisting of a mixture of units with different stock dynamics. Therefore, populations differing significantly in the relative abundance of the same year class – the populations with different stock dynamics (e.g., populations of certain species dwelling in different environmental conditions) – cannot be correctly assessed together, but should rather be considered separately.

The differentiation of macro-regions and regions referred to below is chiefly based on the configuration of the sea, its bottom relief, the regime of inflows of saline and fresh water, the system of density-dependent currents, the character of vertical mixing, chemical differences between the areas and the location of nutrient deposits (cf. Wulff et al. 2000; Figs. 2.12 and 2.13). These units are also related to the occurrence of the units of biota (e.g., populations of herring, sprat or other fish, invertebrates, etc.) adapted to the conditions of a certain sea area and forming functional units with the environment.

Below, the regions or ecosystems are considered on two scales: the macroscale (macro-regions) and the regions and subregions inside the macro-regions (it can be foreseen that the number of subsystems will increase in the future).

2.6.1 *Macro-regions*

Macro-regions differ significantly from one another in climatic, geomorphological, hydrographic, chemical and biological parameters (Table 2.1). Three macro-regions can be differentiated in the Baltic Sea: (1) the Transition Area between the Baltic and North Seas; (2) the open part of the Baltic Sea; (3) large gulfs – the Gulf of Bothnia, the Gulf of Finland and the Gulf of Riga. In each macro-region, smaller units – regions and subregions – can be delineated.

The Transition Area is a large border area between the North Sea of oceanic character and the brackish Baltic Sea. The climate is of the Atlantic type and a continuous halocline exists in the area. Large variations in the dynamics of the water layers and in their vertical and horizontal stratification can occur. Generally, the whole water column from the surface to the bottom is populated by higher forms of life. However, the oxygen concentration in different water layers can substantially vary, and at some places, the bottom layers show short-term oxygen deficiency. From the Kattegat to the Arkona Basin, the ecosystems significantly differ from area to area. Their appearance changes: the predominantly marine characters in the Kattegat are more and more replaced by those of brackish-water.

The open part of the Baltic Sea is the most homogeneous and stable macro-region in the Baltic. The region is under the influence of the Atlantic and also of continental climate systems. A continuous halocline divides the water masses into

Table 2.1 General characteristics of macro-regions, relative importance (%%) of their area, volume and freshwater input

	Transition area	Open Baltic	Large gulfs
Area	15	46	39
Volume	5	58	37
Freshwater input	8	23	69
Surface salinity	7,5–34,5	5,5–8,5	2,5–6,5
Bottom salinity	10,0–35,5	9,0–22,0	3,5–11,0
Peak of prim. prod. (month)	I–IV	IV–VI	VI–VII
Domin. climate type	Marine	Marine/continental	Continantal
Permanent halocline	+	+	–[a]
Domin. biological system	Modified marine	Modified marine/ brackish	Brackish/ modified freshwater

[a]Halocline occurs in limited area of the western Gulf of Finland

the upper and deeper layers, with lower and higher salinity, respectively. In the case of deep-water stagnation causing oxygen deficiency in large water masses, deeper layers can be impenetrable for higher life forms.

The bottom relief with deeps and thresholds between them, as well as separated deposits of biogenes in deeps, accentuates regional differences. The open part of the Baltic Sea does not constitute a single uniform system. Large homogeneous areas well mixed by currents exist in the open part of the Baltic Sea. Regions in the open Baltic differ from one other substantially in water salinity, oxygen concentration and other environmental parameters important in determining the reproduction possibilities, abundance and distribution of organisms. Concerning the distribution of some bottom invertebrates, it is especially important that the depth of the isocline indicating the oxygen concentration < 1 cm ^{3} dm^{-3} (important in distribution of certain invertebrates) differs between the regions. Although differences in the environmental conditions between these areas are significant, the borders separating them are not clear-cut. There is some exchange of water between the systems, which smoothens the transition of salinity, temperature and oxygen from one system to the next.

The large gulfs are an important part of the Baltic Sea. The main environmental conditions (salinity, oxygen conditions, etc.) of the Gulfs of Bothnia, Finland and Riga are rather similar and justify them being grouped as one macro-region. The gulfs have low salinity. From time to time, the gulfs suffer arctic climate and some other harsh living conditions that limit the occurrence of marine organisms. However, oxygen deficiency found elsewhere in the Baltic systems does not occur. Except for in the westerrn part of the Gulf of Finland, the gulfs have no halocline. However, in the summer season, a thermocline occurs. In the coastal zone of the Baltic Sea, especially of the large gulfs, big archipelagos are situated at the border of the fresh and saline water, offering suitable biotopes for sea birds, mammals and freshwater fish species. The coastal zone is, in general, very productive and usually acts as a filter for the water flowing from the mainland to the sea. In the large gulfs, the importance of freshwater species is markedly higher than in the open part of the sea. As the temperature below the thermocline is also low in summer, and usually no oxygen deficiency exists (excluding the western part of the Gulf of Finland), the large gulfs maintain viable stocks of glacial relicts.

2.6.2 Regions and Subregions

Large geographically/hydrographically delimited areas exist in the sea with the conditions broadly similar both in the lower layers (in the deeps) and the surface layers. The existence of regular currents and other hydrodynamic factors are important in levelling the conditions throughout such areas. These areas can be considered regional systems of the Baltic Sea. Hydrological and biological features (including dynamics of production cycles and biodiversity – HELCOM 1996, 2002)

Table 2.2 Parameters of natural regions in the Baltic

Region	Salinity (psu)	Oxygen conditionsat bottom	Ice cover (area, %)	Highest surface temperature in summer, °C	Peak of prim production
Kattegat	11,0–35,5	+/±	20	13–23	Jan–Mar
Danish Sounds	8,0–32,0	+/±	30	14–22	Mar
Arkona	7,5–25,0	+	0	13–22	Mar–Apr
SW part	6,5–22,0	±	0	13–22	Apr
Eastern	6,5–14	–	10	14–21	Apr–May
NWpart	5,5–12	–	40	14–20	May–Jun
G. of Riga	3,5–7,5	+	100	13–21	Apr–May
G. of Finland	3,5–11,0	±	100	11–21	Apr–May
Bothnian Sea	4,5–7,5	+	100	11–19	Apr–May
Botnhian Bay	2,5–4,5	+	100	9–18	Jun–Jul

Area of ice cover is given for the years of average temperature regime; dominating oxygen conditions at bottom (good +; variable ±; poor –) corresponding to the requirements of fish and invertebrates (the minimum concentration of O_2 = 1–2 $cm^3\ dm^{-3}$)

considerably vary between the regions (Table 2.2). Therefore, their response both to the natural factors and human-induced pressure is different. Commonly, transition zones of high variation in external conditions exist in the border areas of the regional units. In the Baltic Sea, ten regions can be differentiated.

Kattegat does not belong to the Baltic Sea, if strictly defined. The Kattegat differs from the other regions in the Baltic Sea. It has markedly higher salinity and considerably greater biodiversity and stability of its biota. Both vertical and horizontal salinity stratification is well-pronounced. The unidirectional barotropic flow dominates, but because of the strong stratification, the two-layer circulation is also important (Fonselius 1995; HELCOM 1996, 2002). Due to the high nutrient loading, oxygen depletion in bottom water can occur. The primary production is higher than in the Baltic Proper. In the frontal areas of different salinity and temperature, it can be up to 25 times greater than in the adjacent waters (HELCOM 1996, 2002). Out of more than 1500 species of marine animals in the North Sea, about 836 are found in the Kattegat (Zenkevich 1963), including some exploitable fish and invertebrate species (Ask and Westerberg 2005).

Belts and the Sound (The Danish Belts) is a shallow area with variable salinity. In this region, normally three different water masses occur: Baltic surface water, Kattegat surface water and Kattegat deep water. Hydrological variations are considerable, being determined by the in- and outgoing water flows, which depend on the wind conditions. The water masses in the Belt Sea are well-mixed, while in the Sound, a strong halocline exists. The Sound should probably be considered as a separate subregion in this region. Conditions vary: the open areas have a good water exchange, while there are also closed coastal areas where local processes are of vital importance. Primary production and biodiversity are high in both pelagic and demersal communities. Oxygen deficiencies with occasional occurrence of hydrogen sulphide have affected benthic animals, especially in the inner areas (HELCOM 1993, 1996, 2002). However, moving from the Kattegat to the Belts and the Sound, the number of marine species declines sharply – to 436 in the Belts (Zenkevich 1963).

The eastern border of the **Arkona region** (and of the macro-region) roughly coincides with the 8 psu surface isohaline (i.e., with the higher limit of the universal salinity barrier), which is of great physiological importance for aquatic animals. The low salinity causes serious osmotic problems for marine organisms and the border is an effective barrier for their migration into less saline water, but in addition, freshwater species cannot penetrate this barrier in the opposite direction. Therefore, the Arkona Deep represents the eastern limit of the distribution of a number of invertebrate (Järvekülg 1979) and fish species. In the Arkona Basin, the variation in biodiversity is high, but the number of marine species decreases to 145. However, the decrease is even greater when moving from the Arkona Basin further east (Zenkevich 1963; HELCOM 1996, 2002; Ask and Westerberg 2005). It is probably reasonable to handle the southern shallow-water part of the area as a separate sub-region.

The climate of the **Southwest (SW) region** is mainly of the Atlantic type and the temperature regime is comparatively mild. The region is under a strong and continuous impact of inert oceanic systems.

Saltwater intrusions varying in strength take place from time to time. After rather short periods of increased salinity and oxygen concentration, stagnation with possible occurrence of hydrogen sulphide may take place. The depth of the upper limit of the halocline can vary between 40 m during the periods of very strong intrusion of saline water and 70–90 m in stagnation periods. Due to the inflow of large freshened water masses of low temperature from the surface layers of the Northwest region in winter, the Southwest region of comparatively limited volume is fully covered by ice earlier than the Eastern region, where ice formation starts earlier. In the Southwest region, the environmental conditions are less variable and much more favourable for marine species than in other parts of the open Baltic. However, the marine fauna and flora are impoverished compared with the Transition Area. The biodiversity clearly depends on the saltwater influxes (Zenkevich 1963; HELCOM 1996, 2002). Together with the Transition Area, the Southwest region constitutes a very important bridgehead for adaptation and invasion of marine species in the Baltic Sea.

In the **Eastern (E) region**, there are several large deeps, the largest being the Gotland Deep and the Gdansk Deep. These are influenced by saltwater intrusion from the Transition Area through the Southwest region. However, whereas many inflows reach the Bornholm Deep, only strong and moderate saltwater inflows can reach the Gotland Deep. This means that an oxygen deficit is more common in the Eastern deeps than in the Bornholm Deep and the periods of oxygen deficit in the deep layers are considerably longer than in the Bornholm Deep. In the Gotland Deep, hydrogen sulphide can be found in the deeper half of the water column. The depth of the halocline can vary from 60–80 to 80–140 m from the surface. The deep layer can be divided into an active deep layer with some production taking place, and a bottom layer where long stagnation periods prevail. The Gotland Basin has the largest reserves of nutrients in the Baltic (Järvekülg 1979). Populations of marine boreal euryhaline eurytherm species dominate in the zooplankton, zoobenthos and fish fauna (Ojaveer and Kalejs 2008). Based on natural conditions, two sub-regions can be distinguished in the Eastern region: the central subregion with the largest deep-water area of the sea, including the Gotland and Fore deeps, and the southeast subregion consisting of the Gdansk Deep. In the Gdansk Deep, a separate circulation system and nutrient pool exist and the hydrological and chemical conditions are different from those in the central sub-region. Both the production processes and the organisms adapted to the regions are somewhat different in these sub-regions (Elwertowski 1957; Popiel 1958; Seletskaja 1970; Järvekülg 1979; etc.).

In the **Northwest (NW) region**, Atlantic and Arctic climate periods alternate. The severest winters for the Baltic Proper can occur. The region is strongly impacted by fresh water discharged into the northern part of the open sea and the Gulfs of Finland and Bothnia. In this region, the changes in vertical stratification and in seasonal and long-term variation of oceanographic conditions are large (Table 2.2). Only very strong saltwater inflows reach the deep layer of the Northwest region and their impact on the oxygen situation is small. Deep waters of this area are usually stagnant and the oxygen concentration can increase chiefly due to the vertical mixing. During long-term intense deep-water inflows consisting of oxygen-poor water pushed here with the current from the Gotland Deep, the salinity increases, the halocline grows stronger, the oxygen concentration decreases and hydrogen sulphide forms in the deep layer. In the periods with larger inflows of fresh water into the Baltic Sea, the halocline may drop much more deeply that in the average situation (70–80 m from the surface) and over large areas, the deep-water layer can disappear (Fig. 2.10). The vertical exchange increases, and with that the oxygen concentration at the bottom magnifies as well (cf. Fonselius 1995). In this region, several sub-regions should be distinguished: the northern sub-region around the Northern Deep and the western sub-regions with the Landsort, Norrköping and Karlsö deeps. In the Western Gotland Basin, currents mainly transport biogenes southwards, including into the Bornholm Deep (Lehmann and Hinrichsen 2000; etc.).

Salinity and temperature have a direct as well as an indirect impact on phytoplankton and primary production, especially in relation to the adaptation of phytoplankton species to concrete conditions in their habitat.

The Gulf of Riga is connected with the open part of the sea by the Irbe Sound, which is of a depth of 25 m, and the Väinameri Archipelago Sea, which is of even more limited depths. Both the Irbe and Väinameri areas are sub-regions of the Gulf of Riga region. After the intrusion of open-sea water into the gulf, some difference between the surface and bottom salinities can occur, mainly in the northwestern part of the gulf (Ojaveer 2014). Biological productivity in the Gulf of Riga is high

(HELCOM 1996). The organisms are distributed from the surface to the bottom. In the central, northern and western parts of the gulf, euryhaline marine zoobenthos dominates and the importance of marine species is rather high in zooplankton, whereas in the coastal areas, freshwater species dominate. Anthropogenic influences are stronger in the southern part of the gulf near the estuaries of large rivers.

No sill exists between the Baltic Proper and the **Gulf of Finland.** The part of the gulf west of 26° E with the existence of a halocline and quasi-permanent hydrological frontal areas (Alenius et al. 1998) should be considered to be a highly dynamic transition zone between the open part of the sea and the central areas of the Gulf of Finland. The western part of the gulf is the northeast end of the deep area in the Baltic Sea where nutrient-rich saline water is pressed to mix with upper layers (Fig. 2.12). The differences between the eastern, central and western areas of the gulf in salinity, temperature and other environmental parameters form the background for the existence of a clear transition from the brackish-marine zooplankton, zoobenthos, macrophyte and fish species in the western parts to the brackish-freshwater communities in the central and eastern parts of the gulf (Järvekülg 1979; Lumberg and Ojaveer 1991; Trei 1991; etc.). The simulation carried out by Andrejev et al. (2004) supports the existence of cyclonic circulation in the gulf and confirms a rapid increase in the salinity in its western part compared to the much more stable situation in the central areas. It has been estimated that in this gulf, human impact has considerably increased the concentration of nutrients. Periodic anoxic conditions can occur in the deeper areas of the gulf (HELCOM 1996, 2002). Several studies (Ostov 1971; Lumberg and Ojaveer 1991; Telesh et al. 1999; etc.) indicate that the Gulf of Finland region should be divided into several subregions. The depth of the sill separating the Baltic Proper from the Gulf of Bothnia is 40 m. The Åland Sea and the Arhipelago Sea, with characteristic ecological systems and high biological productivity (HELCOM 1996, 2002; Ask and Westerberg 2005; etc.), could best be regarded as subregions of the Bothnian Sea.

In **the Bothnian Sea**, a weak halocline occurs at a depth of 50–75 m (Voipio 1981), but the oxygen conditions mostly allow for habitation of the whole water column. For a number of species, the northern borders of occurrence run through the southern part of the Bothnian Sea. The diversity and variation of the benthic macrofauna in the deep soft bottom areas are extremely low compared to the Baltic Proper and the Gulf of Finland (Järvekülg 1979). The fish fauna includes about 25 marine species, of which 10 reproduce there, together with a number of freshwater, migratory and relict fishes (Andreassson and Petersson 1982; HELCOM 1996, 2002).

The ecosystem of **Bothnian Bay** is considerably different from other areas of the Baltic Sea. The conditions in the bay depend on the oceanographic properties of the shallow (up to 25 m) Northern Quark strait controlling the water exchange between the Bothnian Sea and Bothnian Bay. Phosphorus concentrations are considerably lower and nitrogen concentrations higher than in the Bothnian Sea (Fonselius 1995; HELCOM 2002; etc.). Phytoplankton primary production is low, and it is not the dominant energy source like in the other parts of the Baltic Sea. Meiofauna is as important as the macrofauna for biological production. Most of the benthic production is provided by microalgae. Marine bivalves are absent. There is a drastic difference in the energy flow between Bothnian Bay and all other parts of the Baltic, because of the loss of the suspension-feeding pathway in the bay. A number of organisms distributed in the Baltic Sea only reach the southern limit of Bothnian Bay. The short growing period and the movement of the normal phytoplankton maximum towards midsummer are typical of arctic conditions (Järvekülg 1979; Elmgren et al. 1984; HELCOM 1996, 2002; etc.). The fish fauna is poor. The local spring herring population dominates, but freshwater species and some glacial relicts are also abundant (Andreasson and Petersson 1982).

References

Alenius P, Myrberg K, Nekrasov A (1998) The physical oceanography of the Gulf of Finland: a review. Boreal Environ Res 3:97–125

Andreasson S, Petersson B (1982) The fish fauna of the Gulf of Bothnia. In: Müller K (ed) Coastal research in the Gulf of Bothnia. Dr W. Junk Publishers, The Hague, pp 301–315

Andrejev O, Myrberg K, Alenius P, Lundberg P (2004) Mean circulation and water exchange in the Gulf of Finland – a stydy based on three-dimensional modelling. Boreal Environ Res 9:1–16

Anonymous (1921–1959) Bull Hydrogr Copenhagen

Ask L, Westerberg H (eds) (2005) Fiskbestånd och miljö i hav och sötvatten. AB Danagards Grafiska, Odeshog

Baltic Marine Environment Protection Commission–Helsinki Commission (1986) Baltic Sea Environ Proc no 16

Berzinsh V (1995) Hydrology. In: Ojaveer E (ed) Ecosystem of the Gulf of Riga between1920 and 1990. Estonian Academy Publishers, pp 7–31

Dera J (1995) Underwater irradiance as a factor affecting primary production. Inst Oceanol Polish Acad Sci, Sopot. Dissertations and Monographs 7/1995

Elken J, Matthäus W (2008) Baltic Sea oceanography. In: Assessment of climate change for the Baltic Sea basin, Annex A. 1.1. Springer, Berlin

Elmgren R, Rosenberg R, Andersin A.-B, Evans S, Kangas P, Lassig J, Leppäkoski E and Varmo R (1984) Benthic macro- and meiofauna in the Gulf of Bothnia (Northern Baltic). Finn Mar Res 250: 3–18

Elwertowski J (1957) Szprot. Biologia, polowy, przetworstwo. Wydawnictwo Morskie, Gdynia

Fonselius S (1995) Västerhavets och Östersjöns oceanografi. SMHI, Västra Frölunda

HELCOM (Baltic Sea Environment Protection Commission) (1986) Baltic Sea Environ Proc no 16 Helsinki

HELCOM (1990) Second periodic assessment of the State of the Marine Environment of the Baltic Sea, 1984–1988; Background document Baltic Sea Environ Proc no 35B

HELCOM (1993) First Assessment of the State of the Coastal Waters of the Baltic Sea. Baltic Sea Environ Proc no 54

HELCOM (1996) Third Periodic Assessment of the state of the Marine Environment of the Baltic Sea 1989–1993; Background document. Baltic Sea Environ Proc no 64B

HELCOM (2002) Environment of the Baltic Sea Area, 1994–1998. Baltic Sea Environ Proc no 82B

HELCOM (2007) HELCOM Thematic Assessment in 2007. Climate Change in the Baltic Sea Area Baltic Sea Environ Proc no 111

HELCOM (2009) Eutrophication in the Baltic Sea. Baltic Sea Environ Proc 115 (B)

ICES (1932) Rapp P-v Réun Cons int Explor Mer 81

ICES (2006) ICES Advisory Committee on Fishery Management, Advisory Committee on the Marine Environment and Advisory Committee on the Marine Ecosystems. 2006 ICES Advice, Books 1–10, 8, 119 pp
ICES (2011) ICES Advice 2011, Book 8. The Baltic Sea
Järvekülg A (1979) Donnaja fauna vostotchnoi tchasti Baltiiskogo morya (Bottom fauna of the eastern part of the Baltic Sea). Tallinn, Valgus
Kowalewski A, Ostrowski M (2005) Coastal up- and downwelling in the Southern Baltic. Oceanologia 47(4):453–475
Laevastu T (1963) Surface water types of the North Sea and their characteristics. Serial Atlas of the Marine Environment. Folio 4. American Geogr Society, Broadway at 156 th St., New York 32, N.Y.
Lehmann A, Hinrichsen HH (2000) On the thermohaline variability of the Baltic Sea. J Mar Syst 25:333–357
Lumberg A, Ojaveer E (1991) On the environment and zooplankton dynamics in the Gulf of Finland in 1961–1990. Proc Estonian Acad Sci Ecol 3:131–140
Matthäus W (1993) Major inflows of highly saline water into the Baltic Sea – a review. ICES Statutory Meeting, paper CM 1993/C:52
Myrberg K, Andrejev O (2003) Main upwelling regions in the Baltic Sea. Boreal Environ Res 8:97–112
Ojaveer E (2014) Läänemeri. Tead Akad Kirj, Tallinn
Ojaveer EA, Kalejs MV (1974) O nekotorykh okeanologitcheskikh predposylkakh opredelyajustsikh kolitchestvo i raspredeleniye pelagitcheskikh ryb v Baltijskom more (On some oceanological preconditions determining the abundance and distribution of pelagic fish in the Baltic Sea). Okeanologiya 14(3):544–554
Ojaveer E, Kalejs M (2008) On ecosystem-based regions in the Baltic Sea. J Mar Syst 74:672–685
Ojaveer H, Jaanus A, MacKenzie BR, Martin G, Olenin S, Radziejewska T, Telesh I, Zettler ML, Zaiko A (2010) Status of biodiversity in the Baltic Sea. PloS ONE 5(9):e12467
Ostov IM (1971) Kharakternye osobennosti gidrologitcheskogo i gidrokhimicheskogo rezhima Finskogo zaliva kak osnova jego rybokhosjaistvennogo osvojeniya (Characteristic peculiarities of the hydrological and hydrochemical regime of the Gulf of Finland as the basis of fisheries). Reports of NIIORH 76:18–45
Popiel J (1958) Differentiation of the biological groups of herring in the Southern Baltic. Rapp P-v Réun Cons int Explor Mer 143, II:114–121
Seletskaja AV (1970) Morfologitcheskaja kharakteristika kiljki Baltijskogo morya (Morphological characteristics of the Baltic sprat). Rybokhoz issledovaniya v bass Baltijskogo morya 5:76–92
Steemann-Nielsen E (1940) Die Produktionsbedingungen des Phytoplanktons im Übergangsgebiet zwischen der Nord- und Ostsee. Medd Dan Fisk-. o Havunders. Ser Plankton (3)4:1–55
Steemann NE (1975) Marine photosynthesis. Elsevier, Amsterdam/Oxford/New York
Svansson A (1973) Interaction between the coastal zone and the open sea. Merentutkimuslait julk/Havsforskninginst skrift 239:11–28
Telesh IV, Alimov AF, Golubkov SM, Nikulina VN, Panov VE (1999) Response of aquatic communities to anthropogenic stress: a comparative study of Neva Bay and the eastern Gulf of Finland. Hydrobiol 393:95–105
Trei T (1991) Taimed Läänemere põhjal. Valgus, Tallinn
Voipio A (ed) (1981) The Baltic Sea. Elsevier, Amsterdam, Oxford, New York
Wulff FV, Rahm LA, Larsson P (2000) A systems analysis of the Baltic Sea. Springer, Berlin/Heidelberg/New York
Zenkevich LA (1963) Biologiya morei SSR (Biology of seas of the SSSR). Academy of Sciences, Moskva

Chapter 3
Life in the Baltic Sea

Abstract It is probable that the main cause of the poverty of the Baltic Sea in species is its low salinity. Apparently also complicated and variable climatic conditions as well as the anthropogenic impact exert their regionally varying pressure. Regional differences in environment are the main background of variation of life forms in the Baltic Sea. The ecophysiological influence of salinity on the biota is supposedly related with the salinity conditions during the formation and the earlier epochs of development of marine biota when the salinity was likely higher of the universal salinity barrier – 5–8 psu. The influence manifests itself mainly in the changes in osmotic pressure and ion composition of the environment, which affect the exchange of water and salts in organisms and is related with their oxygen consumption, growth and activity. Differences in the osmotic tolerance are genetically determined. The Baltic Sea is composed of a number of natural regions continuously differing in external conditions. Separate populations of a species may produce generations significantly differing in their relative abundance in various environments/areas in the same year. Present stock assessment methods do not allow correct treatment of the data consisting of a mixture of units with different stock dynamics. Therefore, populations differing significantly in the relative abundance of the same year class should be considered separately. For the differentiation of stock units suitable for assessment and management, it is practical to apply the conception of the biological species by Mayr (Systematics and the origin of species. Columbia University Press, New York, 1942): *actually or potentially crossing natural populations which are reproductively isolated from other such groups.* The reproductive isolation of species/groups in the Baltic Sea has evolved in the process of adaptation to the external conditions. The distribution, development, abundance dynamics, reproduction, feeding, growth, diseases, enemies and parasites, mortality and other population parameters as well as management problems of living organisms (phytoplankton and bottom vegetation, heterotrophic microorganisms, zooplankton, zoobenthos, 22 fish species involving marine, diadromous, freshwater and relict species), marine birds and marine mammals are considered in the chapter.

Compared with other seas, the number of species differentiated on the basis of morphological features is relatively small in the Baltic Sea. Considering the brackish-water origin of the Baltic Sea, the great majority of scientists are of the opinion that

E. Ojaveer, *Ecosystems and Living Resources of the Baltic Sea*,
DOI 10.1007/978-3-319-53010-9_3

the main cause of the poverty of the sea in regard to species is its low salinity. It is also apparent that complicated and variable climatic conditions, as well as anthropogenic impact, exert their own regionally varying pressure. Undoubtedly, regional differences in environment are the main background for variation of life forms in the Baltic Sea.

3.1 Salinity-Induced Ecophysiological Problems of Organisms in the Baltic Sea

Most likely, life was born in the sea (Khlebovich 1974; Steemann Nielsen 1975). The type of biota settling water environments – the organisms living in fresh, brackish, marine or hyperhaline environments – is determined by the amount and composition of salts dissolved in the water. The basic types are marine and freshwater biota. The development of the marine biota started in a remote geological epoch. It is thought that at that time, the salinity exceeded the universal salinity barrier, or 5–8 psu (Khlebovich 1974). When this salinity interval is exceeded, some physical and chemical properties, conductivity, etc., of the marine water suddenly change. The changes also concern a number of other biological characteristics on various integration levels. Some macromolecular compounds important in biological processes (e.g., some albumins) can persist under stable conditions only at a salinity higher than 5–8 psu (Khlebovich 1974). The environmental salinity was important for the determination of the direction of the development of the organic world in the remote past and is also important today, as it exerts influence on the adaptation of organisms and the development of corresponding structures.

Alteration in the total concentration of salts appears as a change in the osmotic value of marine water. The ecological–physiological influence of salinity on biota is mainly related to the changes in the osmotic pressure and ion composition of the environment, which affect the exchange of water and salts in organisms and are related to their oxygen consumption, growth and activity. In a number of species, the limits of salinity tolerance of germ cells, larvae and other juvenile stages are lower than the corresponding limits for the adult specimens. In this case, the population cannot reproduce in the full extent of its area. Differences in osmotic tolerance are species specific, and consequently genetically determined. They are related to the variability of other environmental conditions (mainly temperature and oxygen content) important for the organism.

Excluding a comparatively limited number of especially euryhaline species, marine and fresh waters have almost no common species. Despite the pressure from the evolution process over millions of years, fauna of respective marine and freshwater backgrounds have not intermingled and the endemicity of both of them is high. Due to the comparatively short existence of brackish waters, no endemic brackish-water genera and families have eveolved and the number of brackish-water species is limited. The fauna of brackish waters is mainly composed of immigrants from marine and fresh waters. In the Baltic Sea, they have to adapt to an environ-

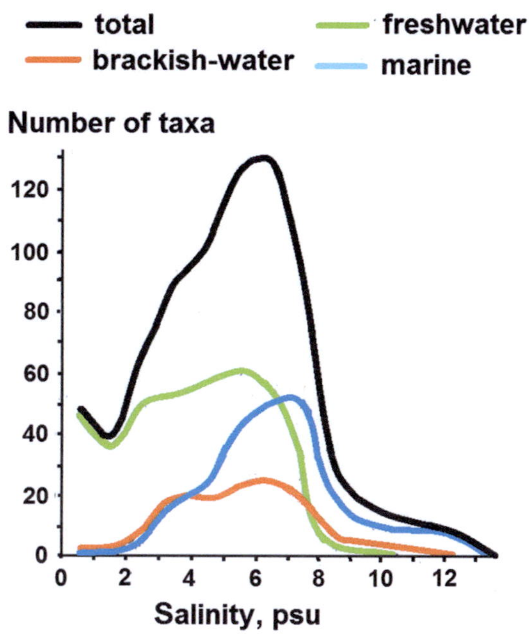

Fig. 3.1 The number of taxa of freshwater, brackish-water and marine bottom invertebrates at different salinities in the Baltic Sea (Järvekülg 1979)

ment which has qualities different from those in their previous area. The reaction of the immigrated species to the water salinity in the Baltic, mediated by the osmotic pressure, is very specific and variable. The species respond adequately to comparatively small variations in salinity. Long-term deviations in the osmotic values of the inner environment exert their effect as metabolic disorders. In the brackish water, the organisms of marine origin have additional energy expenditures to adapt to the changed salinity, therefore their growth rate decreases. Consequently, the adaptation process to the brackish-water environment and to its different versions of salinity, temperature and other conditions is complicated and requires time and energy. Different systematic units of organisms, as well as different populations of the same species, can differ in their osmoregulation, i.e., they can be euryhaline to various rates (Khlebovich 1974).

The above standpoints are substantial for understanding the processes of the formation of the biota in the Baltic Sea. The special importance of salinity is underlined in the formation of areas of the main fauna elements in the Baltic and as the basis for their reproduction and abundance fluctuations.

According to the general graph of the relationship between the salinity of brackish-water environments and their species richness (Remane and Schlieper 1958), the lowest number of species should theoretically be confined to the waters of 'critical salinity' (from 5 to 8 psu). Having studied the bottom fauna in the eastern part of the Baltic Proper, the Gulfs of Finland and Riga over long periods of time, Järvekülg (1979) arrived at the conclusion that in that area in this salinity interval, the number of taxa/species is not minimum but maximum (Fig. 3.1).

It is probable that in the Baltic Sea, this phenomenon is caused by the adaptation of both marine and freshwater species to the brackish environment. Cod, *Pseudocalanus elongatus minutus* and other organisms of marine origin populating the Baltic Sea cannot be equalized with these species in the seas with average oceanic salinity. Generations of these species have long lived at much lower salinities characteristic of the Transition Area, the deep-water parts of the Baltic Proper, etc. Therefore, their reaction to lower salinities is the reaction of a member of a modified population, adapted (at least to a certain extent) to its current environment. The reaction of such a population cannot be adequately compared with the reaction of the same species in eumarine seas. Also, freshwater species have populated and adapted to the coastal and other low-salinity areas of the Baltic Sea for a long time (including some species that came in during the stages of historical development of the Baltic Sea). Therefore, the reaction of the organisms with a freshwater background populating currently brackish areas of the Baltic Sea should also be regarded as the reaction of modified freshwater species.

3.2 Multitude of Ecosystems

The number of typological fish species populating the Baltic Sea is rather small. Many fishes that have immigrated from the North Sea have adapted to much lower salinities than those in their previous area. The species having arrived from freshwater bodies live in the areas of higher salinity than that in the areas of their historical origin. Adaptation to new environmental conditions modifies the body shape, biology and other properties of the immigrants. For the adaptation of organisms to different environmental conditions, expenditure of additional energy is necessary; this does not facilitate an increase in productivity.

Concerning ecosystems, the Baltic Sea is rather heterogeneous within its limits. It is composed of a number of natural regions continuously differing in external conditions. To colonize a region, it is necessary to adapt to the conditions reigning within it, which generally means development of a different population.

Separate populations of a species may produce generations significantly different in their relative abundance in various environments/areas in the same year. Present stock assessment methods do not allow for correct treatment of the data consisting of a mixture of units with different stock dynamics. Therefore, populations differing significantly in the relative abundance of the same year class should be considered separately. In the Baltic Sea, the species of organisms having developed several populations with different abundance dynamics include the most important exploited commercial fishes (cod, herring and sprat) and probably also a number of other species.

Specialists differ in their understanding of the fundamentals of species definition. For solving applied problems, especially for correct assessment of the stock condition and management, the existence of groups of various abundance dynamics in the databases of stock units should be avoided. For the differentiation of stock units suitable for assessment and management, it is practical to apply the conception of biological species by Mayr (1942): *actually or potentially crossing natural populations which are reproductively isolated from other such groups*. As the basis of this conception, the most important precondition – reproductive isolation – is taken. This conception also has the highest importance in the differentiation of stock units for their assessment and management. Below, the populations of the most important fish of the Baltic Sea have been differentiated on the basis of their actual reproductive isolation. The reproductive isolation of species/groups in the Baltic Sea has evolved in the process of adaptation to the ecosystems. Geographical isolation, which has probably relied mainly on currents and other water movements, has played a very important part. Formation of local populations in the Baltic Sea has obviously been influenced by the complicated history of the sea and a new source of pressure – human activity. In spite of the very high scientific and practical importance of differentiating population structure of species (particularly of the exploited fish stocks), genetic, physiological, behavioural and other aspects of species have up to now failed to attract the due attention of scientists. The above-mentioned population conception based on the genetic integration of the unit stocks allows for the best consideration of practical requirements and is concordant with the situation actually occurring in the Baltic.

In the eastern areas of the Baltic Sea and in the Gulfs of Finland and Riga, zoobenthos species, including glacial relicts, have formed a number of local populations differing from each other both in the characteristics of their areas and their population parameters (Järvekülg 1979). For the estimation of the impact of regions of differing environments upon the formation of populations in fish species, the processes around the halocline (probably the most important micro-area of adaptation of marine species to the brackish conditions), and also around the seasonal thermocline and the near-coast mixing zone (related mainly to the adaptation of freshwater organisms to the brackish-water conditions), should be considered.

Of marine fish, the species reproducing in hydrologically (cod, flounder) or both geographically and hydrologically (herring) well-differentiated areas have shown a relatively good adaptive separation. The only important commercial fish species adapted to all regions of the Baltic Sea is spring spawning herring. The population structure of this species corresponds well to the regional structure of ecosystems, and the differences between the populations belonging to various ecosystems are clearly exposed. Cod has formed two populations in the Baltic; one of them (the western population) has adapted to the Transition Area and the other (the eastern population) dwells mainly in the open part (mainly in the SW region) of the sea (Nielsen et al. 2003; Pocwierz-Kotusa et al. 2014).

The occurrence of coastal-spawning flounder in the Baltic Sea (in the Gulfs of Finland, Riga and Bothnia, at the Swedish east coast, Gotland Isle and in the past also on Odra Bank) side by side with the deep-spawning flounder populations of the Sound, Belt Sea, Arkona Deep, Gdansk Deep and Gotland Deep (Strodtman 1918; Mikelsaar 1957, 1984; ICES 2011; etc.), can probably be explained by the repeated colonization of the sea by this species, possibly from different sources.

The embracing of fish and invertebrates of freshwater background into the brackish-water ecosystems takes place continuously in shallow brackish-water coastal areas of higher productivity and near the thermocline. One of the originally freshwater species now included in the brackish-water ecosystem is the pikeperch (*Sander lucioperca* L.) of Pärnu Bay, which differs from freshwater populations of this species in a number of parameters, including the type of development of oocytes (Erm 1981).

In deep layers of the large gulfs, temperature is rather low and the oxygen regime satisfactory. Therefore, coldwater species (including glacial relicts) have found acceptable living conditions in the bottom layers of the Gulfs of Riga, Finland and Bothnia. Järvekülg (1979) has stated that panboreal and boreal species constitute 54% of the benthic invertebrates in the Estonian and Latvian brackish waters. In the deep layers of the central Baltic, the panarctic species are the most important (about two thirds of the number of species in the deep area east of Gotland). However, in the process of occupying new productive areas, some glacial relicts (e.g., eelpout) are rather successfully adapting to the environmental conditions outside their historical area (Ojaveer and Lankov 1997). This indicates intense adaptation of the main components of the Baltic Sea – marine, freshwater and relict organisms – to the present-day environmental conditions and their contribution to the diversification of its food webs.

3.3 Living Organisms

The production system of the Baltic Sea comprises pelagic and littoral trophogenic layers. A clear maximum of light energy inevitably necessary for primary production occurs in summer, as the sea is situated at high latitudes. The duration of the growth period lasts 9–10 months in the southern part of the Baltic Sea, but in the northern areas, it only lasts 4–5 months. Ice cover, important from the point of view of both the development of vegetation and the whole ecosystem, has wide annual variations. Northern areas of the sea are covered with ice about 4–5 months a year. Ice can also occur in the central and southern parts of the Baltic Sea.

Green plants consist mainly of carbon, hydrogen and oxygen. In the organic matter, calcium, potassium, magnesium, sulphur, nitrogen and phosphorus are also necessary. Additionally, in micro-quantities, the organic matter constitutes manganese, copper, zinc, selenium and other so-called trace elements.

Scantiness of vital nutrient salts limits the growth of plants and primary productivity. In the aquatic environment, the growth/productivity is effectively limited by nitrogen as a component of cell proteins and phosphorus occupying an important position in cell energetics. These elements are needed in different proportions, depending on the requirements of the organism concerned. On average, in the organic material of phytoplankton, 16 nitrogen atoms occur per every phosphorus atom. This ratio between phosphorus and nitrogen atoms (the Redfield ratio) characterizes the requirement of these atoms in primary production. In general, such an average relationship occurs in the organic matter of all oceans.

Nutrient salts in the productive zone originate from the following main sources: (1) inflow of natural waters from land; (2) sewage waters; (3) upwellings of nutrient-rich deep waters; (4) wells in the bottom of the coastal sea delivering subsoil water of high nitrogen content, an important source of nutrient salts. In addition, a certain amount of nitrogen is bound from the atmosphere by planktonic organisms (including bluegreen algae).

Benthic heterotrophic organisms depend on the organic matter settling down from the trophogenic layer. The Baltic Sea is a typical detritus-based system. Deposited nutrient salts are again carried with biogeochemical circulation into the trophogenic layer. Since substantial differences occur in hydrological conditions between various parts of the Baltic Sea, such a system also causes differences in the productivity and succession of planktonic and benthic communities between the regions of the sea.

Plants assimilate nutrient salts chiefly as inorganic ions. Nitrogen is used mainly in the form of nitrites, nitrates, ammonium or urea, and phosphorus mainly as phosphates. Thus, nitrogen and phosphorus are involved in the ecosystems in different manners. Nitrogen and phosphorus assimilated by living organisms from the atmosphere or minerals, and later released again from the composition of organisms into the atmosphere or minerals, are circulating in the so-called biogeochemical cycle. Nitrogen is released from water as a result of denitrification processes, while phosphorus passes through the ecosystem in one direction: from soil into surface water and from there into the sea. Therefore, the phosphorus cycle in water is simpler than that of nitrogen.

Planktonic algae and other organic matter unused or not degraded in the water sink to the bottom, decompose and mineralize slowly, due to the activity of meiofauna and bacteria. In this process, phosphates and ammonium are released. Resulting from bacterial nitrification, ammonium is oxidized to nitrites and nitrates. Inorganic nutrient salts return into the water and can once again be assimilated into primary production.

A substantial part of phosphorus is bound (frequently tied to iron) to sediments, which means its removal from the biological cycle. However, depending on the physical and chemical conditions, redox processes, etc., large quantities of phosphorus can also be transferred from sediments into water (e.g., in deep water layers under oxygen-free conditions).

3.4 Vegetation and Primary Production

Phytoplankton In the sea, the phytoplankton is by far the most important vegetation. The amount of photosynthesis going on in the planktonic algae is many times greater than that of all other kinds of marine plants combined (Steemann Nielsen 1975). Research into phytoplankton started in the Baltic Sea at the end of the eighteenth century. At the beginning, it included taxonomy, floristics and ecology. Due to the variety of applied methods, the results of the first studies suffered from poor comparability between the authors. Presently, for quantitative phytoplankton studies (including microplankton), the sedimentation and counting technique developed by H. Utermöhl is applied (accepted also by the HELCOM – Wasmund and Siegel 2008).

Since the second half of the 1970s, satellites have been used for investigation of cyanobacteria. For monitoring phytoplankton blooms from space, oceanographic ocean colour sensors installed on satellites and optical models have been developed.

The growth of the phytoplankton biomass is closely related to the concentration of the growth-limiting nutrients – nitrogen and phosphorus. The importance of these nutrients varies between the seasons and regions of the sea. The importance of phosphorus is especially high in Bothnian Bay and coastal waters. Phytoplankters also depend on the concentration of CO_2 in the water, which is usually present in sufficient amounts in the Baltic Sea. Sometimes other nutrients, such as silica (SiO_2) or trace elements such as iron, can also become limiting (HELCOM 2009).

One of the most important external conditions that can limit phytoplankton growth and abundance is light. Excessive light intensity can depress photosynthesis in phytoplankton. However, changes in light availability (depth-dependent variations, changes in solar radiation, etc.) are reflected as variations in photosynthesis.

It is assumed that the eastern limit of the distribution of marine phytoplankton in the Baltic is the Darss sill. West of the sill, the importance of the marine phytoplankters belonging to the genera *Chaetoceros*, *Protoperidinium*, *Rhizosolenia*, *Coscinodiscus* and *Ceratium* notably increases. In the Arkona Basin and the Belt Sea, the development of phytoplankton is more complicated than in the central and northern parts of the sea. In the southern part of the sea, from the Darss sill to the Gotland Basin, the same species of phytoplankton dominate as in the north of the Gotland Basin. Generally, in the Baltic Sea, the maximum development of phytoplankton and the highest primary production occur in spring. Then, the abundance is high in *Achnantes taeniata*, *Chaetoceros wighami*, *C. holsaticus*, *Melosira arctica*, *Skeletonema costatum* and *Thalassosira baltica*. In early summer, *Dinobrion balticum* and, later on, *Chaetoceros danicus*, *Aphanizomenon flos-aquae* and *Nodularia spumigena* are characteristic. Mainly in the western Baltic, a separate postspring bloom of flagellates has been differentiated (Feistel et al. 2008). In late summer, the importance of the diatoms *Actinocyclus octonarius* and *Coscinodiscus granii* increases.

Species with marine backgrounds (*Ceratium tripos*, *C. fusus*, *Dinophysis acuta* etc.) increase their share in plankton in late summer and autumn. In archipelagos and coastal areas, phytoplankton composition does not substantially differ from that in the open sea. However, in estuaries and eutrophied areas, the character of phytoplankton strongly depends on the regime of freshwater runoff, especially at high concentrations of nutrients. In the coastal zone, phytoplankton is a mixture of freshwater, brackish-water and marine forms. The phytoplankton of isolated areas is formed in dependence on the salinity and eutrophication of the areas in question.

In eutrophied areas of the northern Baltic, *Oscillatoria agardhii* assemblages dominate in summer and at the beginning of autumn. Therefore, after the spring phytoplankton bloom, the low production period is absent. Instead, with the development of an *Oscillatoria agardhii* assemblage, production rises far higher than during the spring maximum. As a result, the maximum production shifts to late summer. The *Oscillatoria agardhii* assemblage can be well developed even in the northern Bothnian Bay, in eutrophied estuarine waters (Voipio 1981).

Seasonal Development In winter, phytoplankton production is low in the central and northern parts of the Baltic and in the Gulfs of Bothnia, Finland and Riga, as the ice and the snow on the ice do not allow for penetration of light sufficient for high production. If snow is absent or its layer on the ice is thin, diatoms will form a yellowish-brown sheet on the lower side of the ice. Frequently, the central part of the sea is ice-free. In winter, the concentration of phosphates, nitrates and silicates in the water may exceed the concentration, limiting production. However, weak illumination and the effective mixing of water layers downwards of the critical depth allowing primary production hinder phytoplankton activity in the production of organic matter. In spring, the temporary mass occurrence of phytoplankton (phytoplankton spring bloom) starts with the improvement of light conditions in early spring, sometimes under the ice (Hällfors and Niemi 1974). The development is regulated by the rise in light intensity and the increase in the stability of water layers after stratification (Kaiser and Schulz 1976). As the south–north range of the sea is large and the illumination and temperature conditions vary correspondingly, the start of the phytoplankton spring bloom (commonly representing the highest productivity) commences in the southwestern areas and shifts step by step northwards.

In the SW Baltic Sea, the phytoplankton spring bloom occurs in late February or in March, usually starting in the southern part of the sea in April. In the central part of the sea and in the Gulf of Finland, where the formation of a thermocline starts later, the phytoplankton mass spring bloom begins in the latter part of April, and in the northern Baltic, at the beginning of May (Fig. 3.2). In Bothnian Bay, it only peaks in June.

Large fluctuations at the beginning of the development of the phytoplankton mass can be related to meteorological conditions. The highest values of phytoplankton biomass coincide with the time of the formation of a weak thermocline. The thermocline delimits the water surface layer, which generally coincides with the euphotic layer, and simultaneously limits the possibility for phytoplankton to fall deeper than the critical depth (i.e., into the deeper water layers where illumination is too weak for primary production). In cases of marked deviations of the environmental situation, the phytoplankton spring bloom has occurred much earlier or later than the normal time (for example, in January of 1997 and 1998 in the Kattegat – Wasmund and Siegel 2008).

Production is started mostly by the marine cold-water diatoms *Achnantes taeniata, Chaetoceros holsaticus, C. wighami, Melosira arctica, Thalassosira baltica, Sceletonema costatum* etc., and the dinoflagellates *Peridiniella catenata*, etc. In the

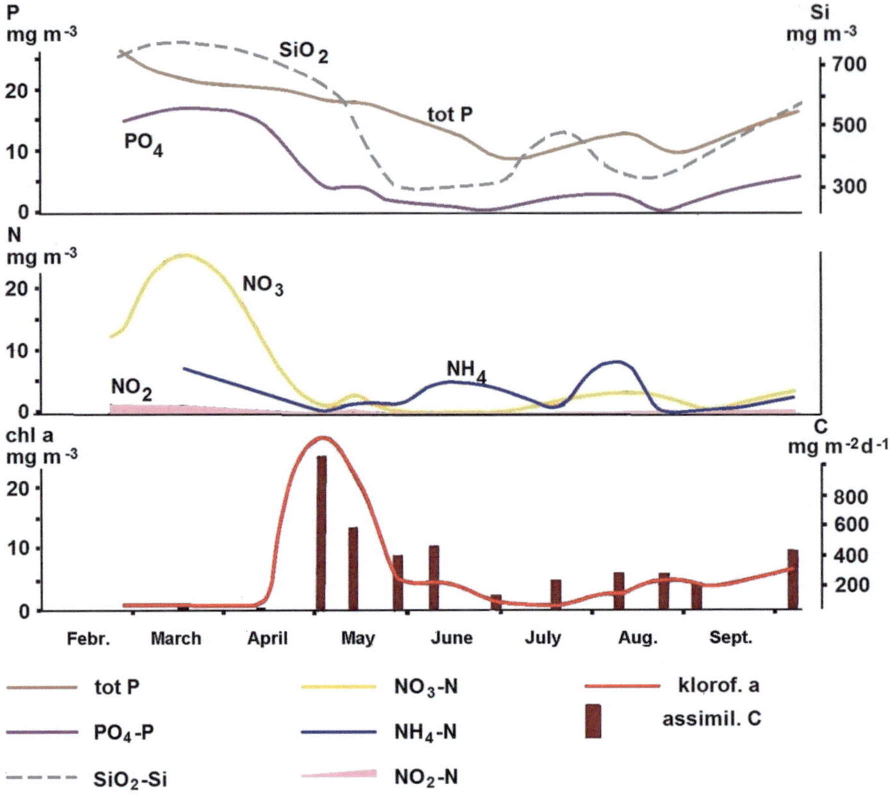

Fig. 3.2 Seasonal fluctuations in primary phytoplankton production, chlorophyll-a, nutrients and hydrographic properties in 1969 at the entrance to the Gulf of Finland (mean values of the euphotic layer) (From Ojaveer 2014 after Voipio 1981)

last stage of this development in coastal areas, the freshwater diatom *Diatoma elongatum* dominates. Up to the early summer, species of the genera Rhizosolenia, Protoperidinium and Geratium also emerge.

The nitrates and phosphates present in the upper, mixed layer are used in phytoplankton development, which continues up to the expiration of nitrates in this layer regardless of phosphates. During the spring production period, the abundance of consumers of this production is rather limited. Therefore, a large part of the production sinks from the productive water layer to the layer of consumers. This is a substantial addition to the benthic system.

Expiration of the spring bloom is related to the nutrient limitation. After the spring maximum, in the stage of low production under the conditions of temperature stratification, only traces remain of phosphates, nitrates and nitrites in the euphotic layer. The limits of this layer generally coincide with the borders of the mixed layer. In summer, Cyanobacteria, procariotic autotrophic bacteria, dominate. At temperatures of 17–18 °C, *Aphanizomenon flos-aquae* and *Nodularia spumigena* can develop extensive blooms. Autumn is characterized by the outcrop of *Ceratium* spp. and the diatoms *Actinocyclus octonarius* and *Coscinodiscus granii*. In the Belt Sea, at the domination of outflowing Baltic Sea water, phytoplankton of the central Baltic is in the majority. In the case of the domination of the Kattegat, water species of marine background prevail. The concentration of siliceous compounds is at its annual minimum, but this probably does not limit primary production. In this period, phytoplankton reuses the biogenic salts involved during the preceding production cycles (which by this time have gotten rid of the mortal remains of planktonic and benthic organisms). In the northern Baltic Proper and in the Gulfs of Bothnia, Finland and Riga, this low level period of primary production occurs in June and the first half of July. At that time, the chief producers are small flagellates. Due to good transparency, the euphotic layer is comparatively deep. At this stage, probably only a limited amount of organic matter sinks to the bottom.

The greater part of phytoplankton consists of diatoms, dinoflagellates, cyanobacteria and a number of other groups. This period of generally low productivity ends in late summer. Beginning in July, *Aphanizomenon flos-aquae*, *Nodularia spumigena* and cyanobacteria capable of binding nitrogen from the atmosphere are in the majority. Their blooms can be seen from July until the early autumn, and even up to October in some years.

Blooms of cyanobacteria usually occur after calm and settled weather. Plankters accumulate in the upper water layers or at the surface. Blooms are supposedly facilitated by upwellings, which are rich in phosphates. However, just before and during the blooms, no phosphates and nitrates have been found in the euphotic layer. The blooms are possible because the species participating in them have the following characteristics: (1) they can accumulate large phosphorus reserves in their cells, which they utilize for growth in case the phosphorus reserves in the environment become exhausted; (2) thanks to the gas vacuoles, in the period of intense growth, they rise to the water's surface; (3) they can bind nitrogen from the air, therefore they do not depend on the mineral nitrogen in the environment. In addition to the above, cyanobacteria are rather poor food for zooplankton, which they avoid if possible.

Part of the nitrogen bound from the air leaves the cells for the environment at the bloom, as does the remaining part afterwards, during the disintegration of the cells. Some phosphorus from below the thermocline is brought to the upper water layers by the cyanobacteria rising into surface layers. Therefore, it is thought that cyanobacterial blooms enrich the surface water layers with nutrient salts.

The addition of phosphates, which may be caused, e.g., by upwellings, enhances cyanobacterial growth. The process also facilitates increasing production in higher levels of the ecosystem.

In October, the phosphorus and nitrogen content in the surface water rises. However, it is only in an especially favourable environmental situation that the autumnal mixing of water layers reaching the halocline (or, in the absence of a halocline, the bottom) gives rise to diatom blooms. In November and later, the primary production in the open sea is very low, as intense mixing of water layers takes place and the light conditions are poor.

In the Bothnian Sea, the maximum of phytoplankton growth starts later than in the Baltic Proper. In Bothnian Bay, the maximum phytoplankton development takes place in June. It consists mainly of brackish-water diatoms (*Chaetoceros wighami*, *Skeletonema costatum*, etc.), freshwater diatoms (*Diatoma elongatum*, *D. vulgare*) and dinoflagellates. Already by October production has considerably decreased. The conditions for primary productivity in Bothnian Bay are markedly differ from other areas of the Baltic Sea (the period is notably shorter than in other parts of the sea, with the maximum occurring in midsummer, which is characteristic of the Arctic conditions). The annual primary production of carbon clearly differs between sea areas. In Bothnian Bay, it is lower than in other parts of the sea, probably due to differences in illumination and other environmental conditions, as well as the low concentration of nutrients, especially of phosphates compared to nitrates (Voipio 1981).

Primary production is created in pigments of a green cell, i.e., mainly in plants distributed in planktic (phytoplankton) and bottom layers (phytobenthos). Productivity is based on photosynthesis: the light energy absorbed by chlorophyll and other pigments of a green cell initiates reactions, which result in the synthesis of organic matter from carbon dioxide and water. As the first step, the light energy is transformed into chemical energy, and thereafter inorganic carbon is converted into organic, first into glucose, and then into various parts of the cell.

As the chlorophyll *a* content in autotrophic and mixotrophic phytoplankton organisms is known, it can be applied as an indicator for the primary phytoplankton parameter. For measuring primary production rates in an aquatic environment, the oxygen method (based on the measurement of the released oxygen) and, since 1952, the ^{14}C method (based on the fixation of inorganic carbon in organic matter, Steemann Nielsen 1975) have been used. The latter, the ^{14}C method, allows for estimation of the size of the primary production with satisfactory precision. It is widely used in standard investigations.

In various areas of the Baltic Sea, a number of primary production estimates have been made. However, their comparison is complicated due to strong variability in insolation, natural patchiness, etc. Based on the data from a reference station near the Askö Laboratory, Elmgren (1989) estimated that from the beginning of the twentieth century up to 1985 primary production in the Baltic Sea increased by a factor of 1.3–1.7, from the estimated original 74–98 g cm^{-2} to 127 g cm^{-2} at the time of his studies. Wasmund and Siegel (2008), estimating the results of a number of authors, indicate that in the total Baltic Sea, a clear increase in primary production took place from the 1970s to the 1990s. Over roughly two decades, primary production doubled (from 84 to150 g c/m^2 $year^1$). This increase has mainly concerned the Kattegat, the Belt Sea and the Baltic Proper.

3.5 Bottom Vegetation

A recent inventory (HELCOM 2012a) showed that in the Baltic Sea, 531 species of bottom vegetation exist altogether. The bulk of the phytobenthos consists of macrobenthic algae Bangiophyceae (Rhodophyceae) – 156 species, Fucophyceae (Phaeophyceae) – 128 species, Tribophyceae (Xanthophyceae) – 12 species, Charophyceae – 16 species and Chlorophyceae – 110 species. This number also includes 74 species of higher (flowering) plants, 7 species of bryophytes, etc. (Nielsen et al. 1995). The richest region in bottom macroalgae is the Kattegat (325 species). Along the declining salinity gradient, the number of species decreases to less than 100 in the Gulf of Bothnia. The coastal areas used for the treatment of bottom vegetation by the BMB Working Group are shown in Fig. 3.3. The greatest decrease in the number of marine algae has been found between subareas 1d (the Sound) and 1e (the area south of the Sound) and 2a and 2b (westward and eastward of the Darss sill). The distribution of species of bottom vegetation follows the regularities of the distribution of other species in the Baltic Sea. Water salinity has the greatest impact. In the southwestern and southern areas of the sea, the species of marine background are in the majority, while in the northern and eastern parts,

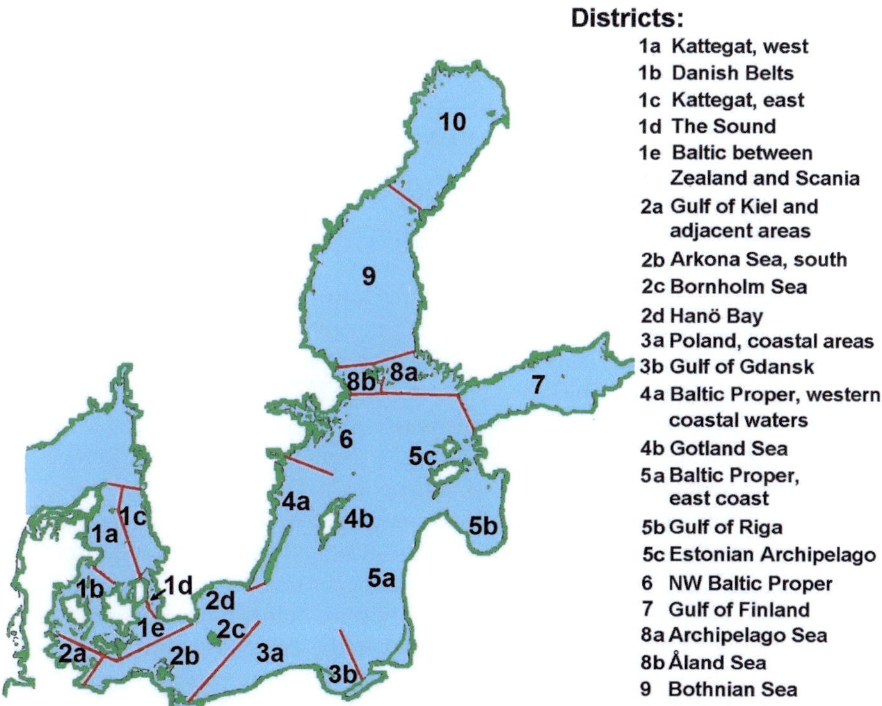

Fig. 3.3 Geographical grouping of bottom vegetation in the Baltic (Ojaveer 2014)

brackish-water and freshwater species dominate. The number of species found all over the Baltic Sea is rather small, including only some filiform chlorophytes. In the western and southern parts of the sea, red and brown algae dominate, as do chlorophytes to the east and north (Ojaveer 2014).

The Baltic Sea is a very complicated environment for benthic vegetation. The north–south and east– west salinity gradients, alternating coast types, variable substrates, changing water transparency related to eutrophication, inflow of fresh water of various properties and other environmental conditions create unique conditions for the bottom plants.

Regularities of the Distribution of Bottom Vegetation in the Baltic Sea Overall, the lower limit of the distribution of bottom vegetation in the Baltic Sea is 30 m. Generally, the lower border of bottom vegetation is 20–25 m; in the gulfs of poor water transparency, benthic plants occur only to a depth of 5–10 m (Trei 1991). Consequently, production of organic matter takes place in a limited depth interval. Outside of it, life can exist only when biological production created by phytoplankton and phytobenthos in the euphotic layer of a rather small vertical range is consumed.

One of the most important regularities followed by the distribution of bottom vegetation in the Baltic is actually followed by aquatic vegetation globally. This is the vertical zonation of phytobenthos. In the Baltic Sea, phytocenoses populating various depth zones are differentiated based on their structure and functions. Three to four vertical zones of phytocenoses are differentiated, also depending on the character of the substrate of their habitat (Martin 2000).

(a) The lowest depths of the coastal slope belong to the zone of green algae. The vegetation is composed of filamentous green algae of a rapid growth rate. In summer, the most common inhabitants of this zone are *Cladophora glomerata* and *Ulva intestinalis*. Depending on the trophic level of the sea area and its salinity, mass occurrence of other species of green algae is also possible. Generally, this zone reaches the depth of the lowest level of sea surface. In cases of soft bottoms, this zone is rather poor in vegetation. In the bights closed to waves, this depth zone is usually occupied by reeds.
(b) The zone situated just below the previous zone hosts brown algae on hard substrates. There, the vegetation of this zone can be very rich in species. In the central and southern Baltic with salinities over 3–4 psu, bladder wrack and other perennial species dominate simultaneously. The depth distribution of this zone is very much related to the local conditions, especially to water transparency. Also, the depth reach of this zone can be influenced by the presence or absence of hard substrate. On the soft bottoms of this zone, the vegetation can be composed of flowering plants or some species belonging to Charophyceae.
(c) The zone of red algae is situated deeper than the zone of brown algae. In this zone, perennial red algae dominate. However, species of brown algae and even some species of green algae can also occur there. In the case of acceptable salinity conditions in the central Baltic, *Furcellaria lumbricalis* or other perennial species of red algae with strong thalli are in the majority. On soft bottoms, this

zone is, in most cases, free of vegetation. On some exceptional occasions, the species characteristic of this zone can occur as loose-lying forms (e.g., loose-lying *Furcellaria lumbricalis*).

Some authors are of the opinion that on hard bottoms under the zone of brown algae, one more zone should be differentiated. The lower limit of bottom vegetation is situated there. The zone is named after *Mytilus edulis*.

The Baltic Sea is situated in the climatic zone characterized by a clear impact of seasonal changes on environmental conditions. This exerts an influence on the development of bottom vegetation. A number of important environmental conditions controlling the development of bottom vegetation vary with seasons: temperature in the upper water layers, the amount and quality of light energy reaching the sea bottom, ice conditions, wave regime, etc. This also causes seasonality in association with bottom vegetation. Seasonality is expressed chiefly in shallower water (in the zones of green and brown algae) and is associated with the mass development of some opportunistic ephemeral species during certain periods. These variations are usually not related to perennial, long-lived species, although in different seasons, the quantitative structure of associations can substantially vary. Such species of seasonal occurrence are a number of filiform brown algae like *Pilayella littoralis* and *Ectocarpus confervoides*, some red algae like *Ceramium tenuicorne*, etc. Also, occurrence of the only endemic algal species in the Baltic Sea – *Monostroma balticum* – just after the disappearance of ice in early spring is related to seasonality.

Environmental Factors Affecting the Distribution of Bottom Vegetation Bottom vegetation is simultaneously influenced by a large number of environmental factors. Their importance varies depending on the temporal and spatial scale of the conditions. For the ecosystems of the whole sea, the most important are water salinity and the factors related to climate. Due to the salinity gradient, the stress on the species of marine origin increases towards the north and the stress on the species of freshwater origin towards the south. Of the factors related to climate, temperature has a very important effect, manifesting itself in the intensity of metabolism, duration of the ice period with its diversity and impacts, duration of the growth period, etc. The growth period decreases from south to north. The ice period is about half a year in the Gulf of Bothnia, whereas in the central, southern and southwestern Baltic, it is much shorter, and in some years may even be absent. These conditions primarily limit the living conditions of the annual plants. In the Gulf of Bothnia, annual macroalgae of freshwater origin dominate. The share of perennial and marine species markedly increases towards the south. These circumstances are the most important in the formation of the frequency of occurrence of marine/freshwater and annual/perennial aquatic plants in various areas of the Baltic Sea (Kautzky 1988).

For the distribution of bottom vegetation on a regional scale, the general aspects of environmental conditions (e.g., salinity) can be regarded as background information, while other conditions that are important on a local scale are of vital importance. These include

(a) factors influencing light availability: depth of the habitat, transparency and content of suspended matter in water, depth of the photic zone, the level of eutrophication, etc.;
(b) character of the attachment substrate: the nature of the substrate (hard/soft, stable/mobile), the amount of loose sediments on the hard substrate (sedimentation), oxygen content or its absence in the soft sediments. The absence of suitable substrates for attachment restricts the depth range of bottom vegetation in the areas where mobile soft bottoms prevail. On the most sheltered muddy shores, the vegetation is constituted by reeds, rhizophytes (*Potamogeton* spp., *Zostera* spp. etc.) and also charophytes. Strong wave action excludes the vegetation from sediment bottoms due to the instability of the substrate. In the areas of bedrock coasts, the substrate for the attachment of algae is good and supports vegetation even in the areas of the outer archipelago;
(c) mechanical activity of marine water: openness to the wave actions, velocity of near-bottom currents, impact of ice damaging the bottom mechanically or inducing seasonal oxygen deficit. Strong wave action excludes the vegetation from sediment bottoms due to the instability of the substrate. The algal vegetation of rocky shores is affected mainly with respect to the relative abundance and morphology of the species. The species composition is less variable. The biomasses tend to be inversely related to the intensity of the wave action;
(d) biological interrelations between organisms involving a large number of effects, beginning with competition for the attachment substrate, light, nutrient salts, etc.

The majority of the above factors are mutually related and the resulting effect of their impact is very difficult to describe or predict. The situation is complicated by anthropogenic eutrophication and toxic pollution, which, in addition to their direct impact, alter the total effect of various simultaneous natural factors.

Figure 3.4 illustrates the formation of bottom vegetation of characteristic structure and biomass on the bottoms of variable characteristics in the Gulf of Riga, accounting for different conditions of human impact (Martin 1999). In group (1) at a depth of 1–2 m, filiform green and brown algae of high biomass dominate. The area of the group is confined to the relatively eutrophied areas of the coastal zone north and northeastwards of the Daugava River, and in Pärnu and Kuressaare Bays, which all have soft bottoms and waters of limited transparency. Species diversity in this group is limited: one species mainly dominates. Group (2) occurs at the eastern and southwestern coasts of the gulf, at the southwestern coasts of Saaremaa Island and between Kihnu Island and the mainland on hard and mixed-type bottoms. In this group, the species diversity is notably higher than in the previous group, and the biomass and coverage down to a depth of 5–6 m are more or less uniform, although lower than in the former group. The area of group (3) mainly embraces mixed-type bottoms at the west coast of the Gulf of Riga and most of the southern coast of Saaremaa Island. Anthropogenic impacts are relatively weaker than in the areas of the previous groups. Biomass and coverage are higher than in the other groups. In the upper part of the littoral, *Fucus vesiculosus* is in the majority, with maximum

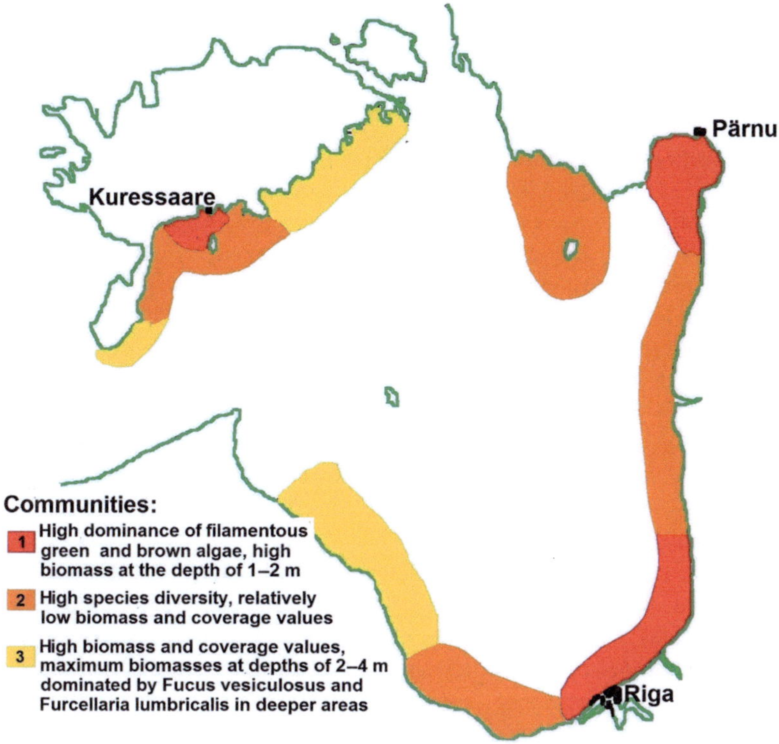

Fig. 3.4 Impact of substrate and human activity on the formation of assemblages of bottom vegetation in the Gulf of Riga (Ojaveer 2014)

biomass in the depth interval from 2 to 4 m, while *Furcellaria lumbricalis* dominates deeper.

The communities of bottom vegetation are simultaneously influenced by a number of various environmental factors, which all participate in the formation of the community during certain periods. Here, the temporal factor is also of importance, as some species are annual, and others perennial, with complicated reproduction mechanisms/cycles.

Trei (1991) draws attention to a characteristic peculiarity of the Baltic Sea: in addition to the habitat of the main biomass of bottom vegetation, certain habitats substantially differ from them, for example, boddens and haffs in the southern part of the sea and ponds of some hundreds of square metres on rocky coasts and islands at the northern coasts of the sea. In the Darss–Zingst area at the northern coast of Germany, in the chain of shallow boddens with soft bottoms and very unstable external conditions, only some species of ecologically extremely plastic aquatic plants are able to endure the existing conditions, among them charophytes, the green algae *Enteromorpha* and *Cladophora*, and also some rhizophytes.

In the habitats separated from the open Baltic by the Curonian, Vistula and Odra sand spits, environmental conditions (including salinity) are somewhat different, but their aquatic vegetation is rather similar. In the Vistula Lagoon, the marine element is composed of green algae. Other vegetation consists of reed, *Potamogeton* spp., etc. This is probably related to the characteristic chemical regime in the areas and the absence of rocky bottoms necessary for the attachment of vegetation.

The Most Important Species of Bottom Vegetation Bladder wrack (*Fucus vesiculosus*) is one of the most widely spread species of aquatic plants in the Baltic. It is thought to be the key species of littoral communities. It originates in the tide zone of the North Atlantic rocky coasts, but has occupied a rather different habitat in the Baltic – the area deeper than the limit of the lowest water level. The species is distributed throughout the entire Baltic Sea, settling in moderately opened habitats with hard substrates at salinities of 3–4 psu. The plant can grow to a length of one metre and have a dry weight of 1.5 kg. Different authors have estimated its highest age from 5 to 25 years. Bladder wrack is a relatively tenacious species, capable of standing short-term worsening of light conditions, deficit or oversaturation of nutrients, epiphytes, and rather strong mechanical stress from crushing waves or ice. However, the species is sensitive to long-term changes in the quality of environmental conditions and has deserved much attention related to the notable decrease in its population in a number of areas of the Baltic Sea at the end of the twentieth century. The latest data, however, refer to a partial restoration of the communities at the Åland Islands and in the northern Gulf of Riga, etc. (Torn et al. 2006).

Furcellaria lumbricalis is the key species of bottom vegetation on hard bottoms below the zone of brown algae. Two ecologically separable forms of this species occur in the Baltic Sea. The fixed form, *F. fastigiata*, is very common, settling hard substrates in the depth interval between 3 and 15 (20) m. Rather frequently, *F. fastigiata* populates the lower depth limit of the bottom vegetation. The loose-lying form of *F. lumbricalis* is rare, and its areas of occurrence can be found in the Väinameri area. In Kassari Bay (depth 5–9 m, salinity 6–7 psu), Väinameri, *F. lumbricalis* with other species of red alga (mainly *Coccotylus truncatus*) constitutes a unique community stretching like a continuous carpet over more than 200 km^2 of bottom. The 3–4 cm to 10–15 cm thick and sporadically even thicker algal mass, estimated at about 140,000 tonnes (Trei 1991), continuously moves in circular currents. Presently, *F. lumbricalis* is the only commercially used macroalgal species in the Baltic Sea (both loose-lying and fixed forms are used).

Zostera marina is the only widely distributed and also the best known marine rhizophyte in the Baltic. The species grows on sandy and gravelly sites on moderately opened bottoms at depths of 1–4 (5) m. In the southwestern part of the Baltic, its length may be up to 1 m, and in the central part of the sea, 20–30 cm. The assemblages of *Zostera* constitute habitats for a number of bottom invertebrates and epiphytes. Also, a number of fish use the species as their spawning substrate. In the northern Baltic, the species reproduces only vegetatively. Due to its sensitivity to eutrophication and vegetative propagation, the species is rather endangered today. During recent decades, its distribution has diminished in the Baltic Sea. The species is limited first of all by the decrease in water transparency and the mechanical disturbance of bottoms.

Charales (*Chara aspera*, *C. baltica*, *C. canescens* etc.) is a special characteristic brackish-water order of mainly freshwater algal phylum. They are mainly connected with unspoilt environments. Therefore, species of this order are used in assessments of coastal environments as indicators of the good quality of the sea water.

Utilization of Macrovegetation of the Baltic Sea In marine ecosystems, plants are of unique importance as suppliers of oxygen, sources of energy and producers of organic matter. However, marine plants are also directly used in everyday life. The brown alga *Furcellaria lumbricalis* is used for manufacturing agar-agar. Danagar has been manufactured from the raw material taken from the central Kattegat beginning in 1946. In Poland, agar-agar has been manufactured from drifting *Furcellaria lumbricalis* from Puck Bay since 1960. Starting in 1966, a gelatinous substance has been produced from the agar-agar manufactured from *Furcellaria lumbricalis* harvested in Kassari Bay, Estonia. It is used for reasons confectionary, culinary, etc., and also as a nutrient substratum in microbiology and medicine (Trei 1991). Large amounts of bladder wrack thrown up on the coast during gales are used for the fertilization of fields, but could also be applied to the production of alginates necessary for manufacturing foodstuffs, textiles and medicine. *Zostera marina* is, to some extent, used in the furniture industry.

3.6 Heterotrophic Microorganisms

In the Baltic Sea, microbiological studies were started at the end of the nineteenth century (Yurkovska et al. 1983). Nevertheless, compared to other living systems, data on microorganisms, including heterotrophic microorganisms, are limited, especially the long-term data both from the sediments and the water of the open sea. However, the role of both the planktonic algae treated as microorganisms and the heterotrophic microorganisms is extremely substantial in the ecosystems of the Baltic Sea. They constitute a very important level of the secondary production utilizing dissolved organic material in such amounts and concentrations impossible on other levels, and break down organic compounds difficult to change. This is due to the high level of their metabolism, intense multiplication and conspicuous ability for adaptation to the environmental changes. Microorganisms are able to assimilate and transform very variable organic compounds (Apine 1984; Platpira 1995; Yurkovska et al. 1983, etc.). They can serve as food for organisms of higher trophic levels, accumulators of pollutants, and sources of diseases for other organisms. They also can be used as indicators of ecosystem status. Microorganisms take part in eutrofication processes together with primary producers at the beginning of the food chain.

The most essential functions of microbes in the ecosystem are (1) regeneration of mineral compounds of biogenes terminating the circulation of matter in the ecosystem; (2) biodegradation of pollutants resulting in full or partial purification of the water body from them; (3) execution of nitrification, denitrification, mercury methylation, reduction of sulphates and other important biochemical processes; (4) utilization of various dissolved organic compounds (including excretions of plants and animals) and rendering them accessible to zooplankton. This facilitates effective consumption of primary and bacterial production by organisms of higher trophic levels. Microbes participate in the regulation of inorganic nutrient resources and in forwarding them to the heterotrophic organisms. Therefore, the activity of microorganisms is important both in pelagic and demersal food chains.

The importance of microorganisms as indicators of pollutants is related to their wide adaptation abilities, large variation in physiological properties and high biochemical activity. The microbial indication of the state of a marine environment is based on the relation of the physiological activity of microflora and the character of pollutants. This is an important means for early diagnostics of changes in the chemical regime.

The variation in the production and biomass of bacterioplankton should be treated as components of changes in the common inorganic and organic stock, which reflect the state of the aquatic environment. Generally, phyto- and bacterioplankton react similarly to changes in abundance and biomass due to the addition of inorganic nutrients. In coastal areas, this reaction to the changes in environmental conditions may differ, as the part of the nutrients originating in the mainland may cause alteration of only the production and biomass of bacteria.

Because of large gaps in the knowledge of microbiological processes, we do not have acceptable databases for the estimation of the characteristics of biogeochemical transformation, biodegradation of pollutants, the rate of mineralization of organic matter and a number of other processes. Therefore, for the characterization of ecosystems, indirect methods, that is, the total abundance of microorganisms, abundance of specific groups of organisms having the ability to oxidize certain substrates etc., should be considered.

In the Kiel and Mecklenburg bights, the Gulfs of Riga, Finland and Bothnia, and also in some open sea areas, more and more complete microbiological investigations have been carried out. Therefore, some generalizations can be made. HELCOM (1996) data indicate that the annual average values of bacterioplankton production in different regions of the Baltic Sea are substantially different rising from the northern areas to the southwest (in gC m^{-2}): Bothnian Bay 13, coastal areas of the Bothnian Sea 17–21, open part of the Bothnian Sea 22, coastal areas of the Gulf of Finland 9–19, coastal areas of the northern Baltic Proper 29, the Belt Sea 22–43.

Regional Differences: Open Baltic Parts of the Baltic Sea differ in the annual level of phytoplankton and also in bacterioplankton productivity. The distribution of bacterioplankton is very uneven: in summer, the biomass can vary from the level of a gram of carbon per cubic metre of water in the southern parts of the sea to hundredths of a gram in the north (Yurkovska et al. 1983). On the other hand, it has been shown that in large areas, bacterioplankton has only little variation both annually and seasonally. This may indicate its dependence on comparatively stable factors (control by predators, dependence on bacterioplankton carbon sources in the food web). Under the conditions of the existence of a thermocline and temperature stratification, the largest bacterioplankton assemblages are situated in the upper 30 m water layer. Below the thermocline, the concentration of bacterioplankton sharply diminishes, but rises again towards the bottom. Both the vertical distribution of bacterioplankton and the annual differences in the structure of microorganisms are notable (Yurkovska 1990).

Bacterioplankton is substantially influenced by seasonal changes in the environment. The lowest concentration of bacterioplankton has been found in February and March. The highest concentrations at the same place have been observed in September–October. In the central and northern parts of the Baltic Proper, seasonal regularities have been found in the reproduction of microorganisms: in autumn, reproduction is estimated to be twice, and in winter, 4–5 times slower than in summer. In the scientific cruises from the northeast to the southwest Baltic, it was noted that in spring, phytoplankton developed patchily and bacterioplankton developed after phytoplankton with a time shift of some days. In summer, the growth of phytoplankton and also of bacterioplankton is limited by the deficiency of nutrients. The growth of bacteria primarily depends on the phytoplankton carbon supply, and the abundance of bacterioplankton cells is in significant correlation with the amount of chlorophyll *a* in phytoplankton. The lower part of the microbial food web, the pico- and nanoplankton, dominates the carbon and energy flow during the late summer season in the open parts of the Baltic Sea (HELCOM 1996).

The distribution of physiological groups of bacteria (saprophytes, petroleum-oxidizing bacteria, including paraffin- and xylol-oxidizing bacteria, lipolythic and solar-oil-oxidizing bacteria) depends on area, season and depth. Saprophytes – ammonifiers – are widely distributed throughout the entire Baltic and they are also plentiful in the central part of the sea. In the northern Baltic, their average number is about 10^3 cells cm^{-3}, in the southern Baltic, about 10^4 cells cm^{-3}. The most widespread taxonomic groups are *Pseudomonas* and *Micrococcus* (Yurkovska et al. 1983).

In all the areas investigated in the Baltic Sea (but not at all depths), bacteria using petroleum hydrocarbons as their source of energy and carbon have been found. The abundance of bacteria oxidizing hydrocarbons has varied in the study areas between 0 and 10^6 cells cm^{-3}. These and lipolythic bacteria were found over the entire studied area of the Baltic Sea, with their mean abundance being 10 cells mL^{-1} (Yurkovska 1990).

Occurrence of the xylol-oxidizing bacteria in the Baltic was detected in 1976, and since then, they have been found at ¾ of sampling stations. They constitute 0.1–0.01% of the saprophytic microbes, 10–1000 cells cm^{-3}. In the coastal zone, their abundance increases to 1000–10,000 cells cm^{-3}. Lipolythic bacteria, which take part in the treatment of petroleum products, occur mainly in the upper 0–30 m water layer. Their abundance constitutes 0.1–10% of the saprophyte numbers or 0.1–1% of the total numbers of bacteria (Yurkovska et al. 1983).

Coastal Zone Microbiology of the coastal zone of the sea has been rather unsatisfactorily studied. According to Yurkovska et al. (1983), the biomass of bacterioplankton in the eastern part of the open Baltic between Baltiisk and Kolka varies from 0.02 to 0.86 mg dm^{-3}. Saprophytes, the most sensitive indicators of the easily oxidizing organic compounds, are distributed unevenly in this area. Their highest density (1000–20,000 cells cm^{-3}) occurs in the vicinity of ports and settlements, at a distance of 2–3 km from the coast. In other areas, the abundance of saprophytes is comparatively low. The number of petroleum-decomposing lipolythic bacteria reaches up to 100,000 cells cm^{-3}, of xylol-oxidizing bacteria 10–1000 cells cm^{-3}. The proportion of these bacteria in the total number of microorganisms of this area is rather low.

Bottom sediments of the coastal zone are composed mainly of fine-grained yellow sand and fine gravel. The total number of bacteria in the surface layer of these sediments is rather small: 36.3–159.6 mill cells in a gram of sediment. The concentration of saprophytes constitutes 0.07–0.84% of the total abundance of bacteria. Anaerobic cellulose-decomposing bacteria occur in rather low numbers (10–1000 cells g^{-1}). The maximum concentration of petroleum-oxidizing bacteria in bottom sediments of the coastal zone is found in the harbour areas at Baltiisk and Klaipeda.

Kiel and Mecklenburg Bights High seasonal variability in bacterioplankton production, with maximum values during summer and/or autumn, has been recorded in all localities. Regression analysis between the rate variables of phyto- and bacterioplankton show a tendency towards an increasing correlation along a gradient from coastal to offshore stations with the best fit at the open sea stations.

This suggests that bacterioplankton measures are not fully phytoplankton-dependent and that bacterioplankton carbon production should be used in assessing the organic production of the Baltic Sea (HELCOM 1996).

Gulf of Riga In the southern Gulf of Riga, microbiological studies were started at the end of the 1960s, chiefly in estuarine areas of large rivers. Their main goal was the collection of data for evaluation of the sanitary state of the area and also assessment of the participation of microorganisms in the process of transformation and destruction of organic compounds. Afterwards, the studies were extended over the whole gulf.

The bacterioplankton of the Gulf of Riga includes various morphological forms, mainly bacilli and cocci. The majority of stems are represented by simple bacilli, whereas cocci and sporiferous microbes are rare. Single bacteria are the most numerous; yeast-like microbes also occur. Bacterial micro-colonies are found relatively seldomly. Investigations into the biochemical properties of bacterial neuston and bacterioplankton cultures have shown that the microorganisms isolated from neuston have the highest activity.

Long-term investigations have shown the total count of bacteria in the Gulf of Riga to vary between 0.4 and 5.4 mill cells cm^{-3} (Yurkovska et al. 1983). In 1988, the abundance of bacterioplankton varied from 0.6 mill cells cm^{-3} in the central part of the gulf to 3.2 mill cells cm^{-3} in the coastal areas, with the average for the gulf being 1.7 mill cells cm^{-3}. Bacterial abundance is the highest in surface layers and decreases towards the bottom.

Regionally, the highest bacterial abundances have been found in the southern part of the gulf against the mouths of the Daugava and Gauja Rivers and in the coastal zone, while the minimum values have been registered in the Irbe Sound (Apine 1984). The biomass of microbes in the Gulf of Riga is variable, depending on areas, depths, etc. High values have been found against the Daugava River (the average biomass 0.90 mg dm^{-3}). This is supposedly connected to the organic matter carried to the gulf by river water containing large quantities of municipal and industrial wastes and its positive impact on the development of microflora. The temporal dynamics of bacterioplankton was investigated in the neighbourhood of estuaries of the large rivers Daugava, Gauja and Lielupe in the Jaunkemeri-Skulte area in 1988–1989. In winter, the abundance of bacteria was 0.04–1.02 mill cells cm^{-3}, characterizing the area as oligotrophic. In April, the number of microorganisms somewhat increased, and in May, in the period of freshet, the number of bacterioplankton rose to 4.6–8.1 mill cells cm^{-3}, which is characteristic of strongly polluted and eutrophied waters. In June, July and August, the numbers of bacteria varied between 1.88 and 5.09, 2.01 and 5.47 and 0.24 and 7.25 mill cells cm^{-3}, respectively. In October, November and December, the concentration of bacteria decreased, but still amounted in most samples taken at the beginning of November up to 3–5 mill cells cm^{-3}, indicating that the coastal waters were polluted and eutrophied.

These dynamics can be explained by large amounts of organic matter formed from the phytoplankton on the spot and brought in with rivers. Thus, the amount of bacterioplankton depends primarily on the mass of organic material, and not so much on temperature. Monthly surveys have shown that in the seasonal dynamics of bacterioplankton, two maxima occur: the summer maximum related to the optimum temperature as well as plankton development, and the early autumn maximum related to the elimination of phyto- and zooplankton (Fig. 3.5).

In the period from1977–1989 saprophyte bacteria, important in the destruction of labile organic matter, notably increased their abundance in the gulf. The number of ammonifying microorganisms varied between 10 and 10^4 cells cm^{-3}. In the layer just below the surface, it constituted 10^2–10^4 cells cm^{-3}. Investigations into the total abundance of saprophytes and ammonifiers allow for a conclusion that in the Gulf of Riga, bacterioplankton is distributed rather unevenly.

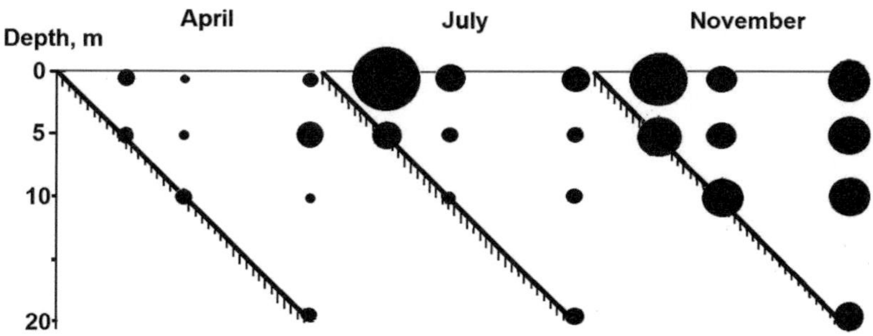

Fig. 3.5 Distribution of paraffin-oxydizing microbes at the mouth of the Daugava River in the Gulf of Riga in April, July and November 1989. In figure the area of the circle is proportional to the number of cells in ml^{-1}. The lowest sampling depth was 0.5 m below the surface (Platpira 1995; Ojaveer 2014)

Heterotrophic bacteria actively participate in the transformation of various compounds containing both natural and synthetic hydrocarbons. In the evolution process, microorganisms have developed fermentation systems for the biodegradation of hydrocarbons and the utilization of them as the growth substrate. Microorganisms making use of these compounds are widely distributed, including in marine systems. The occurrence of hydrocarbon-oxidizing microorganisms mainly indicates the intensity of self-purification processes in the marine environment and, to a lesser extent, the degree of pollution by petroleum hydrocarbons.

The number of paraffin-oxidizing microbes varies in summer between 250 and 2500 cells cm^{-3}, with their maximum concentration confined to the upper 0.5 m water layer. Spatial distribution of this group of microorganisms hints at the stability of substrates and self-purification processes in the Gauja, Daugava and Lielupe mouths at the 5 m depth isolines. In autumn, the amount of paraffin-oxidizing microorganisms varies within limits of 950–2000 cells cm^{-3}. The distribution of the group is rather homogeneous and their amount has been rather stable.

In the destruction of intermediate products of *n*-paraffins, the leading role belongs to the lipolytic microflora. In the Gulf of Riga, lipolytic microflora is abundant in the neuston adapted to the existence in the border layer between the hydrosphere and the atmosphere (Platpira 1995). This layer is characterized by large temperature fluctuations, instability of its salt regime and a high concentration of organic matter and oxygen. Towards the central part of the gulf, fluctuations in the abundance of lipolytic microflora increase considerably ($250–2.5 \times 10^6$ cells cm^{-3}). The near-bottom layer of the gulf is also rich in lipolytic microflora ($10^4–10^6$ cells cm^{-3}) (Platpira 1995). Consequently, the processes of microbial transformation of the compounds of lipid origin are most active in the areas of river mouths. Off the coast, intensification of the processes has been observed in the bottom layer. The amount of microorganisms in bottom sediments depends primarily on the content of organic matter in them. Transformation of the matter of lipid origin is most intense in the neuston at depths up to 20 m near the estuaries of the Daugava and Lielupe Rivers. In offshore areas, such processes dominate in the near-bottom water layers. Occurrence of paraffin-oxidizing bacteria in the water column indicates the self-purification capacity of the area of estuaries. Long-term data allow for the conclusion that the abundance, as well as the spatial and temporal distribution patterns, of paraffin-oxidizing bacteria has been rather stable. Toxic aromatic hydrocarbons have been found to affect microbial cenoses. The main function of microbes in the coastal water is the transformation of unstable organic matter.

Silt deposits represent a specific ecological zone in all water bodies. In the processes going on in bottom sediments, microorganisms have an important role. Development of bacterial populations depends on bottom characteristics and sediment properties. It has been shown that in water bodies, only those organic and mineral compounds that are situated in the surface layer of bottom sediments participate in the recirculation process. The destruction of organic matter in bottom sediments, including those in aerobic conditions, is a multistage process. The first stages of mineralization of organic matter take place under the influence of nonsporiferous microflora. With the degradation of easily accessible matter, sporiferous bacteria start to dominate. In the river-mouth areas, their abundance maxima are confined to the 0–2 cm and 2–4 cm layers. The maximum development of sporiferous bacteria in the Gulf of Riga has been found in the 9–11 cm layer in the pelite muds at a depth of 40 m. In summer, the abundance of aerobic saprophytes constitutes 900–312,000 cells g^{-1} (Platpira 1995). The distribution of saprophyte bacteria in the bottom deposits depends on the availability of labile organic compounds. It has been established that anaerobic microflora and sporiferous forms of these bacteria exist everywhere. The occurrence of anaerobic saprophyte microflora is connected with reducing conditions that have been revealed in the surface layer of bottom deposits. The relationship between nonsporiferous and sporiferous forms indicates that bottom sediments in the southern part of the Gulf of Riga are eutrophied (Platpira 1995).

Under the conditions of oxygen deficit usually resulting from the destruction of organic compounds, sulphates are reduced to hydrogen sulphide. This takes place with the participation of anaerobic microorganisms – the sulphate-reducing bacteria. The process of sulphate reduction up to hydrogen sulphide with the participation of living organisms takes place, for instance, in dissimilative reduction, e.g., in sulphate respiration resulting in the excretion of hydrogen sulphide. The main producers of hydrogen sulphide are sulphate-reducing bacteria. Their development is positively influenced by the reducing conditions in the bottom sediment. The distribution of the sulphate-reducing bacteria, constituting 51–700 cells $g^{-1}$of wet sediment, shows that the highest abundance of that microbial group is confined to the 2–4 cm layer of the deposit. The development of bacterial populations in the bottom layers largely depends on sediment properties.

Putrefactive bacteria produce hydrogen sulphide from albumen residues. It is known that these bacteria cannot maintain a high intensity of sulphur circulation in water bodies. The distribution of putrefactive bacteria in bottom deposits indicates the existence of seasonality in microbial development. In the mesotrophic areas, their numbers amount to 10–100 cells g^{-1} and their distribution in the 0–2 cm, 2–4 cm and 9–11 cm layers is rather uniform. In eutrophied areas, their maximum concentration (24,000 cells g^{-1}) is confined to the 0–2-cm layer, whereas in deeper layers, their numbers decrease considerably (Platpira 1995).

Gulf of Finland According to Telesh et al. (1999), at the beginning of the twentieth century, the whole Neva estuary was oligotrophic and the state of its ecosystem was determined by natural processes. By the early 1920s, glacial relicts were generally replaced by Oligochaeta and Mollusca, which are more resistant to organic pollution. The pollution of the Neva estuary continued and anthropogenic pollution accelerated.

Microbiological investigations in the Gulf of Finland started at the beginning of the twentieth century, but remained episodic. Studies during the 1970s–1980s showed that the concentration of bacteria in this gulf (0.3–6.4 mill cells cm^{-3}) is higher than in the water of the coastal zone of the Baltic Proper. The most conspicuous is the concentration of bacteria in the Neva estuary: 3.5–6.4 mill cells cm^{-3}. On 80% of occasions, the maximum concentration of saprophytes is situated at the depth of the distribution of bacterial neuston, indicating intense pollution of surface layers of the sea. The highest bacterial concentrations in sediments (120–263 mill cells g^{-1}) occur in sandy silt, while in fine-grained grey sand, the corresponding number is smaller (20–86 mill cells g^{-1}) (Yurkovska et al. 1983).

The variation in physiological groups of bacteria in bottom deposits in the Gulf of Finland is far wider than in the Gulf of Riga or in the coastal part of the open sea. The number of saprophytes in the surface layer of sediments amounts from tens to millions of cells per one gram of sediment and constitutes 0.1–10.4% of the total number of bacteria. This indicates active participation of microflora in the decomposition of organic matter. The bottom sediments of the Gulf of Finland are characterized by the prevalence of anaerobic microflora. Sulphate-reducing bacteria have been found in all sediment samples, their numbers amounting to 100–170,000 cells g^{-1}. The activity of mineralization of sulphates has been found to be the highest in the areas in the neighbourhood of large cities/ports (Yurkovska et al. 1983).

In the late 1990s, the development of the situation in the eastern Gulf of Finland and Neva Bay was classified as alarming, due to the progressing eutrophication and the excess of primary production of the decomposition of organic matter (Telesh et al. 1999). Considering the Gulf of Finland as a region of the Baltic Sea, HELCOM (1996) estimated that the development of its eutrophication, which was evident in the open and coastal waters during the 1970s and 1980s, had seemingly stopped.

Gulf of Bothnia The data collected during 1984–1992 indicate an increase in bacterial growth until 1992, after which the values started to decrease. The decrease during 1992–1994 was similar to the trend of **chlorophyll** *a* for that period. Within each year, the bacterioplankton show two intense growth periods: spring and summer. The pooled spring and seasonal data show higher spring than summer averages at the measuring stations. No clear evidence for the changes in the average annual bacterial biomasses have been found. The main biomass component of the heterotrophic bacteria in Bothnian Bay is chemoorganotrophic bacteria. This group is responsible for more than half of the secondary plankton production. Therefore, the main part of the carbon cycle and oxygen demand are related to the activity of bacterioplankton. Based on bacterial growth and carbon consumption, the food webs in the Bothnian Sea and Gulf of Bothnia are clearly different (HELCOM 2002).

3.7 Zooplankton

The largest exploitable resource in the Baltic Sea is pelagic fish – herring and sprat. The studies on the link of the food chain just before the pelagic fish – zooplankton – were started in the last decades of the nineteenth century. In the Southwestern Baltic, the first well-known zooplankton studies with corresponding methodical work were done by V. Hensen and K. Möbius, in the Transition Area, by C. Aurivillus, and in the northern areas of the sea and the Gulfs of Finland and Bothnia, by K. M. Levander and O. Nordquist. Along with the development of investigations, since the end of the nineteenth century and the first decade of the twentieth century, the attention of marine scientists was captured by the dependence of the composition, distribution, abundance and biomass dynamics of zooplankton in the Baltic Sea on the temporal and regional variability of environmental conditions.

In the Baltic Sea, the main food animals of pelagic mass fish are *Pseudocalanus minutus elongatus*, *Temora longicornis*, *Centropages hamatus*, *Eurytemora hirundoides*, *Acartia longiremis*, *Limnocalanus grimaldii*, *Evadne nordmanni*, *Podon polyphemoides* and *Bosmina coregoni maritima*. The annual planktonic food of a sprat weighs on average 89–180 g, that of a Baltic herring 135–200 g. In addition to herring and sprat, during certain periods, all fish species (including demersal fishes, relicts, migratory fishes – cod, flounder, turbot, smelt, salmon, sea trout, etc.), and also benthic invertebrates, live on zooplankton. Zooplankton is particularly important for fish larvae, being their principal food.

Composition Depending on the history and the complicated environmental conditions of the Baltic Sea, the number of zooplankton species in this sea is comparatively low. Ackefors (cited in Voipio 1981) has estimated that, presently, about 40–50 planktonic species and pelagic larvae of benthic invertebrates reside in the Baltic Sea. Adding to them the number of fish having pelagic larvae, the total number of species occurring in zooplankton rises to about 60. Copepods constituting the majority of holoplankton species are represented in the Baltic by only eight species. Yet it should be added that the production and biomass of some copepod species are comparatively high in this sea.

Only three brackish-water zooplankton species are thought to be endemic to the Baltic Sea: *Bosmina coregoni maritima*, *Keratella quadrata platei* and *K. cochlearis recurvispina*. A relict species *Limnocalanus grimaldii*, very important in the Baltic concerning food chains of the sea, dwells in comparatively cool and low-salinity habitats in deeper parts of the Gulfs of Bothnia, Finland and Riga.

Considering the requirements set to the basic environmental parameters, the animal plankton, like other groups of animals in the Baltic Sea, can be divided into marine, freshwater and brackish-water zooplankton. The species of marine background can propagate and dwell constantly only in high-salinity deep water, but the euryhaline marine organisms adapted to the brackish conditions live at lower salinity and also lower temperature.

In the zooplankton of the Baltic Sea, protozoans are mainly represented by marine infusorians. *Tintinnopsis tubulosa* is encountered in coastal waters from spring to autumn, *Mesodinium rubrum* with the highest abundance in spring, mostly in central parts of the sea. Up to now, the microzooplankton of the Baltic Sea has been studied rather incompletely.

Coelenterates include comparatively few species. *Aurelia aurita* with a diameter of up to10–20 cm primarily populates coastal areas of the southern parts of the Baltic. In the northern parts, this species is found infrequently and not every year. Also, the year-round areas of the cold-water *Cyanea capillata* of marine background (diameter in the Baltic Sea up to 30 cm) are mainly deep-water layers of higher salinity in the southern parts of the Baltic, but the species has also been encountered northwards, up to the central Gulf of Finland. The medusa of up to 16 cm height *Perigonimus yoldia-arcticae* is found in the water layers with a salinity of at least10 psu, mostly in the southern areas of the sea, from January to August. However, solitary finds of this species reach up to the Gulf of Finland. The medusa *Melicertum octocostatum* (diameter up to 20 cm) has continuously been found in the deep layers of the Bornholm Deep (Hernroth and Ackefors 1979).

Of Ctenophora, the only species found in the Baltic Sea is *Pleurobranchia pileus.* It populates areas of salinity of at least 7 psu in the open part of the sea and large gulfs.

The most important zooplankton species dwelling in the Baltic Sea belong to euryhaline eurytherm marine and euryhaline marine species whose optimum temperature is low and brackish-water species whose optimum temperature is relatively high. The bulk of zooplankton of the Baltic Proper is constituted by rotifers (*Keratella quadrata quadrata, K. quadrata platei, K. cruciformis eichwaldi, K. cochlearis recurvispina, Synchaeta baltica* and *S. monopus*), cladocerans (*Bosmina coregoni maritima, Evadne nordmanni, Podon leuckarti, P. intermedius* and *P. polyphemoides*) and copepods (*Pseudocalanus minutus elongatus, Centropages hamatus, Temora longicornis, Acartia bifilosa, A. longiremis, A. tonsa, Eurytemora hirundoides, Limnocalanus grimaldii* and *Mesocyclops leuckartii*). *Paracalanus parvus* of marine background lives mainly in the southern areas of the sea, *Oithona similis* both in its southern and central parts (Hernroth and Ackefors 1979; Kostritškina and Laganovska 1983).

Sagitta elegans baltica, which belongs to the chaetognaths, occurs at salinities higher than 10 psu, in the deeps of the southern areas of the sea. The species may constitute a significant share in both zooplankton biomass and food for fish.

Tunicates are represented by Appendiculariae, e.g., some- millimetre-long *Fritillaria borealis* populating both the open part of the sea and large gulfs. Young stages of bottom animals – annelids, cirripeds, molluscs and bryozoans – may be of great importance in the zooplankton.

Compared to the eumarine seas, in the Baltic Sea, the trophic structure of zooplankton is simple. In the planktonic coenoses, the following trophic levels have been differentiated (Kostritškina and Laganovska 1983): (1) primary producers and sarcophagi – phytoplankton and bacterioplankton; (2) phytophags or herbivores – *Pseudocalanus minutus elongatus, Temora longicornis, Eurytemora hirundoides,* nauplii and younger copepodites of all copepods, *Bosmina coregoni maritima, Fritillaria borealis,* Bivalvia larvae, etc. Important components of their food are particles of organic matter with dissolved organic matter and bacters stuck to them; (3) euryphags or omnivores: *Acartia* spp., *Centropages hamatus, Oithona similis* (copepodites), *Synchaeta similis, Balanus* (larvae). Their main food consists of phytoplankton, detritus and microzooplankton; (4) carnivores (*Pleurobranchia pileus, Sagitta elegans*). Their main food is zooplankton, other invertebrates and fish larvae. The most important of them is phytophagous zooplankton.

On the basis of very preliminary calculations, the average annual zooplankton production in the central part of the open Baltic has been estimated at about 100 g m^{-2} $year^{-1}$ in wet weight (equivalent to 5 g Cm^2). Microzooplankton is not included in the calculations. Taking a rough estimate of primary production at 100 g Cm^2 a^{-1}, the relationship between the primary and secondary production outside the coastal waters is 20:1. In the coastal areas, zooplankton production is relatively high (Hernroth and Ackefors 1979).

Distribution In the Baltic Sea, the composition, distribution and production of zooplankton differ by region, as the environmental conditions regulating the relevant processes have a regional character. Temperature and food security are considered the primary external conditions defining the abundance and biomass of zooplankton throughout its adapted salinity range. Reliability of feeding conditions gains the highest importance in the period of low levels of nutrient salts and phytoplankton in summer. These external conditions and illumination also form the background for the vertical distribution and diurnal migration of zooplankton.

Vertical stratification of zooplankton is expressed throughout the year. The Baltic, as a sea of the temperate zone, is characterized by sharp seasonal zooplankton dynamics, caused first of all by the temperature regime and the dynamics of food organisms. In the 1960s–1970s, the latter was influenced above all by phosphorus and nitrogen compounds, which limit primary production.

In winter, when chiefly copepods occur in plankton, the biomass is the highest in deeper water layers. The winter season lasts from December to March. Due to low temperature and weak illumination, reproduction, feeding and other biological processes are limited. Older copepodites of *Pseudocalanus m. elongatus* and *Temora longicornis*, as well as species of the genus *Acartia*, have a prominent position in the zooplankton. The winter zooplankton is quite poor, being richest in deep areas. Due to low temperature, cladocerans and rotifers are absent. Warm-water species descend deeper and cold-water ones ascend to higher water layers. In the eastern part of the sea, two zooplankton maxima (at depths of 0–25 and 50–75 m) form in cold winters, but in warm winters, only one (at 0–25 m) is formed. In particularly cold winters, a single zooplankton maximum is formed in medium water layers. In the Baltic Sea, a number of zooplankton species (chiefly rotifers and cladocerans) do not overwinter in their definite shape, but mainly as 'resting eggs'. In such species, an explosive development of eggs and hatching of nauplii takes place, usually in February–March, considerably increasing the abundance of the species.

Along with the increase in biological production, the progress of the reproduction period and the increase in the abundance of warm-water species in spring, the biomass of zooplankton shifts closer and closer to the surface. In midsummer, the importance of the upper 25-m water layer, compared to the deeper layers, is high (Fig. 3.6). The spring season, being dependent on the phytoplankton spring bloom, starts in different parts of the sea at different times. The maximum zooplankton reproduction follows a couple of weeks after the phytoplankton bloom. In the south-western and southern Baltic, the rapid increase in zooplankton biomass begins in April–May, in the eastern part of the sea, in May–June. The reproduction of zooplankton takes place in the upper 25-m layer of the sea. In spring, the distribution of zooplankton clearly depends on water temperature. Copepod nauplii, *Pseudocalanus* spp. and *Synchaeta* spp. dominate in the zooplankton in spring.

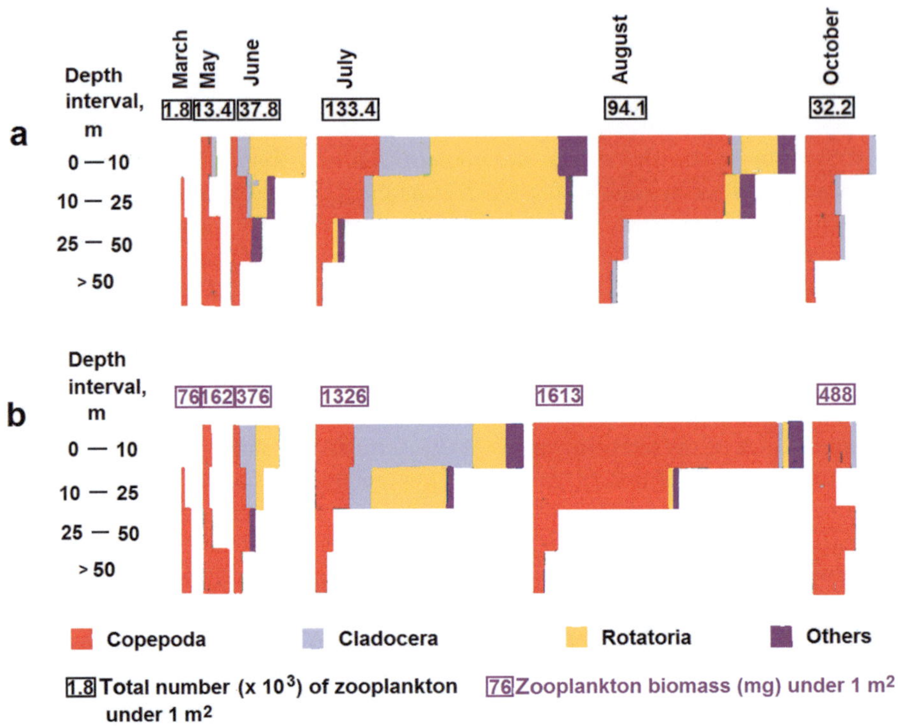

Fig. 3.6 Vertical distribution of zooplankton abundance (**a**) and biomass (**b**) in the central Gulf of Finland in March, May, June, July, August and October 1974. Above monthly average values of zooplankton total abundance (**a**) and biomass (**b**) under 1 m^2 are shown. The number and biomass of zooplankton groups is proportional to the area indicated in the figure (Ojaveer 2014)

Zooplankton is richest in summer. Its distribution depends on temperature, but also on primary productivity, which is associated with the ascent of nutrient salts from deeps to euphotic water layers.

In autumn, zooplankton abundance and biomass significantly decrease. Zooplankton biomass rather rapidly drops up to November, mainly due to the decline in cladocerans. Thereafter, zooplankton is strongly dominated by copepods, with the genera *Acartia*, *Pseudocalanus* and *Temora* well represented. Zooplankton is distributed in two belts: 0–25 m and 50–100 m. The lower belt is clearly dominated by *Pseudocalanus* spp.

Zooplankton inhabiting various depths cannot undertake long migrations, and therefore relocation takes place chiefly together with water movements. However, planktonic organisms, especially copepods, also perform diurnal migrations characteristic of a number of pelagic species. At sunset, together with the weakening of daylight, planktonic organisms rise into surface layers. They spend the entire night there (mainly feeding) up to daybreak, when they begin the descent to their daily habitation. Considering the minute body length of the plankters, the range of the migration (up to 20–50 m) is, for them, extremely long. The vertical reach of the migration, as well as its upper limit, differs by species. This depends on lighting conditions, season, temperature, nutritional conditions, etc. A number of marine biologists are of the opinion that the diurnal vertical migration has evolved into a feeding–hiding migration, i.e., plankters rise into the surface layer of the sea to feed there in favourable conditions (food richness, favourable temperature), being simultaneously inconspicuous (due to faint light) to enemies. Strengthening daylight induces the descent of plankters into deeper and darker water layers to become even less conspicuous to their enemies.

Baltic Proper In correspondence with hydrological zones, three ecological zooplankton complexes can be differentiated in deep regions: (1) warm-water plankton of the upper hydrological zone (*Temora longicornis*, *Acartia longiremis*, *Centropages hamatus*, *Bosmina coregoni maritima*, *Synchaeta monopus* etc.); their habitat is the water layer from the surface to the thermocline, which warms up in summer; (2) plankton of the cold intermediate layer (*Pseudocalanus minutus elongatus*, *Fritillaria borealis*, young stages of *Mysis mixta* etc.), and (3) deep-water (below the halocline) plankton (*Oithona similis*, *Sagitta elegans* etc.). In shallow areas (depending on corresponding coastal zone, rivers and hydrological situation), one or two plankton complexes with dominating marine or freshwater organisms exist. Zooplankton of the open Baltic generally occurs in the boreal water with a salinity of 7–8 psu.

The dynamics of zooplankton abundance and biomass depends on the long-term dynamics in salinity and temperature in the Baltic Sea, based on the alternation of climate periods. Naturally, the preconditions for primary production are solar energy and availability of nutrients. Phytoplankton production represents the energy source for both microzooplankton and herbivorous mesozooplankton. Another important food source is detritus composed of organic and inorganic parts. The production of pelagic bacteria is based on the solution of organic matter from phytoplankton. Bacteria represent a main nutrition source for microzooplankters, which have an important position in the diet of some species of carnivorous zooplankton (genus *Synchaeta*, etc.).

In connection with eutrophication, which began in the late 1960s, an increase in zooplankton production was observed. Compared to the 1960s, in the 1970s, zooplankton abundance increased in the Bornholm area by 1.4–1.7 times, in the Gotland area, 1.3–1.8 times, and in the Fore Deep area, 1.1–1.6 times. During this period, the zooplankton biomass increased in the Gotland basin in summer by 1.7 times and in the Fore Deep by 1.5 times. Zooplankton production has augmented in the southern Baltic since the second half of the 1960s, and in the eastern and northeastern areas of the sea after 1971. Thereby, the abundance of different species has changed in different ways. A decrease in the abundance was noted in the species depending on higher salinity (*Sagitta*, *Oithona* etc.). These changes brought along changes in the trophic structure of zooplankton in the Baltic Sea. The importance of herbivorous zooplankton increased from 65% in the 1960s to 74% in 1971. Also, changes occurred in the importance of depth zones as habitats of zooplankton (Kostritškina and Laganovska 1983).

In the first half of the 1980s, a significant decrease in the most important component of zooplankton – marine copepods – began. This decrease was preceded by an increase in the freshwater runoff since the second half of the 1970s and the accompanying decrease in salinity in the marginal parts of saline deep-water areas, e. g., in the Gulf of Finland (Fig. 3.7). The influence of this very important event soon extended to the whole Baltic Sea, inducing a serious drop in the growth rate of a number of fish species (herring, sprat, etc.), which has lasted up to the present. The economic impact of this change in the environment has been immense.

Large Gulfs A constant halocline and below it water of comparatively high salinity exist only in the western part of the Gulf of Finland. In the Gulf of Bothnia, the halocline is weak and unstable, while in the Gulf of Riga, it is absent. In large gulfs, the importance of freshwater organisms is significantly greater than in the open Baltic. In the gulfs, brackish-water euryhaline species dominate (*Eurytemora hirundoides, Acartia bifilosa, Bosmina coregoni maritima, Podon polyphemoides, Synchaeta monopus* etc.), but *Leptodora kindtii*, *Keratella quadrata*, *Mesocyclops leuckartii*, etc., of freshwater background, marine euryhaline and eurytherm species

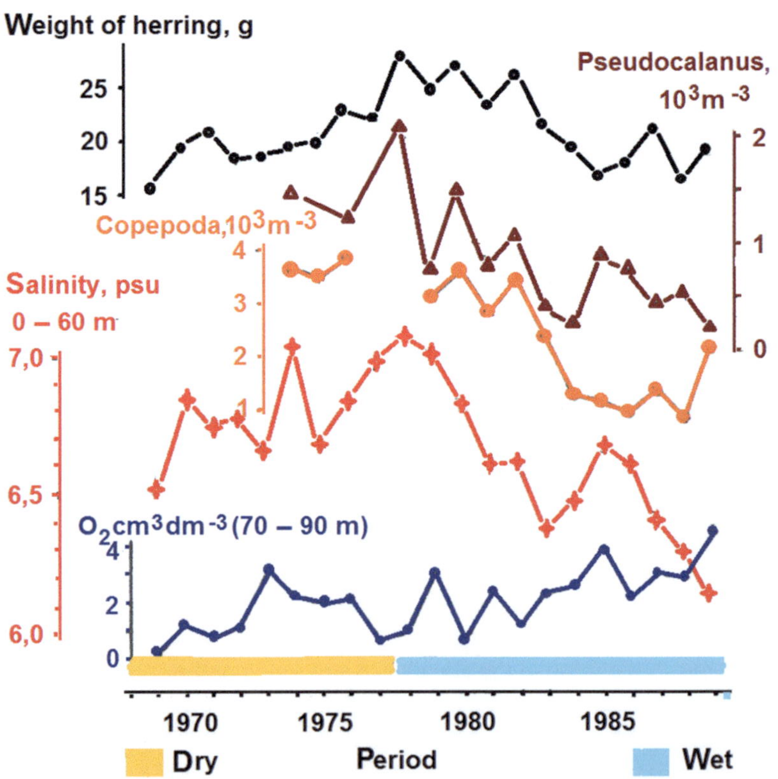

Fig. 3.7 Dependence between the average salinity, oxygen content, abundance of copepods and the weight of the 2-year-old herring in the Gulf of Finland in the period 1969–1989 (Ojaveer 2014)

(*Pseudocalanus minutus elongatus*, *Temora longicornis* etc.) and larvae of marine euryhaline species *Balanus improvisus* and *Fritillaria borealis* are also important. These species develop especially violently in the estuarine areas of eutrophied rivers. Freshened bights and lagoons mainly host euryhaline brackish-water species together with freshwater ones (*Keratella cochlearis*, *Daphnia cucullata*, *Chydorus sphaericus*, *Cyclops* spp., etc.). Compared to the Baltic Proper, the temperature regime in large gulfs is generally more continental. Due to differences in the geographical location, water exchange, climate and other important conditions, large gulfs clearly differ from one another in the composition and dynamics of zooplankton.

Gulf of Riga In the open part of the gulf, the number of zooplankton species is smaller (about 50) than in the river estuaries (about 132 species). The difference is chiefly due to the higher number of freshwater organisms in estuaries. The central areas of the gulf are dominated by brackish-water copepods of wide distribution *Eurytemora hirundoides*, *Limnocalanus grimaldii*, *Acartia bifilosa*, cladocerans *Podon polyphemoides*, *P. intermedius*, *P. leuckarti*, *Bosmina coregoni maritima*, *Evadne nordmanni*, etc., and rotifers *Synchaeta baltica*, *S. monopus*, *S. fennica*, *Keratella cochlearis*, *K. cruciformis*, *K. quadrata*, etc. Freshwater species *Mesocyclops oithonoides*, *M. leuckarti*, *Daphnia longispina*, *D. cucullata*, *Sida crystallina*, *Kellicottia longispina*, etc., populate particularly estuarine areas. The species of marine background *Pseudocalanus minutus elongatus*, *Temora longicornis*, *Acartia longiremis*, *Centropages hamatus*, etc., are of smaller importance. They can chiefly be found in the northern and western parts of the gulf, carried there with the open sea water. Coldwater plankters are usually absent, but they can occur in the coastal zone in the cold season, rarely in the case of coastal winds in summer.

Considering the relation of zooplankton organisms to the variation in salinity and temperature, mesozooplankton in the Gulf of Riga can be divided into the following complexes (Line and Sidrevitš 1995):

1. euryhaline eurytherm brackish-water species – *Erytemora hirundoides*, *Acartia bifilosa*, etc.;
2. thermophilic brackish-water species *Synchaeta monopus*, *S. baltica*, etc.;
3. brackish-water species of high temperature optimum – *Bosmina coregoni maritima*, *Podon intermedius*, larvae of *Balanus improvisus* and of molluscs.
4. freshwater species *Keratella quadrata*, *K. cochlearis*, etc.;
5. euryhaline eurytherm marine species *Pseudocalanus minutus elongatus*, *Temora longicornis*, *Evadne nordmanni*, *Podon polyphemoides;*
6. brackish-water glacial relicts – *Limnocalanus grimaldii*.

The Gulf of Riga belongs to the water bodies where big seasonal changes in the regime, including clear rearrangements in zooplankton, take place. The winter season usually lasts from the middle of December to the middle of March. However, depending on the thermal situation, both the beginning and the end of the season can shift. The abundance and biomass of the zooplankton are low and annual variations are considerable. Zooplankton consists mainly of copepods, notably *Acartia bifilosa* (during 1960–1987, this species made up on average 58%, in severe winters, up to 68% of the zooplankton). *Eurytemora hirundoides* has the second highest abundance, with its average proportion varying between 17 and 35%. Winter zooplankton consists mainly of older copepodites and adult specimens, but nauplii are also represented. It is comparatively uniformly distributed both horizontally and vertically. However, generally, the zooplankton aggregations in the central part of the gulf are denser than at the coasts.

Depending on water temperature, reproduction of copepods begins after mild winters in March and after cold winters in May. Reproduction of *Limnocalanus* takes place in March–April. Most of the spring generation of *Acartia* and *Eurytemora* mature after mild and medium winters from the end of May/the beginning of June. *Limnocalanus* terminates its development by July–August. After cold winters, reproduction of copepods shifts to a later time and the spring generation matures in late June or early July. Consequently, after the winter minimum, a new increase in zooplankton abundance begins in April and the height of the augmentation depends on the temperature regime in winter and spring. In addition to copepods, rotifers also increase their abundance and biomass in spring. The annual maximum of *Synchaeta* spp. occurs in June, when warm-water rotifers and cladocerans also make their appearance. In spring, the spatial variation of zooplankton in the Gulf of Riga is notable. Before the water in the coastal zone warms up the zooplankton abundance is larger in the open part of the gulf, and after the water at coasts has warmed, the amount of plankton there grows larger.

In summer (July–August), the abundance of zooplankton sharply increases and reaches its annual maximum. At that time, the amount of zooplankton is mainly determined by the amount of food and the water temperature. In summer, the relative impact of the feeding of fish upon zooplankton is the highest. In the summer period, in the zooplankton of the Gulf of Riga, *Eurytemora hirundoides*, accounting for up to 70–80% of the total abundance, is of greatest importance. In surface layers (0–20 m), younger copepodites dominate, in deeper layers, older copepodites and adults. Already in late spring, cold-water *Limnocalanus* descend into the bottom layers. Characteristic of the summer zooplankton is an explosion-like increase in the importance of cladocerans. The abundance of *Podon* spp. rises in July, and of *Bosmina* later (its abundance in a cubic metre can constitute up to 100,000 individuals – Line and Sidrevitš 1995). Also, the abundance of *Keratella* spp. grows in summer, reaching the yearly maximum of 80,000 ind.m^{-3} in the upper 40 m water layer. However, towards the end of summer, the abundance of rotifers rapidly diminishes. In addition to zooplankton species, a large number of larvae of bottom invertebrates (mainly of *Balanus*) occur in summer. The most vital conditions primarily determining the zooplankton numbers in summer are the amount of food and water temperature.

In autumn, cladocerans disappear from zooplankton, especially the genus *Bosmina*. In September, the summer generation of copepods multiplies, giving birth to the autumn generation. The genus *Acartia* of this generation matures in December, and *Eurytemora* in February–March of the following year. As the *Limnocalanus* population is composed of adult specimens at that time, their abundance is rather limited. *Synchaeta* are comparatively abundant, but zooplankton total abundance is 3–4 times lower than in summer. The decrease is most rapid in the coastal zone where water temperature falls earliest. In autumn, the difference in vertical and horizontal distribution of zooplankton in water layers is even greater than in summer.

In various parts of the gulf, neither zooplankton structure nor its development is uniform. In spring, the development starts in the western part of the gulf. Zooplankton is constituted primarily of euryhaline brackish-water and also marine species there. In the southern part of the gulf, in the estuarine areas of the large rivers Daugava, Lielupe and Gauja, planktonic fronts are situated. The freshwater–brackish-water zooplankton of these areas transfers into the euryhaline brackish-water– marine plankton coenosis in the ordinary coastal zone. In the eastern part of the gulf, the highest water layer mostly consists of fresh water discharged by large rivers. In spring, this part of the gulf is usually the longest covered with ice and therefore zooplankton development is hindered there. In this part of the gulf, the number of marine species is smallest. Due to higher salinity in the northern part of the gulf, brackish-water marine zooplankton organisms dominate there. In the deepest and thermally stratified central part of the gulf, organisms can have problems with oxygen conditions.

Plankton investigations have indicated that in the period 1952–1986, fluctuations of both zooplankton abundance and biomass were notable (Line and Sidrevitš 1995). In the 0–40 m water layer, the dominating trend was increasing abundance in May, October and most clearly in August (Fig. 3.8). The background of this process was created with the increase in the numbers of copepods (especially *Eurytemora* spp.), cladocerans (primarily *Bosmina* spp.) and also rotifers (first of all *Keratella* spp.). This process was favoured by the rise in temperature and also eutrophication. The increase was most expressed in warm-water species. Simultaneously, the tendency of decreasing abundance and biomass of *Limnocalanus grimaldii*, preferring cold and clean water, was obvious (Line and Sidrevitš 1995). However, in the Gulf of Riga, the sudden drop in the quantity of copepods, discovered in the open Baltic and the Gulf of Finland in the 1980s, was not stated.

Gulf of Finland Based on environmental conditions and the structure of biota, Kostritškina and Laganovska (1983) divided the Gulf of Finland into five subareas: (1) The Neva estuary – the richest area in the Gulf of Finland considering zooplankton species composition. Water salinity in this area is below 3 psu. Rotifers make up 87% of the zooplankton species. The biomass of zooplankton, constituted, for the most part, of freshwater species, is rarely more than 50 mg m^{-3}. (2) In the transition area with a salinity of 3–5 psu, situated westward of the previous subarea, oligohaline freshwater zooplankton is replaced with mesohaline marine species. Trophic relations, thermal conditions and intense water mixing create good conditions in some areas for rich zooplankton production (up to 1000 mg m^{-3}). The occurrence of estuarine plankton fronts results in large differences (up to 5–6 times) in zooplankton biomass between neighbouring areas. (3) In the shallow brackish-water area, salinity may reach up to 6 psu from time to time. Brackish-water species dominate, but in the estuarine area, freshwater species occur. They include both warm-water

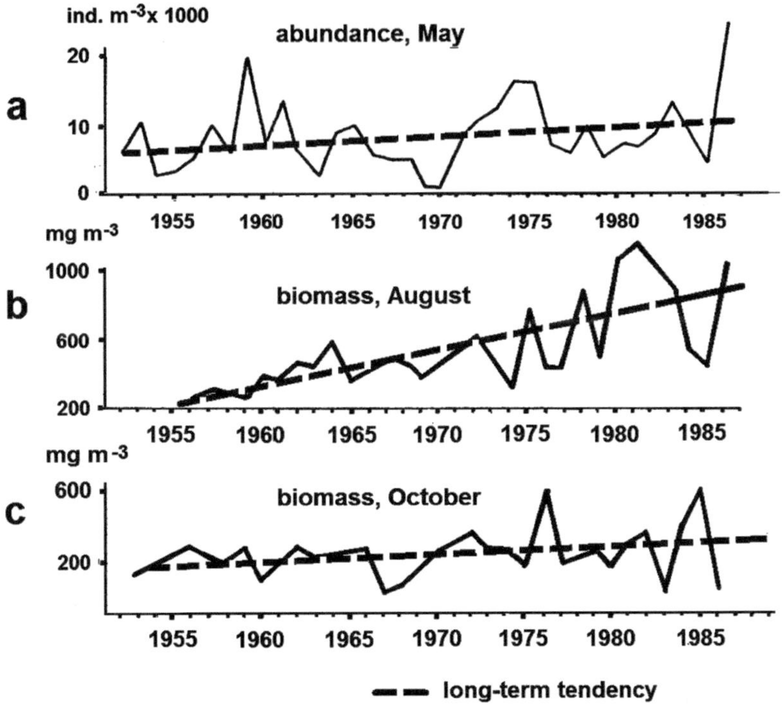

Fig. 3.8 Long-term dynamics of zooplankton in the 0–40 m water layer in the Gulf of Riga in 1952–1986. (Line and Sidrevitš 1995)

and cold-water organisms, while in certain places, marine species are also important. During the vegetation period, zooplankton biomass varies between 100 and 1000 mg m^{-3}. (4) In the eastern deep-water area, salinity rises up to 6–8 psu, resulting in an increase of brackish-water and marine species instead of freshwater ones. The zooplankton biomass varies between 200 and 500 mg m^{-3}. (5) The western deep-water area differs rather slightly from the previous subarea in species composition, abundance and biomass of zooplankton. However, the importance of marine species is higher. The zooplankton biomass ranges from 200 to 1000 mg m^{-3}.

After a severe winter, the abundance and biomass of a number of zooplankton organisms are low. However, in this case, the abundance and biomass of the relict species *Limnocalanus grimaldii* are the highest. Rotifers and cladocerans winter mostly as resting eggs. In spring and early summer, the abundance and biomass of cladocerans and rotifers rise considerably. However, because of the small size of rotifers, their biomass does not constitute a large percent of plankton biomass. The importance of rotifers and cladocerans is the highest and that of copepods the lowest in the most freshened eastern part of the gulf, whereas copepods dominate in the western areas. In the eastern part of the gulf, the importance of *Limnocalanus grimaldii* (with the upper salinity limit of 6 psu), *Eurytemora hirundoides* and *Pseudocalanus minutus elongatus* is high in all seasons. Westwards, the importance of *P. elongatus*, *Temora longicornis* and *Acartia bifilosa* continuously increases. Independent of the season, the density of zooplankters diminishes from lower to higher depths. The phenomenon probably depends on the clinal change of environmental parameters. It can also be related to the peculiarity of the distribution of plankton: the age and size of organisms increase with distance from the coast (Lumberg and Ojaveer 1991).

In winter, almost all zooplankton in the Gulf of Finland is distributed deeper than 25 m. In spring, with the warming of surface layers, most of the plankters ascend to the surface layers for feeding and reproduction (Fig. 3.6). Domination of copepods increases from summer to autumn. In the eastern part of the gulf with larger variations in temperature, seasonal differences in zooplankton distribution are much more obvious than in the western areas.

Plankton samples taken in various parts of the Gulf of Finland indicate considerable variation between the plankton sampled in different stations in a year and the plankton sampled in different years from the same station (Fig. 3.9). However, in the eastern part of the gulf, the zooplankton is of more freshwater (high abundance of freshwater rotifers), and in the western part, of more marine character (higher abundance of marine copepods), despite the patchiness in plankton distribution masking the differences somewhat.

In the Gulf of Finland, both seasonal and regional variations in the structure, abundance and biomass of zooplankton are clearly expressed. The general reason for such notable dissimilarity in the salinity and temperature regime is the influence of the large freshwater input by the Neva River into the eastern part of the gulf and the impact of the saline water under the halocline in its western part. The differences are most obvious in summer as reproduction, which is strictly controlled by environmental conditions, takes place then. The results of mesozooplankton studies that were regularly carried out in 12 stationary stations in May, August and October–November 1963–1991 showed that zooplankton abundance and biomass were relatively high in 1963–1967 and diminished thereafter. A new increase occurred in 1974. The augmentation of the total plankton biomass lasted up to 1982, mostly in warm summers when the abundance of the warm-water *Bosmina coregoni* and rotifers was high (Fig. 3.9). However, a drop in the important zooplankton component *Pseudocalanus minutus elongatus* had already started in 1978, after an increase in river discharge and a decrease in salinity of the 0–60 m water layer. The coeffect of the decrease in salinity and variation in temperature and oxygen conditions caused a substantial decline in the abundance of copepods in the entire Gulf of Finland up to the early 1990s.

Gulf of Bothnia In the Bothnian Sea, the difference in salinity between the surface and demersal layers is 2–3 psu, in Bothnian Bay, about 1 psu. As the halocline is unstable, it is supposed that the chief hydrological factor influencing zooplankton distribution in the Gulf of Bothnia is the thermocline.

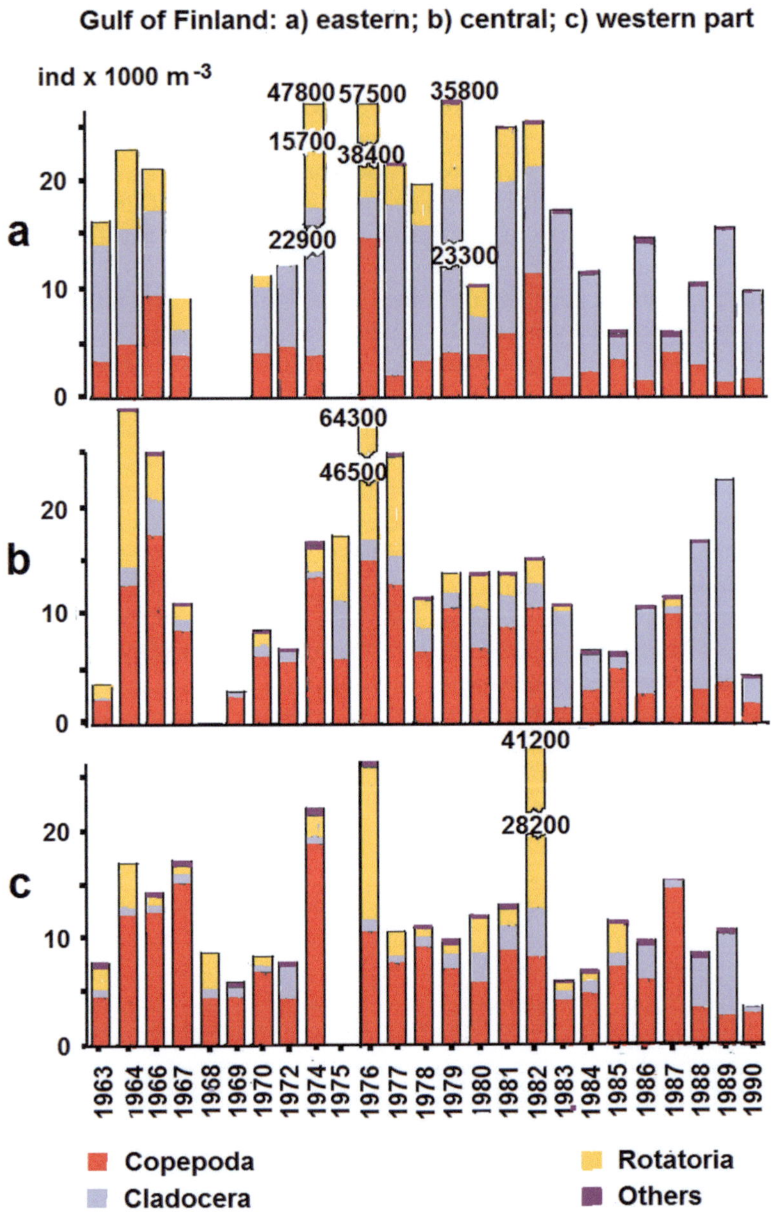

Fig. 3.9 Average abundance of zooplankton in the eastern (**a**), central (**b**), and western (**c**) parts of the Gulf of Finland in August 1963–1990 (Ojaveer 2014)

An overview of the zooplankton of the Gulf of Bothnia has been composed by O. Sandström (1982), who summarized the data of earlier studies beginning with the pioneer expedition by Oscar Nordquist in summer 1887. He also presented the results of his own studies carried out in the1970s (sampled with plankton nets from surface to bottom). His work indicated that in the Gulf of Bothnia, rotifers prefer the surface layer and do not undertake migrations through the thermocline. Also, the most abundant cladocerans *Bosmina coregoni maritima*, *Evadne nordmanni* and *Pleopsis polyphemoides* spend all summer in the warm surface layer. The stenothermic cold-water species *Limnocalanus grimaldii* lives in deep layers of the Gulf of Bothnia, while in surface layers, it has been found only on a few occasions. This has happened in Bothnian Bay in autumn, when the thermocline is weak and cannot influence the vertical distribution of this species. Young stages of this species dwell in the surface layers, and therefore it is thought that their preference for low temperature develops with age. The nauplii of the species live in the surface water in most cases and are common in shallow coastal water in summer. Copepods of this species are the only zooplankton group preferring the thermocline. In summer, *Acartia* spp. and *Eurytemora* spp. are limited to surface water, although adults have also numerously been found in the thermocline. *Pseudocalanus minutus elongatus* dwells without exception below the thermocline. Nauplii of this species have been found closer to the surface, but their main area is also below the thermocline. Larvae of *Harmothoë sarsi* mostly dwell in deep water, but they have been found incidentally in the thermocline. The larvae of *Macoma baltica* generally occur in the surface layers, but they have also been found deeper. In its diurnal migration, *Limnocalanus macrurus* penetrates into the thermocline.

The zooplankton of the Gulf of Bothnia can be divided into two parts: (1) in the warm surface layer, rotifers, cladocerans and, from copepods, *Acartia* spp. and *Eurytemora* spp. occur. In addition, young stages of *L. macrurus* and larvae of *Macoma baltica* may use this habitat; (2) in Bothnian Bay below the thermocline older stages of *L. macrurus* dominate. In the Bothnian Sea, deep-water fauna is richer. In deep layers, in addition to *L. macrurus, Pseudocalanus minutus elongatus, Pleurobranchia pileus* and larvae of *Harmothoë sarsi* can also be encountered (Sandström 1982).

The zooplankton of the Gulf of Bothnia has been investigated since 1887, but not so regularly as in other areas of the Baltic Sea. However, significant changes are well registered and related to environmental parameters. In the first half of the twentieth century, the importance of marine species was rather limited in the Gulf of Bothnia, especially in its northern part. This has been related to low salinity: in the 1930s, up to 5.7 psu in the bottom layers of the Bothnian Sea, and in its deep layers, up to 3.8 psu. In the 1950s, the salinity in these areas rose to 6.7 and 4.2 psu, respectively. This situation enabled marine fauna to shift in and forced freshwater fauna to retreat. In the 1920s, the rotifer *Keratella quadrata platei* was relatively sparse in Bothnian Bay and concentrated in the coastal areas in the Bothnian Sea. Fifty years later, this species, generally occurring in late summer, was common in the entire Gulf of Bothnia, although more frequently in the south. In Bothnian Bay, *Keratella cochlearis recurvispina* has reached an abundance of 85,000 ind. m^{-2}, and in the Bothnian Sea, 132,000 ind m^{-2}. The density is higher in coastal areas. *Keratella cruciformis eichwaldi* is present, but relativey scanty in the Bothnian Sea. The species of the genus *Synchaeta* are distributed throughout the entire Gulf of

Bothnia. In 1934, representatives of this genus were only occasionally found in Bothnian Bay. During the last five decades, its abundance has notably increased in the Gulf of Bothnia. Species of this genus are found throughout the summer (probably various species at different times) up to the first week of November. *Kellicottia longispina* is a rotifer of freshwater background, which has mostly been encountered in the coastal zone of the Bothnian Sea.

The cladoceran *Bosmina coregoni maritima* is a very important zooplankton species for the whole Gulf of Bothnia. Its highest densities occur in the coastal zone of southern areas. It can be found from early spring, but is a typical species of late summer, with its abundance maximum in August–September. *Evadne nordmanni* is one of the two most important cladocerans of the Gulf of Bothnia. Its maximum abundance in Bothnian Bay occurs in late July, in the Bothnian Sea, somewhat earlier. Other cladocerans found in the Gulf of Bothnia are *Podon leuckarti*, *Daphnia cristata*, *D. cucullata*, *Chydorus sphaericus*, etc.

Copepods have strengthened their position in the Gulf of Bothnia. At the beginning of the twentieth century, the Åland Sea was the distribution limit of *Temora longicornis*. In the 1950s, it penetrated into the Bothnian Sea and survived there. The importance of *Pseudocalanus minutus elongatus* in the Gulf of Bothnia has clearly risen during the last decades. It is mainly distributed in the deeps of the Bothnian Sea, but is found also in Bothnian Bay. *Acartia longiremis* and *A. bifilosa* have recently been limited to the southern part of the Gulf of Bothnia. Of the most important plankton species in the Gulf of Bothnia, *Limnocalanus grimaldii* is outstanding mostly because of its body dimensions. It dominates in the deep areas of the gulf, but compared to the 1920s, its abundance has diminished. Of other copepods in the Gulf of Bothnia, the abundance of species of the genus *Eurytemora* (*E. hirundoides, E. hirundo* and *E. affinis*) is probably the highest in the Bothnian Sea.

The most important larval bottom invertebrates in the Gulf of Bothnia are *Harmothoë sarsi*, *Macoma baltica*, etc. (Sandström 1982).

3.8 Zoobenthos

Zoobenthos includes animals of very different systematic groups living on or dug into bottom sediments, and also the organisms living in water layers but closely connected to the bottom. Bottom invertebrates of the Baltic Sea have been investigated by a number of scientists for a comparatively long time.

In German waters, the study of macrozoobenthos dates back to the end of the eighteenth century and the systematic observations to the second half of the nineteenth century (Zettler et al. 2008). In the Northern Baltic, the first mention of demersal invertebrates in scientific literature goes back to the year 1769, and studies in both the Gulf of Finland and the Gulf of Bothnia were started from the middle and the second half of the nineteenth century (Leppäkoski 2001).

The majority of scientists have stressed that the bottom fauna of the Baltic Sea is notably poor in species. This is obviously related to the development of the sea and its biota in the climatic conditions during the Pleistocene glaciation/interglaciation periods with continuous changes of the most important environmental conditions in the sea: connection with the World Ocean, water salinity, thermal regime, composition of biota, etc., but also with the ecological conditions that have reigned later and the ever-strengthened anthropogenic influences during recent decades.

An important part of the contemporary bottom fauna stems from previous stages of the development of the Baltic Sea. In contemporary ecosystems of the Baltic Sea, an outstanding position is occupied by the crustaceans *Saduria entomon*, *Mysis relicta, M. mixta, Monoporeia affinis* and *Pontoporeia femorata*, but also the mollusc *Astarte borealis*. In the contemporary Baltic Sea, with its habitats of very variable conditions, they have found a favourable niche. However, probably due to the short time of development and continuous changes in the environment, no endemic species have developed in the zoobenthos of the Baltic Sea.

Järvekülg (1979) estimated 35 years ago, before the last mass immigration of non-indigenous species, that the Baltic Sea hosted altogether 490 species of bottom animals 48.8% of them freshwater, 35.3% marine and 14.3% brackish-water. Zettler et al. (2008) state that in the Southern Baltic, 667 macrobenthos species representing 29 higher taxonomic groups are present, and in a comparable Danish study, 510 taxa have been found. In German data, polychaetes, gastropods and amphipods (155, 90 and 75 species, respectively), and in the Danish areas, annelids, arthropods and molluscs are the most abundant groups.

The zoogeographical composition of the bottom invertebrates is rather variable. This is supposedly related to the complicated history of the evolution of the biota in the Baltic Sea and also with the present-day ecological conditions, mostly with the thermal regime. In the zoobenthos of the Baltic Sea, the proportion of boreal (28.4%) and panboreal (28.9%) species is almost equal. In the central part of the sea and the Gulfs of Finland and Riga, the most important groups are relict panarctic species (in most cases distributed in the elittoral of the Gulf of Finland and the lower horizon of the elittoral in the Gulf of Riga) and panboreal species (distributed mostly in the central part of the sea, in the sublittoral of the Gulf of Finland and the upper horizon of the elittoral of the Gulf of Riga). The remaining species belong to the panarctic-boreal, panarctic-panboreal, etc., zoogeographical groups. The number of taxa is highest in polychaetes, crustaceans and molluscs.

The organisms dwelling in the bottom water layers or having bored themselves into bottom sediments are generally only capable of limited migrations. Therefore, they are much more dependent on changes in the environmental conditions in their habitat than pelagic organisms. No mass colonization of the whole Baltic Sea by demersal species of marine background has occurred up to now. In the North Sea, 2600–2700 species of benthic marine invertebrates occur, in the Kattegat, 1200–1300, and in the central Baltic and large gulfs, only 120–130 species are found. The number of marine species declines from the Transition Area towards the North East in parallel with lessening water salinity. A number of classes of benthic marine invertebrates are totally absent from the Baltic Sea (Echiuroidea, Sipunculoidea, Cephalopoda, Loricata, Holothuroidea, etc.), and each of the classes Porifera, Anthozoa, Ophiuroidea, etc., are represented by only one species (Järvekülg 1979).

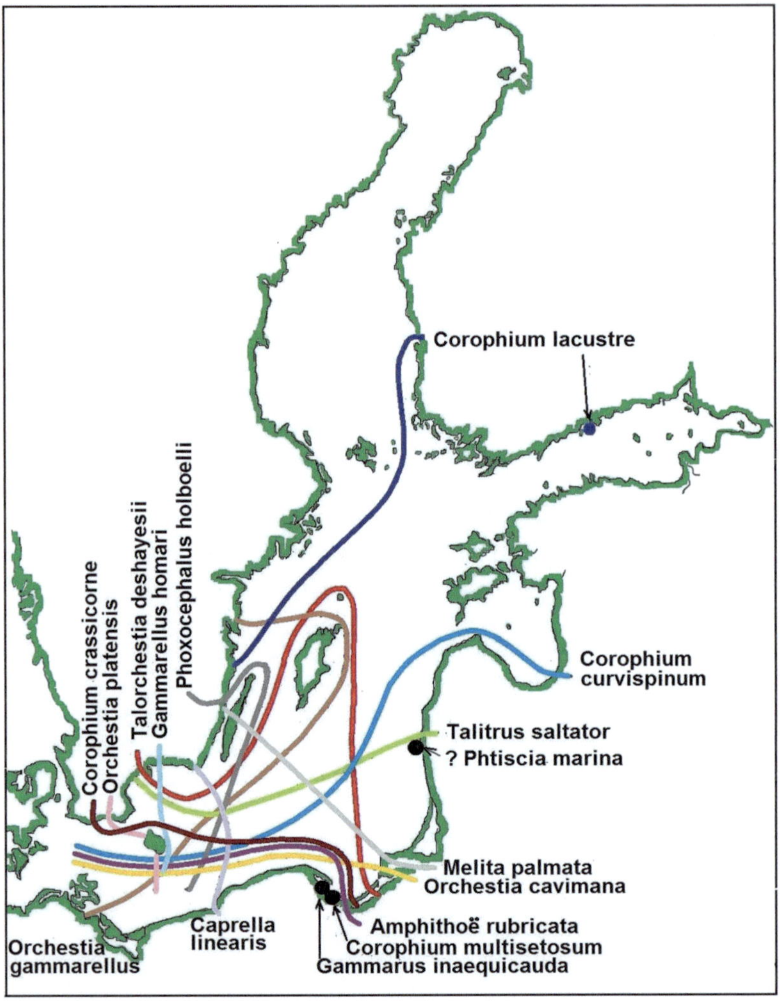

Fig. 3.10 The limits of distribution of some demersal invertebrates in the Baltic Sea (Järvekülg 1979)

In Fig. 3.10, one can see distribution borders of some nektobenthos species in the Baltic Sea. These borders simultaneously show the success of some marine species in adaptation to environmental conditions of various areas in the Baltic. The distribution limits of marine species proceed notably in the areas where environmental conditions substantially change – at the border between the southwestern and southern regions or at the borders between the Baltic Proper and the large gulfs, and also in these gulfs. In addition to salinity, low winter temperatures and too short warm periods for normal reproduction of warmer-water species have probably had an additional effect in the elimination of species from the southwest to the northeast.

The richness of marine biota depends on the concrete complexes of ecological conditions. One of the most important conditions is the amount of available nutrient salts. The main food of zoobenthos is detritus from the surface and upper part of bottom sediments and seston from the bottom layers of water. Phytoplankton is the source of the greater part of the detritus, while a smaller part comes from zooplankton. In the coastal zone, macroflora is an important source of detritus formation. Consequently, development of zoobenthos is closely related to the general biological productivity of seas, especially to the productivity of phytoplankton, but also of zooplankton and macrovegetation (although the importance of phytobenthos in the food of zoobenthos is comparatively low). Rich phytoplankton can develop only with plentiful availability of nitrates, phosphates and other nutrient salts in the trophogenic layer of the water body.

In the coastal sea, production is formed with the participation of the nutrients brought there with river waters from the land. Accounting for the salinity stratification of the Baltic Sea, the nutrients contained in the river runoff play a vital role in determining the bioproductivity in sea areas. In the open sea, the distribution of detritus depends largely on currents and the relief of the sea bottom. Detritus generated in the coastal areas may be carried with currents into other places and accumulate far from the area of its origination, in most cases, in deeps. Waters in which upwellings lift nutrients through the halocline and thermocline into the upper illuminated layers where they are embraced into primary production are productive.

Like the production of zooplankton, the average biomass of zoobenthos is comparatively moderate (about 30 g m^{-2}) in the Baltic. Järvekülg (1979) shows that the maximum zoobenthos biomasses decrease with increasing distance from the Danish Sounds (or with decreasing salinity): in the Southern Baltic, >3000 g m^{-2}; Central Baltic, ~1800 g m^{-2}; Gulf of Finland, ~600 g m^{-2}; Gulf of Riga, ~450 g m^{-2}; in the southern part of the Gulf of Bothnia, ~150 g m^{-2}; in the northern part of the Gulf of Bothnia, <100 g m^{-2}. In addition to higher salinity, as well as higher temperature, earlier beginnings and longer growth periods in southward areas of the Baltic Sea (the Belts, slopes of Arkona, Bornholm and Gdansk deeps) provide a substantial advantage for the formation of large primary production. However, the distribution of benthos biomass in the Baltic is in no case a smooth transfer between neighbouring areas. It is rather like oases of high productivity in large low-productivity regions (Fig. 3.11). These oases are situated in quite limited areas where, due to the bottom relief and existing currents, continuous upwellings take place, fertilizing the photic zone with nutrient salts. In addition to high productivity, these areas are also remarkable for species diversity (Demel and Mulicki 1959; Järvekülg 1979).

The density of zoobenthos settlement and biomass vary considerably by season, related to the seasonal course of temperature, reproduction of organisms regulated by temperature, mortality of generations and also predation by fish and other consumers. In shallow areas, the biomass increases in summer and decreases in winter. Annual differences are related to the amount of river discharge or saltwater influxes, winter severity, etc. Variations in the zoobenthos of the Baltic coastal areas may be induced by the periodic changes in salinity or temperature, and in the pseudoabyssal, by the periodic changes in the gas regime in the bottom layers of deeps.

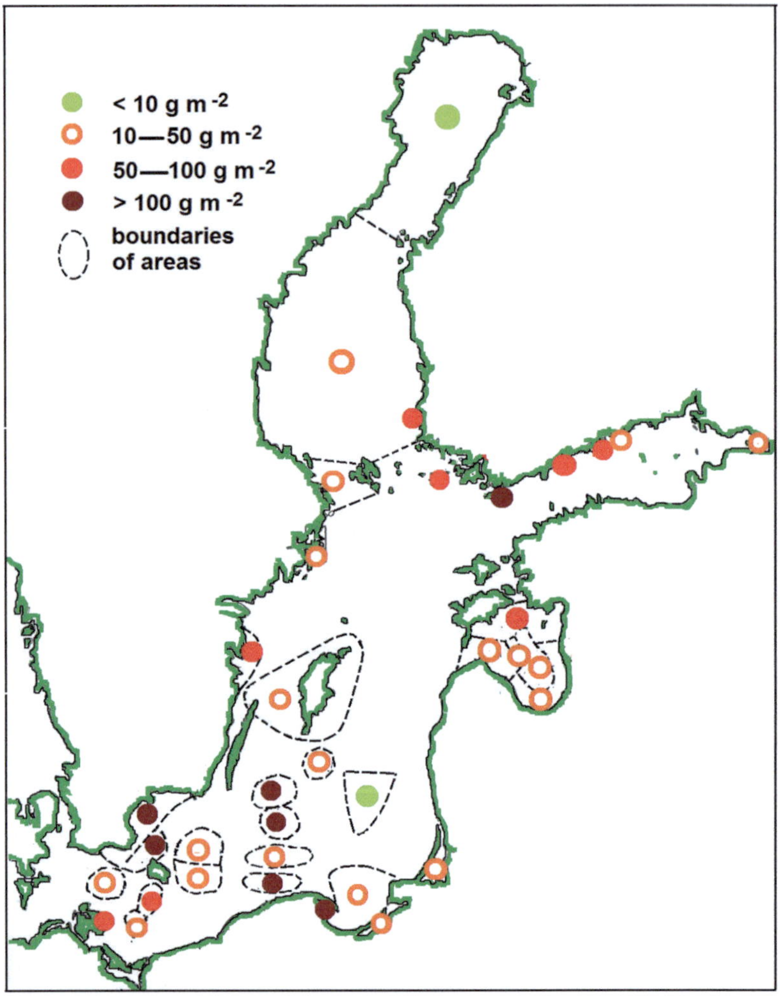

Fig. 3.11 Average biomass of bottom invertebrates in some areas of the Baltic Sea (Ojaveer 2014)

Structure and Distribution Zoobenthos structure depends on salinity, depth of water, character of bottom sediments, gas regime, water pollution and other abiotic variables. Species diversity is wide or very wide, especially in the sublittoral (at depths of 0.5–20–30 m) with a diverse bottom and the presence of bottom vegetation. In the elittoral (30–70–80 m) of the open Baltic, the bottom fauna varies by area and depends on concrete communities. Zoobenthos of the pseudoabyssal (>70 m) is characterized by very limited species diversity, obviously because of the unfavourable gas regime reigning in deeps. The zoobenthos of straits is generally of high biomass and very diverse, more variable than in the open part of the sea. In general, the zoobenthos of small bays is of high abundance and biomass.

In the central Baltic, the areas of maximum zoobenthos biomasses are situated on the coastal slope, in the southwestern, southern and western parts of the Gulf of Riga, and in some places in Pärnu Bay and Väinameri.

The importance of species in zoobenthos abundance varies more than in biomass, notably in the open sea. In southern areas of the central Baltic, east of the Gogland Isle, in Narva Bay in the Gulf of Finland and in the southern and northwestern areas of the Gulf of Riga, the most abundant zoobenthos species is *Monoporeia affinis*, preferring rather low temperature (2–6 °C), not very high salinity and sandy or sandy–muddy bottoms. In the northern areas of the central Baltic, it dominates together with *Mytilus edulis*. In the central Baltic, by depth zones, the follpwing species dominate: at 10–29 m, *Mytilus edulis*, 30–59 m, *Monoporeia affinis*, 60–79 m, *Pontoporeia femorata*, 80–89 m, *Pontoporeia femorata* and *Scoloplos armiger*, over 90 m, *Scoloplos* spp. In the estuarine part and at the southwestern coast of the Gulf of Finland, generally cold-water psammophilic *Heterocyprideis sorbyana* dominates. By depth zones, the following species dominate: at 0.5–9 m, oligochaetes, 10–19 m, oligochaetes and *Macoma baltica*, 20–29 m, *Macoma baltica*, 30–70 m, coldwater relicts *Monoporeia affinis*, *Heterocyprideis sorbyana*, *Paracyprideis fennica* and *Pontoporeia femorata*. In the Gulf of Finland, in the pseudoabyssal (depths over 70 m), the composition is rather variable and no stable dominant is present. In the Gulf of Riga, at 0–0.4 m, larvae of Chironomidae, at 0.5–19 m, oligochaetes, at 20–29 m, oligochaetes and *Monoporeia affinis*, at 30–60 m, *M. affinis* dominate.

In the straits and bights of the eastern central Baltic, an important role in the benthic coenoses is played by *Corophium volutator* of amphipods, oligochaetes, larvae of Chironomidae, *Hydrobia ulvae*, etc.

Impact of Environmental Conditions on the Distribution of Zoobenthos Zoobenthos, with increasing depth, grows both qualitatively and quantitatively poorer in all seas. Together with depth changes, the following occur in some very important conditions of existence: hydrostatic pressure (this factor influences water density, ionic balance and dissolution of gases in water), thermal and gas (frequently also salinity) regime, lighting conditions, development of macroflora, character of water movements, bottom properties, amount of detritus and seston, etc. Generally, with increasing depth, living conditions of zoobenthos constantly deteriorate. Therefore, with the deepening of the settlement area, average biomass of zoobenthos, population density, mean number of taxa, their total count and species diversity diminish.

Salinity The main obstacle hindering marine zoobenthos from colonizing the areas northeast of the Danish Sounds is too low salinity and osmotic problems related to the adaptation of marine species. Salinity has an especially great importance in the distribution of organisms in brackish water. Zoobenthos organisms are highly sensitive to large and aperiodic fluctuations in salinity. This becomes evident in river estuaries, where fluctuations in the freshwater quantity discharged can seriously influence living conditions of benthic organisms.

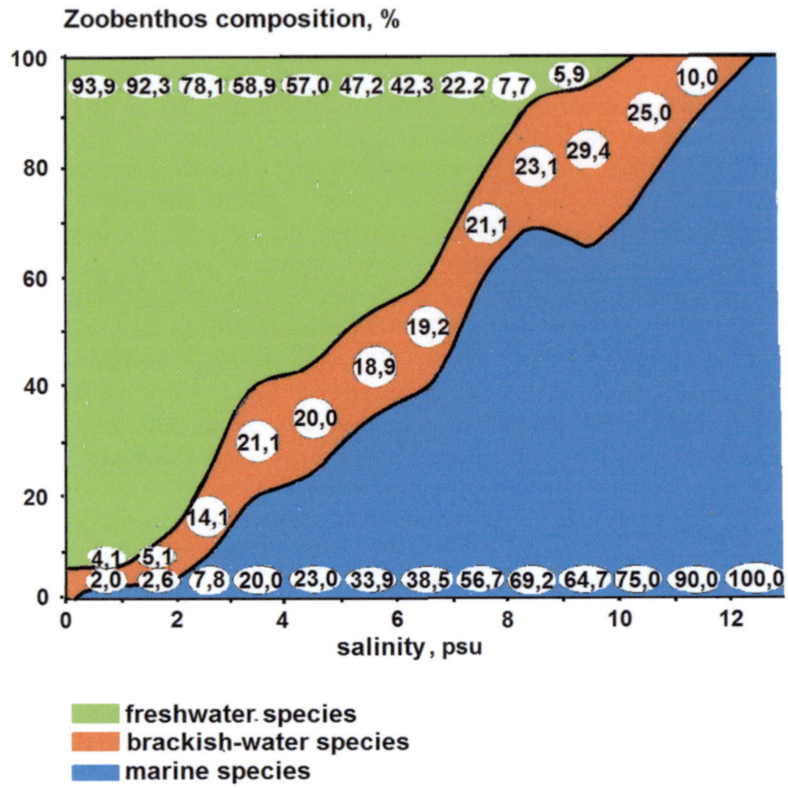

Fig. 3.12 Occurrence of bottom invertebrates of freshwater, brackish-water and marine background at various salinities (Järvekülg 1979)

The ecophysiological aspect of salinity changes through deviations in osmotic pressure and ionic composition of environment has been treated above. The dependence of the distribution of bottom invertebrates of freshwater, brackish-water and marine background on the salinity of the environment is illustrated in Fig. 3.12.

Another important obstacle hindering the increase in marine taxa in the Baltic is too low winter temperature and a too short warm period in summer. In addition to the combined effect of salinity and temperature, the distribution of zoobenthos depends on the gas regime, usually complicated by a low oxygen content of deep water layers in the sea.

In addition to the problems related to the osmotic pressure, adaptation to lower salinity causes extra expenditures of energy, reflected as a decrease in activity, slower growth rate, lessening dimensions, etc. Comparison of the dimensions of the eurybiotic mollusc *Mytilus edulis* in the Baltic and North seas shows that in Narva Bay of the Baltic Sea, the length of *Mytilus edulis* is ten times shorter, and in the central areas of the Baltic, three times shorter than in the North Sea. The length of the largest specimens of *Mya arenaria* in the Gulfs of Finland and Riga is three times shorter, and in the central Baltic, two times shorter than in the North Sea.

Therefore, the productivity of many organisms of marine background in the Baltic Sea is much less than in seas of normal salinity. However, differently from the species mentioned, no notable difference of measurements between the North and Baltic seas has been found in *Macoma baltica* and certain other organisms.

Compared to eumarine seas, in the coastal zone of the Baltic Sea, more species of freshwater background can be encountered. This is probably related to the history of the settlement of the Baltic area with organisms after the last glaciation. Freshwater species that immigrated into the Baltic Sea from lakes and rivers have adapted to the salinity of this sea. Some of them have increased their abundance and become important links in the brackish-water food chain. Towards the north, the importance of freshwater species, especially insects, but also oligochaetes and molluscs, increases. Freshwater organisms constitute an appreciable fraction in the bottom invertebrates of the Baltic Sea, and a part of them populate environments of rather high salinity (5–10 psu). However, such salinity constitutes a condition for them limiting their abundance. In addition to salinity, the distribution of a number of bottom invertebrates is limited by severe temperature conditions. Therefore, the abundance of freshwater species that have immigrated into the Baltic Sea cannot compensate for the low number of species of marine background in this sea.

Bottom Character and Macrovegetation Zoobenthos is very closely related to the character of the bottom, frequently more than with depth. The most important attributes in the bottom characters are granulometric composition, density of sediments, stability and the content of organic matter. The feeding of organisms is very variable and different: among bottom invertebrates, one can find detrivores, sestophags, herbivores, and omnivores, as well as carnivores. For them, bottom sediments are only for support or as a source of food. The character of bottom sediments defines their possibilities for attachment or hiding. Sand, gravel, mud or clay fit for different bottom residents. Too soft sediments do not offer enough support, while too dense ones may be too durable to enable digging. Organic matter occurring in bottom sediments generally favours zoobenthos. Unstable bottom sediment types have a negative influence on zoobenthos. Such a situation may be connected to currents, waves or estuarine processes. Bottom invertebrates depend on demersal macrovegetation, which is important as substrate and the source of detritus, rather than just as direct additional food. Therefore, in the areas rich in bottom vegetation, generally abundant and diverse zoobenthos has developed.

Interspecific Relations Composition and distribution of the biota of a water body are determined not only by environmental conditions and the historical background of the biota, but also by interspecific relations. As a result of the struggle for life, certain organisms are eliminated. This is compensated for by reproduction. Natural selection acts through elimination. Not all species are equally successful in interspecific struggle. However, changes in environmental conditions (e.g., environmental pollution) may generate corrections and significantly influence the final result. The forms of interspecies struggle can conventionally be divided into three main categories: (1) competition; (2) the predator–prey relationship; (3) the parasite–host relationship.

Interspecific Competition Mostly, in marine zoobenthos of the same trophic level, primary consumers dominate through abundance of organisms. Competing species are understood as the species that simultaneously need the same vitally important but limited resources. Competition is the least aggressive struggle between species. It is a matter of warding each other off in the struggle for food, habitat or place of reproduction. Competition usually results in the elimination of a certain portion of individuals. Under relatively invariable conditions, interspecific competition leads to the domination of the species that is best adapted to the environment. The competition of close species of the same feeding type and populating the same niche leads, in most cases, to the extrusion of the less adapted species. Interspecific food competition is of substantial importance in the distribution of a number of bottom invertebrates in the Baltic Sea. For instance, it was earlier thought that the distribution of the amphipod *Bathyporeia pilosa* was related to its high particularity to substrate (dwells only on silver sand) or salinity. Later, it turned out that in a number of places in the Baltic, the species populates areas of rather low salinity, on various bottoms, and tolerates noticeable pollution. The species turned out to be incapable of resisting the competition of other zoobenthos species and could live only in such areas where the abundance of other zoobenthos species was low.

Interspecific competition for living space is frequently interwoven with the competition for food; therefore, their differentiation is complicated. Because of low salinity, the interspecific competition is small in the Baltic. This has allowed some species of moderate success at the normal eumarine salinity to widen the depth of their area in place of other species in the Baltic Sea (*Harmothoë sarsi*, *Pygospio elegans*, *Macoma baltica* etc. – Järvekülg 1979).

Predator–Prey Relationship In the case of this form of relationship, the specimens of one species will be eaten by the members of the other species. Typically, the prey animals are the main food for the beasts of prey. In this case, both the predators and prey should have such morphological, ecological, etc., features which allow their co-existence in the same biotope. Also, abundance fluctuations of the prey population should exert a (belated) impact on the predator population. In some areas of the Baltic Sea rich in benthic food (e.g., in the southern Gulf of Riga), zoobenthos biomass is abnormally low, because in this area, the omnivorous *Saduria entomon* dominates, eating large amounts of amphipods and other benthic invertebrates. The most important diet of *Saduria* is *Monoporeia affinis*, mostly those of its slowly moving young stages. Therefore, the population of *M. affinis* can maintain the stability of its population only through intrapopulational migration. Investigations into the predator–prey relationship have also helped us to understand the reasons why there is no *Macoma baltica* east of Gogland Isle in the Gulf of Finland and in the southern Gulf of Riga. It turned out that the newly settled young stages are eaten up by the panarctic crustaceans *Pontoporeia femorata* and *Saduria entomon*. Obviously, this is also the cause of the absence of oligochaetes, ostracods, *Pygospio elegans* and some other benthic invertebrates in the same areas (Järvekülg 1979).

Zoobenthos Biocenoses Under natural conditions, bottom invertebrates exist as mutually related communities. In the differentiation of communities/bioceonoses, both quantitative and qualitative attributes are taken into account. Biocoenosis is the body of organisms settling a biotope (a comparatively limited area of a conventionally uniform environment). Every biocenosis has its structure formed during its geological history by the qualitative and quantitative species composition and the relationship of organisms of various ecological and zoogeographical appearance and trophic type.

In zoobenthos biocenoses, the number of species is commonly rather large, but mostly only one species dominates through abundance. Such a situation has obviously developed, because in the deep areas of the sea bottom settled by organisms, large areas of stable and uniform ecological conditions reign. In such biotopes, interspecific competition results in the formation of a biocenosis where one species, the best adapted to the concrete conditions, dominates.

The most important relations influencing the structure of biocoenoses are feeding relations. The trophic structure of demersal biocenoses in the sea has the following peculiarities: (1) On the main part of the sea floor, the importance of producers is comparatively negligible. This biotope receives its food from outside: dead organisms sinking from the upper water layers, the food brought with currents, etc. (2) The largest part of the abundance and biomass are composed of the consumers of the same trophic level – detrivores, herbivores, etc. The importance of carnivores is commonly low. (3) The consumers of various feeding types get their nourishment from different ecological niches: sestophags from near-bottom water layers, detrivores from the surface of bottom sediments, etc.

Biocenoses are differentiated on the basis of biomasses and named after the dominating taxon. They can be distinguished by their bathymetric features, relation to salinity, edaphic (sediment type), trophic, zoogeographic and other attributes.

Communities distributed in the pseudolittoral and upper horizons of the sublittoral (depths from 0.5 to 10–20 m) occur chiefly in shallow isolated bights (e.g., the community of *Bithynia tentaculata* and larvae of Chironomidae). In the sublittoral (depths from 0.5 to 20–30 m) of the central Baltic, the community of *Mytilus edulis* is widespread. In the elittoral (depths from 30 to 70–80 m), the *Saduria entomon* and *Monoporeia affinis* communities are notable. In the elittoral and pseudoabyssal (depth over 70–80 m), large areas are occupied by the *Pontoporeia femorata* community. In the pseudoabyssal of the central Baltic, the *Scoloplos armiger* community is one of the most noteworthy ones. Out of the rather numerous zoobenthos communities in the Baltic, only a few are of notable size.

In the southwestern part of the sea, the *Cyprina islandica*, *Astarte borealis* and *Macoma calcarea* biocenoses (Fig. 3.13) are characteristic with regard to high biomasses and species diversity. The macrobenthos diversity is high in the Kiel and Mecklenburg bights (average salinity 15–20 psu, maximum 28–30 psu) with approximately 400 species present. The biodiversity is highest at the entrance to the Great Belt off Fehmarn Island. The areas of high biodiversity are mainly linked to a good oxygen supply, a well-structured bottom surface and continuous recruitment due to appropriate currents (Zettler et al. 2008). The abundance and diversity of zoobenthos decrease noticeably from the southwest to the north, with deteriorating environmental conditions.

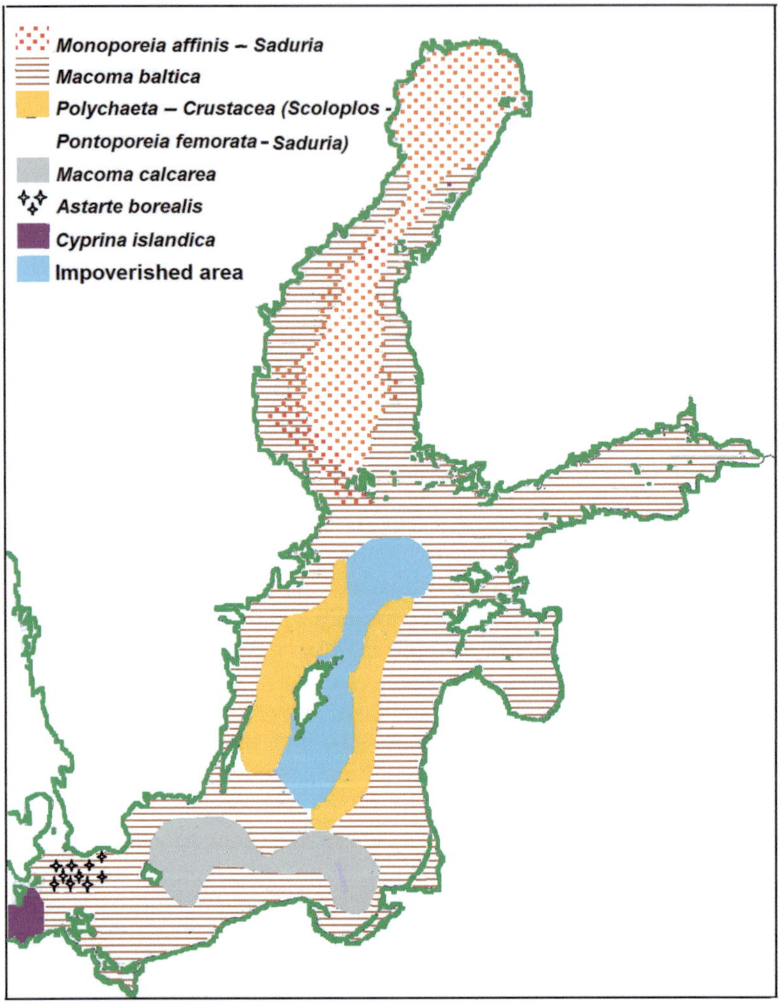

Fig. 3.13 Distribution of the main cenoses of bottom invertebrates of the Baltic Sea (Järvekülg 1979)

The *Macoma baltica* polytypic biocenosis, which is very widely distributed in the Baltic Sea, is divided into modifications differing from one another in composition, abundance parameters and ecology. The western border of this biocenosis is situated east of the Gulf of Mecklenburg and reaches into the comparatively shallow southern part up to the Gulf of Gdansk. The biomass of the biocenosis is relatively high. The share of some other species (*Cardium edule*, *Mytilus edulis*, *Mya arenaria*, etc.) is rather considerable. *Astarte borealis*, *Macoma calcarea* and *Syndesma alba* are less often encountered. Worms are represented by *Scoloplos armiger*, *Nereis diversicolor*, *Harmothoë sarsi*, *Nephthys ciliata*, and *Halicryptus spinulosus*, crustaceans by *Diastylis rathkei*, *Pontoporeia femorata*, and, at low abundance,

Saduria entomon. The last two species have their westward limit west of the Macklenburg Bight and they reach high abundances much farther north in the Baltic. However moving eastward, the share of *Mya*, *Cardium* and *Mytilus* shrinks. In deeper areas of the Southern Baltic, the role of *Macoma baltica* falls (its vertical distribution is limited by oxygen concentration) and its place is occupied by *Scoloplos armiger*, *Halicrypyus spinulosus*, *Priapulus caudatus* and the crustaceans *Diastylis rathkei*, *Pontoporeia femorata*, etc. In the Gulf of Gdansk, *Macoma baltica* dominates up to a depth of 100 m. Deeper, *Scoloplos armiger*, *Saduria entomon*, *Pontoporeia femorata*, etc., become dominant.

The Isle of Gotland divides the area of the *Macoma baltica* biocenosis conditionally into two parts. At these latitudes, *Mytilus*, *Mya* and *Cardium* have not yet lost their importance and *Saduria entomon* and *Pontoporeia femorata* have not achieved the importance they enjoy northwards. The Isle of Gotland is surrounded by a zone with impoverished bottom fauna. With increasing depth, first of all molluscs and after them worms disappear, with the last to be eliminated being crustaceans. After *Pontoporeia femorata* and *Terebellides stroemi* have dropped out, the polychaete *Scoloplos armiger* still remains on the mud bottom smelling of hydrogen sulphide. Under the conditions of oxygen deficiency or absence, the number and diversity of benthic organisms correspondingly decreases up to the full extinction of the bottom fauna in certain areas. Following improvement of the oxygen conditions, the bottoms of the area are recolonized, but the new composition of demersal biocenoses may be basically different from the ones that reigned before the extinction of the previous zoobenthos (Leppäkoski 1975).

In the Åland Sea area, the decrease both in salinity and the organisms depending on higher salinity (*Cardium edule*, *Nereis diversicolor*, *Terebellides stroemi*, *Harmothoë sarsi*, *Halicryptus spinulosus*, *Pontoporeia femorata*, etc.) continues. The abundance of *Monoporeia affinis*, *Saduria* and Chironomidae grows. In the Åland Sea up to the depth of 40 m, *Macoma baltica* dominates, but deeper, the biomass of zoobenthos considerably diminishes and *Monoporeia affinis* becomes prevalent. *Saduria entomon*, *Monoporeia affinis*, *Pontoporeia femorata* and, supplementing them, polychaetes – on shallower soft bottoms, *Nereis diversicolor*, and deeper, *Harmothoë sarsi* – compose the bulk of the nourishment of demersal fish in the northern areas of the Baltic Sea. In the central part of the sea, an important role in fish food is also played by *Halicryptus spinulosus* and *Pygospio elegans*, which prefer sandy bottoms.

In the shallower coastal areas of the Bothnian Sea, zoobenthos biomass constitutes 30–40 g m^{-2}, but with increasing depth, it decreases several times. In the zoobenthos of Bothnian Bay of very low salinity, *Monoporeia affinis* dominates among demersal invertebrates. *Saduria entomon*, *Macoma baltica* and oligochaetes are of lesser importance. *Macoma baltica* occurs up to the salinity of 3.5 psu and disappears at lower salinities. In the coastal areas of Bothnian Bay, the bottom fauna is richer, mainly due to the presence of larval chironomids and oligochaetes. Northwards, both the abundance and biomass of zoobenthos constantly decreases. *Monoporeia affinis* dominates over a huge area. In Bothnian Bay, the average weight of *Saduria entomon* and its total biomass are smaller than in the Bothnian Sea.

In some shallower parts of the Baltic Sea – small bays, lagoons, etc. – drastic changes in zoobenthos have occurred due to changes in environmental conditions. Following the regulation and cutting off of the inflows of the Vistula and Nogat rivers into the Vistula Lagoon in 1914–1916, the average salinity in the lagoon increased to 3–5 psu. The change was estimated as insignificant (Ezhova et al. 2005). However, it initiated a number of basic changes during the twentieth century in the zoobenthos of the lagoon with an area of 861 km^2 and an average depth of 3.1 m, important both for the Baltic Sea and the coastal states. After a strong anthropogenic eutrophication and the beginning of the invasion of alien species, the composition of macrozoobenthos in the lagoon drastically rearranged. Since 1914, mass eliminations of many freshwater mollusc species have taken place and thick layers of shells from *Anodonta*, *Unio*, *Dreissena*, *Lymnea*, *Bithynia* and other representatives of freshwater species could be seen on the bottom of the lagoon. The number of freshwater species was seriously reduced and the introduction of alien species further contributed to the change in the zoobenthos composition and the relationships between species in the ecosystem. This resulted in an increase in the sediment depth penetration by zoobenthos and also in an increase in benthos biomass. Mostly connected to rapid anthropogenic eutrophication beginning in the 1960s–1970s, the lagoon can be estimated as being highly eutrophied. Anthropogenic eutrophication redoubled the fauna impoverishment caused by the salinity increase at the beginning of the twentieth century and provided conditions for the successful naturalization of new species (Ezhova et al. 2005).

As in a number of other groups of animals of the Baltic Sea, it has been found that bottom invertebrates, glacial relicts to begin with, are divided into local populations differing in their qualities and generally spatially separated. Brackish-water *Saduria entomon*, *Monoporeia affinis*, *Paracyprideis fennica*, and also euryhaline marine *Halicryptus spinulosus* and *Heterocyprideis sorbyana*, have each formed a number of local populations. Differences between these populations have evolved in the past (probably in the stage of the postglacial Yoldia Sea) when groups of these species immigrated into the Baltic Sea by different routes. Differences between these populations were evidently deepened with the survival of their ancestors in the almost fresh water of Ancylus Lake or their habitation during that time in saline water in the present-day Skagerrak–Kattegat area. The distance in genetic differentiation between populations can be varied, depending on concrete peculiarities in adaptation and natural selection. Due to ecological and physiological differences, local populations can develop and colonize new territories and ecological niches to secure species from possible elimination. During the complicated history of the Baltic Sea, it has been populated by a number of freshwater, brackish-water and marine populations. Today, they are adapted to the conditions prevailing in certain parts of the sea and are continuously living in these areas. As bottom invertebrates generally do not undertake long migrations, corresponding local populations can occupy both large or rather limited areas and can be situated relatively close to each other. In Fig. 3.14, populations of *Saduria entomon* differentiated by Järvekülg (1979) in the northeastern Baltic are shown. Unfortunately, no corresponding genetic studies have been carried out.

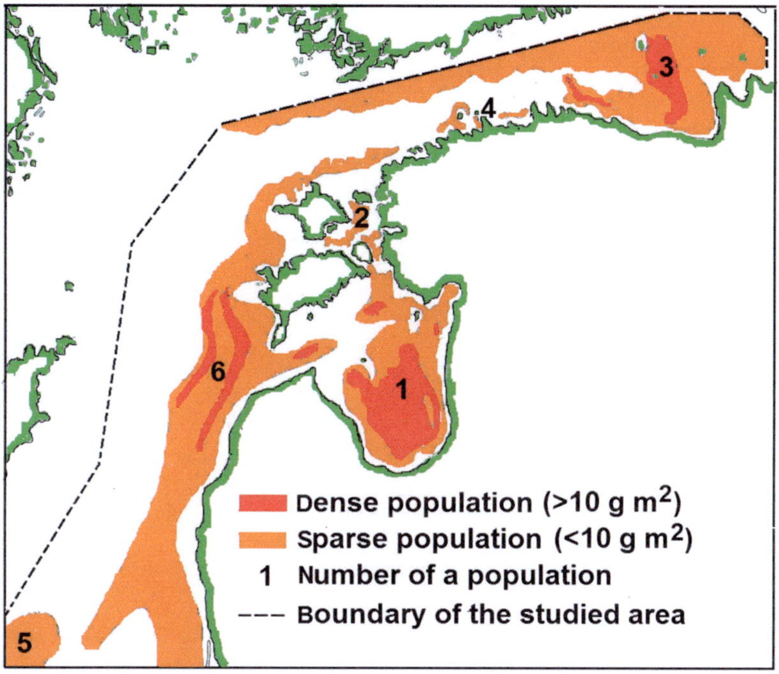

Fig. 3.14 Local populations of the *Saduria entomon* in the NE Baltic and distribution of its biomass (Järvekülg 1979)

Among the zoobenthos of the Baltic Sea, no species occur that are widely used for human food. Because of their slow growth rate, molluscs such as *Mytilus* spp., *Cardium* spp., etc., are of small dimensions, and therefore their utilization does not give substantial profit. Up to the present, the possible profit related to zoobenthos has been connected with benthos-eating fish (cod, flounder, plaice, turbot, eelpout, herring, etc.) for whom demersal invertebrates constitute an important part of their food.

Zoobenthos is an important part of the ecosystems of the Baltic Sea. Studies on benthic ecosystems have lasted more than a century and will also be carried out in the future. The results are important in improving the management of ecosystems of the Baltic Sea. The state of benthic ecosystems is also especially important as an indicator for the assessment of water pollution.

3.9 Fish

There are about 100 fish species (excluding in the Kattegat) that are indigenous and more or less adapted to the Baltic Sea ecosystems (Ojaveer et al. 2010). Baltic Sea ecosystems differ notably from one another and from the ecosystems in the seas of

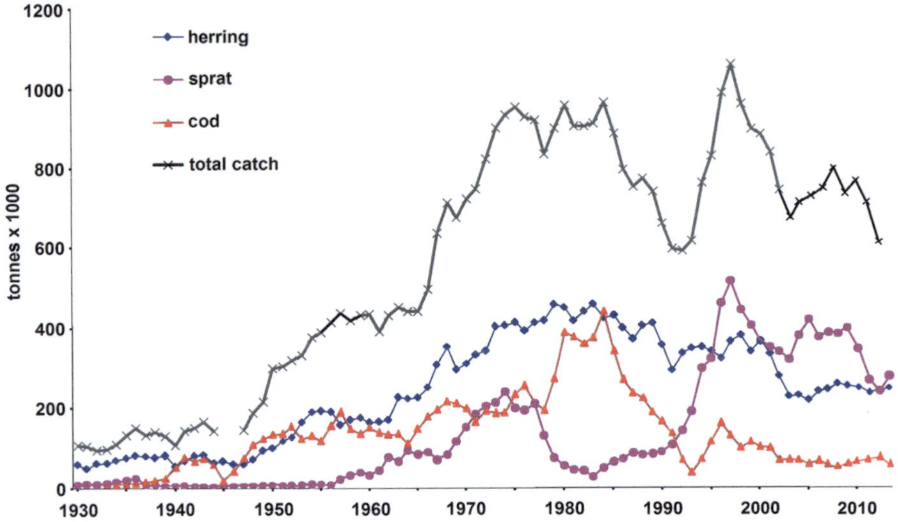

Fig. 3.15 Fish catches in the Baltic Sea since 1930 (Ojaveer 2014; ICES 2013, 2015, etc.)

normal (33–37 psu) salinity. The most important reasons for the limited size of its ecosystems are the moderate dimensions of the Baltic Sea, as well as the present-day environmental parameters which have developed during their complicated history. Therefore, the adaptations of the fish populations in the Baltic Sea are variable and their abundance rather moderate.

In the present Baltic Sea, all the members of ecosystems are newcomers that have settled in this sea during some recent years, after disappearance of the ice. Some populations of wide ecological amplitude of herring, sprat, cod and flatfish species of marine background are successfully adapting in the Baltic Sea. The adaptation has been accompanied by important changes in their population parameters. For instance, compared to their predecessors, the spring spawning herring, well adapted to the present Baltic Sea conditions, has changed one of its most basic features – the salinity requirements at reproduction. Presently, certain populations of the Baltic spring spawning herring cannot reproduce in water of oceanic salinity (Ojaveer 1981).

Considering distribution and feeding biology, very marked phenomenon can presently be seen in relict eelpout. In the Gulf of Riga, the species has two ecological groups which can easily be separated by depth distribution and feeding biology, as well as by the resulting growth pattern and otolith structure (Ojaveer and Lankov 1997). As a result of adaptation, the species is accepting a new feeding strategy, leaving its historical cold-water areas and entering into more productive food chains in the warmer parts of its present area.

The fish fauna of the Baltic Sea strongly differs by region. The southwestern and southern regions have richer species composition of fish fauna, as the higher salinity and temperature favours adaptation of species with a marine background. Northern regions of the sea are populated by the species adapted to low salinity and severe temperature conditions. Only a few marine fish species can reproduce there. Therefore, the composition of fish shoals there is different – the importance of freshwater and relict species is much higher than in the southern half of the sea. No fish species populating the Baltic Sea is adapted to the contemporary high influence of human activity on the quality of the environment as well as directly on the abundance limitation/overexploitation of the species.

There have been dramatic changes in the coastal ecosystems and fish communities over the twentieth century, and this has resulted in an increased focus on this component of the Baltic Sea systems (Adjers et al. 2006; HELCOM 2012b; etc.). Therefore, variations in the structure and abundance of coastal fish stocks should also be assessed and managed. Due to the very important role of fish stocks both in the ecosystems of the Baltic Sea and for the communities around the sea, for the assessment and management of Baltic fish stocks, the best knowledge accumulated up to now should be used.

Based on practical importance, fish fauna of the Baltic Sea is introduced below in the following ecological groups:

Marine pelagic species
Marine demersal species
Diadromous species
Freshwater species
Relict species

As a matter of fact, for management of the ecosystems and resources in the Baltic Sea, all these fish groups should be assessed. The populations of the marine pelagic (spring spawning and autumn spawning herring and Baltic sprat) and marine demersal species, which have key importance in commercial fish landings, have been assessed since the 1970s in an international collaboration by the ICES Working groups using the techniques of the virtual population method (ICES 1995; ICES 2000, 2015b, etc.). Salmon stocks in the Baltic have been assessed using Bayesian methodology (Kuikka et al. 2014). Conventional methods used for open sea stocks are not appropriate for estimating coastal fish resources. It is complicated and costly to sample necessary data for the assessment of freshwater and relict species. In developing the resource assessment system for Baltic coastal fish stocks, Thoresson et al. (1996) suggest starting with standardized test fishing with gill nets and further pertinent treatment of data to get information for stock assessments. This should naturally be performed on the basis of biological and ecological specification of the species in these groups (shortly presented below).

3.9.1 Marine Pelagic Fish

The most substantial group of fish stocks all over the Baltic Sea, both in terms of marine ecosystems, food requirements of the coastal inhabitants and commercial importance, have been the pelagic species with marine background: spring spawning herring, autumn spawning herring and the Baltic sprat. These species are assessed and managed by international bodies based on scientifically valid methods (ICES 1995, 2000, 2015b).

3.9.1.1 Baltic Herring *Clupea harengus membras* L.

Stock Structure Baltic herring is usually treated as a subspecies of the Atlantic herring (*Clupea harengus* L.), which populates areas from Spitzbergen, Novaya Zemlya, the Kara Sea and the White Sea to the Gulf of Biscay in the eastern part of the Atlantic Ocean and from Greenland and Hudson Bay to Cape Hatteras in its western part.

The interrelations between herring groups vary from the status of reproductively related local populations overlapping to a certain extent, to the reproductively isolated units. The most important units are spring spawning herring and autumn spawning herring, which had already been discriminated by Gisler in 1758. Blaxter (1958) showed that spring spawning herring and autumn spawning herring do not mix reproductively and that they should be treated as sibling species. Hence, the species *Clupea harengus* L. should be treated as a superspecies. The mentioned sibling species – spring and autumn spawning herrings – are also present in the Baltic Sea. They have probably immigrated from the Northern Atlantic. C. Linnè named the herring populating the brackish Baltic Sea *Clupea harengus* var. *membras* L. It was shown that in the Baltic Sea, spring spawning herring and autumn spawning herring should be treated as subspecies of different sibling species.

The experiments arranged under natural conditions on herring spawning grounds in the Gulf of Riga (eggs and sperm taken from the spring spawning herrings having arrived at the spawning grounds at the end of the spring spawning season were used in cross-fertilization with the sperm and eggs of the autumn spawning herrings that had arrived at their spawning grounds in the first spawning shoals of the autumn herring) showed no normal descendants hatched from the spring herring eggs fertilized with autumn herring sperm or autumn herring eggs fertilized with spring herring sperm (Fig. 3.16). This experiment indicated that spring and autumn spawning herrings should be treated as reproductively isolated groups.

Later, the levels of genetic divergence between spring and autumn spawning herring in the Baltic Sea was assessed by using two types of DNA markers, microsatellites and Single Nucleotide Polymorphisms, and the obtained results were compared with those of the autumn spawning North Sea herring.

Temporarily replicated analyses reveal clear genetic differences between ecotypes, and hence support the standpoint that spring and autumn spawning herrings are reproductively isolated (Bekkevold et al. 2015).

One of the most basic differences between the spring and autumn herrings is the optimum temperature of their embryonic development. This character is rather conservative in evolution. For spring herrings in the Baltic, in the Pacific and probably also in the Atlantic Ocean, this temperature is about 7 °C, but is noticeably higher for autumn herrings.

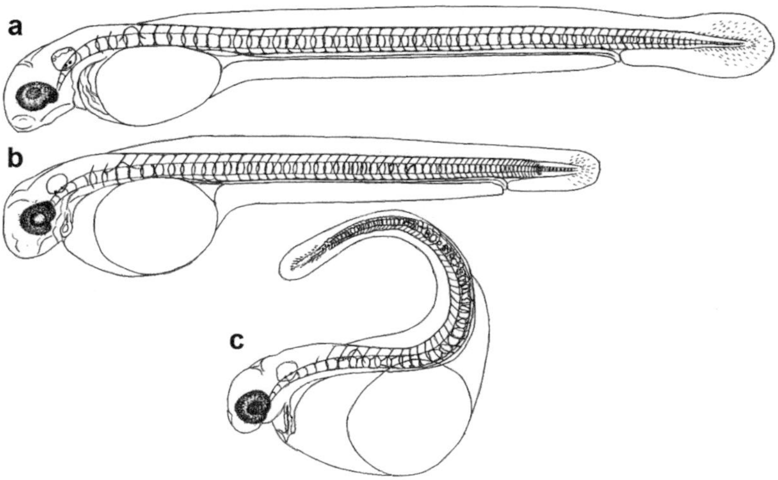

Fig. 3.16 Newly hatched spring spawning gulf herring larvae developed at constant temperature (17 °C) and hatched 1650–1690 degree-hours after fertilization: (**a**) the normal larva of spring spawning gulf herring hatched in June; (**b** and **c**) abnormal larvae developed from the sperm of the autumn spawning herring and the eggs of the spring spawning herring or from the sperm of the spring spawning herring and the eggs of the autumn spawning herring caught on the spawning ground of autumn herring in September. Abnormalities include deviations in pericardium, internal ear, notochord, in many morphological characteristics; the swimming ability is limited or absent (Ojaveer 2014)

Differences between spring and autumn herrings also concern their behaviour, reproduction, and population dynamics, as well as morphological and biological parameters. Compared to the spring herrings, in the autumn herrings of the Gulf of Riga, the number of vertebrae, pyloric caeca and gill rakers are higher, but the relative length and height of head, eye diameter, the maximum body depth and the preventral distance are smaller.

Spring- and autumn spawning herrings can be almost fully separated based on the structure of their otoliths (Fig. 3.17). The life history of a fish is stored in the structure of its otoliths. Otoliths of the spring spawning herring can be separated from the otoliths of the autumn spawners chiefly by their clearly smaller central field, and also by the angle between the rostrum and anterostrum and the relative size of their summer zones. The first growth zone in autumn herring otoliths is almost exclusively wider than in spring herring otoliths. For discrimination of spring- and autumn spawning herrings, otoliths are chiefly used. Otoliths can also be applied for the discrimination of marine herrings from the populations inhabiting the Gulfs of Riga, Finland and Bothnia.

However, as a rule, spring- and autumn spawning herrings cannot be safely recognized by their external appearance. Still, in certain areas, some part of spring and autumn spawning herrings can be discriminated by their characteristic body shape. In the Gulf of Riga, the body of the spring spawning herring is wedge-shaped, with a comparatively large head and eyes and usually featuring a blue back. The body of

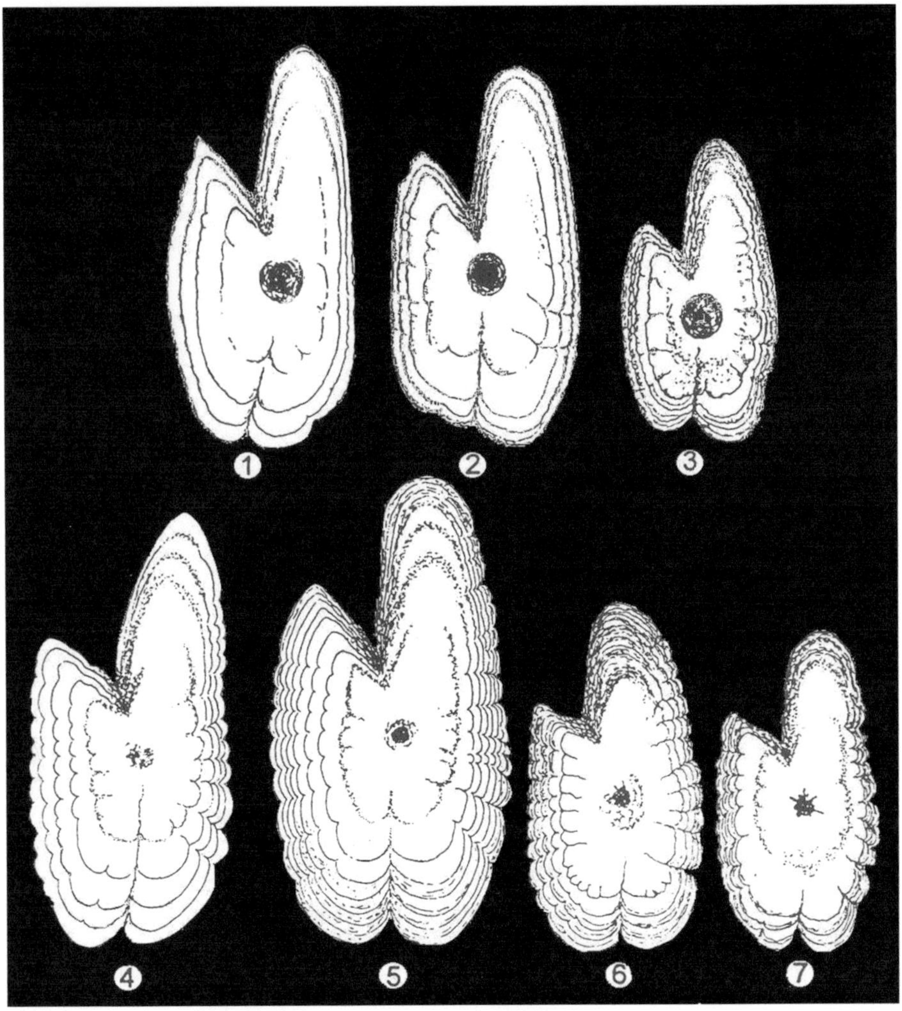

Fig. 3.17 Otoliths of autumn spawning sea herring (*1*, *2*), autumn spawning gulf herring (*3*), spring spawning sea herring (*4*, *5*) and spring spawning gulf herring (*6*, *7*) (Ojaveer 1988)

the autumn herring is commonly spindle-shaped, with relatively smaller head and eyes (Fig. 3.18) and a back that is commonly grey.

It has been known for a long time that in the Baltic Sea, both spring spawning and autumn spawning herrings do not compose a uniform intermingling (panmictic) population. Differences between various herring groups in morphology, growth rate, migrations, abundance dynamics, etc., have induced closer investigations that started as early as the eighteenth century.

It has been found that in the row of herring populations between the NE Atlantic and the open part of the Baltic Sea, a population exists that can be taken as a population of the subspecies *Clupea harengus membras* L., but which has some features

Fig. 3.18 Characteristic difference between the body shape of the spring spawning (*top*) and autumn spawning (*bottom*) herring in the Gulf of Riga (Ojaveer 2014)

characteristic for the populations living in the eumarine environment. This is Rügen herring. During the long period of adaptation, herring populations have obtained new characters and lost some of their previous attributes. The populations living in the eastern part of the Baltic have lost their ability for embryonic development in the water of oceanic salinity (Ojaveer 1981).

3.9.1.2 Spring Spawning Herring *Clupea harengus membras* L.

Extensive marking experiments carried out in Denmark, Sweden, Poland, Germany, Finland, etc., show that in the Baltic Sea, persistent local groups of this species exist. In general, mass mixing of these groups on their spawning grounds has not been stated. This has created favourable conditions for the differentiation of local herring populations inhabiting their areas that clearly differ in environmental conditions.

Summarizing the findings of a number of herring scientists over at least one and a half centuries (Kupffer 1877; Heincke 1898; Hessle 1925; Kändler 1942; Ehnholm 1951; Otterlind 1962; Weber 1971; Biester 1979 etc.) in studies of herring morphometric characteristics, migrations, reproduction, fecundity, growth rate, parasites and diseases, and other population parameters, the following populations of this species can be differentiated (Fig. 3.19): (1) Rügen herring spawning mainly in the area of Rügen Island. The spring spawning herring of the Rügen type also spawns in the vicinity of river estuaries along German and Polish coasts, in Belts, etc. (2) herring spawning in the Gulf of Gdansk, the Wistula lagoon and at the Lithuanian coast; (3) the coastal population of Hanö Bay; (4) Swedish east coast herring; (5) Gulf of Riga herring; (6) sea herring of the NE Baltic; (7) Gulf of Finland herring; (8) herring of the Bothnian Sea; (9) herring of Bothnian Bay.

Below, the herring populations will be treated according to the knowledge of the location of their reproduction and nursery areas. Populations of marine herring inhabiting the open part of the Baltic with higher salinity and milder climate differ from the populations living in the Gulfs of Riga, Finland and Bothnia. Gulf herring live at salinity below 7 psu (but avoid salinities less than 2 psu). In their area, water temperature varies more than in the area of marine herring. Adult marine herrings enter water of salinity below 7 °C for a rather short time – mainly for spawning. After a certain period of larval and postlarval development, young marine herrings leave for the deeper sea. In the life of sea herrings, migrations are much more important than for the gulf herring populations.

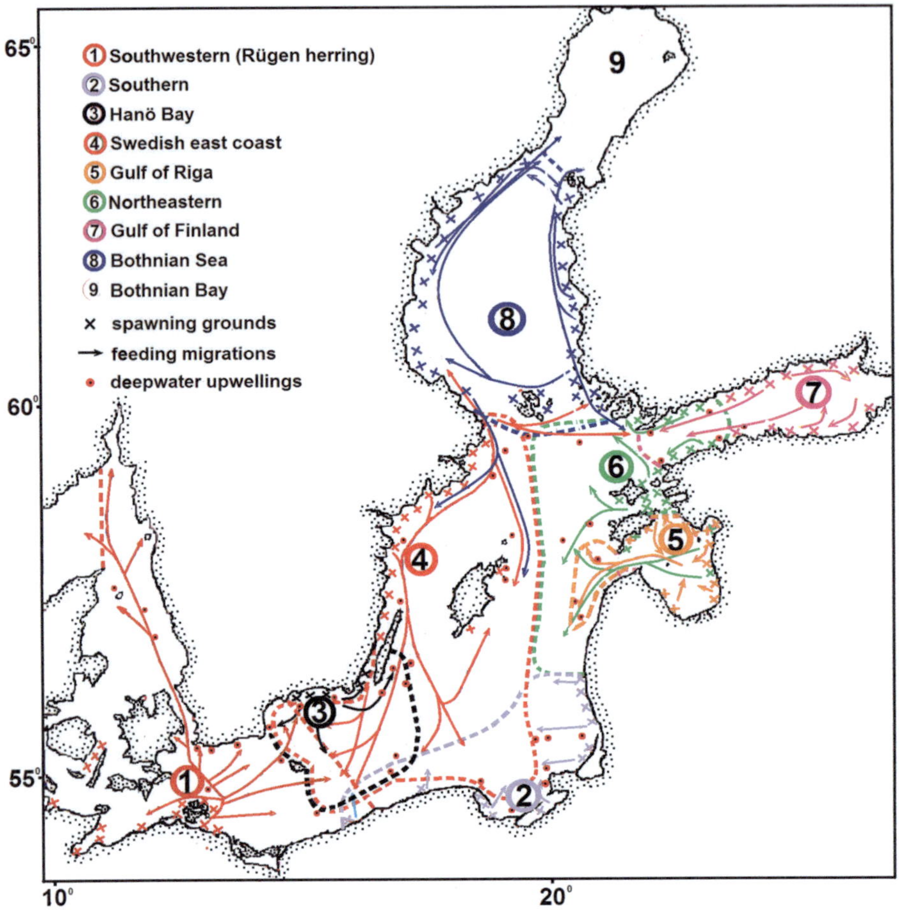

Fig. 3.19 Spawning grounds and feeding migrations of populations of spring spawning herring in the Baltic Sea. The area of the main body of the population is surrounded by *dotted line* (Ojaveer 2014)

Distribution and Habitat Distribution, behaviour and migrations of herring shoals in the Baltic Sea depend on the biological cycle (reproduction, feeding and wintering) of the fish, its physiological condition and environmental influences, including human impact. Herring lives in all parts of the Baltic Sea, at various salinities. In earlier sources, it has been stated that herring can enter into river estuaries and live in freshwater lakes. It is known that herring has developed an endemic population of stunted growth in the coastal water body Windebyer Noor with a salinity of 3–6‰ that separated from the Eckenförde in 1874 (Neb 1970). On the other side, even the herring population in the eastern Gulf of Finland, adapted to comparatively very low salinity, avoids the areas where salinity is very unstable or continuously below 2 psu.

The reaction of the Baltic herring to light considerably changes during its life. The larvae have a positive reaction to light and live in daylight in surface water layers. Also, herring whitebaits and young herring populate well-illuminated upper water layers. However, with increasing age, the depth of the daily distribution of herring constantly increases. In feeding and wintering periods, older herring rises to higher water layers only at night. Still, it is a complicated thing to differentiate the reaction of herring to light from its reaction to nourishment, as the main herring food organisms strictly perform diurnal migration (i. e., at night, their main concentrations are found in surface layers).

Seasonal Distribution In winter, the depth of herring shoals by day can reach 120 m. The upper limit of the shoals is determined by the acceptable temperature (2–6 °C) and the lower limit by the oxygen content (at least 1–2 $cm^3\ dm^{-3}\ O_2$, Fig. 2.22). In the deeper part of the fish layer, older herring dominates. Whitebaits and younger specimens gather mainly at the coastward edge of the fish layer where the temperature is lower but the oxygen concentration higher. In the southern part of the sea, herring winters at higher temperature than northwards (in the Bornholm Deep area, usually at 4–6 °C, in the Gotland Deep, at 2–5 °C). In the Gulf of Riga, the Archipelago Sea and the eastern Gulf of Finland, herring winters in bottom layers at depths of at least 20 m at 0–3 °C, but in good oxygen conditions. Wintering concentrations of pelagic fish are usually distributed in one layer, which is vertically rather compressed, in severe temperature conditions. In milder conditions, the vertical range of the layer is bigger, and in some cases, two-layered disribution of pelagic fish (in the deeper layer, mainly older herring, higher sprat and younger herring) can occur.

In spring, the dispersing of wintering aggregations begins with the increase in temperature of the surface layers up to 0.5–1 °C. Younger, mainly immature herring and sprat ascend into the warmer surface water or originating thermocline, where the concentration of their prey – the warm water zooplankton – is densest. Larger herring keep to the bottom layers, preferably in the mixing zone at the halocline. Spawning migration is usually started by the older age groups with their gonads in the most advanced stages of development, as they usually winter in the most favourable temperature conditions. Herring enter the coastal zone after the breakage of ice, commonly together with the water of open sea origin. After spawning, herring returns to the deeper water layers to start internse feeding.

In summer, herring are very mobile. They are distributed in three layers (Fig. 3.20): the whitebaits keep to the warm surface layer near the coast, young herring and sprat dwell in the thermocline by day, and older herring feeds intensely in bottom layers, chiefly in the area of the densest aggregations of invertebrates at halocline.

In autumn, the fatness of herring has increased and its activity decreased. After the disappearance of the thermocline, pelagic fish shoals and those feeding at the coastal slope dissipate. Herring gather for wintering in warmer water in offshore areas.

Diurnal Vertical Migration During the feeding period (at temperatures >2 °C and also in good oxygen conditions in the wintering period), herring, like its main food animals, perform diurnal vertical migrations (Fig. 3.21). At sunset, the herrings having distributed by day through bottom layers (older fish) or at the thermocline, ascend over 30–40 min into surface layers and stay there all night. The descent of

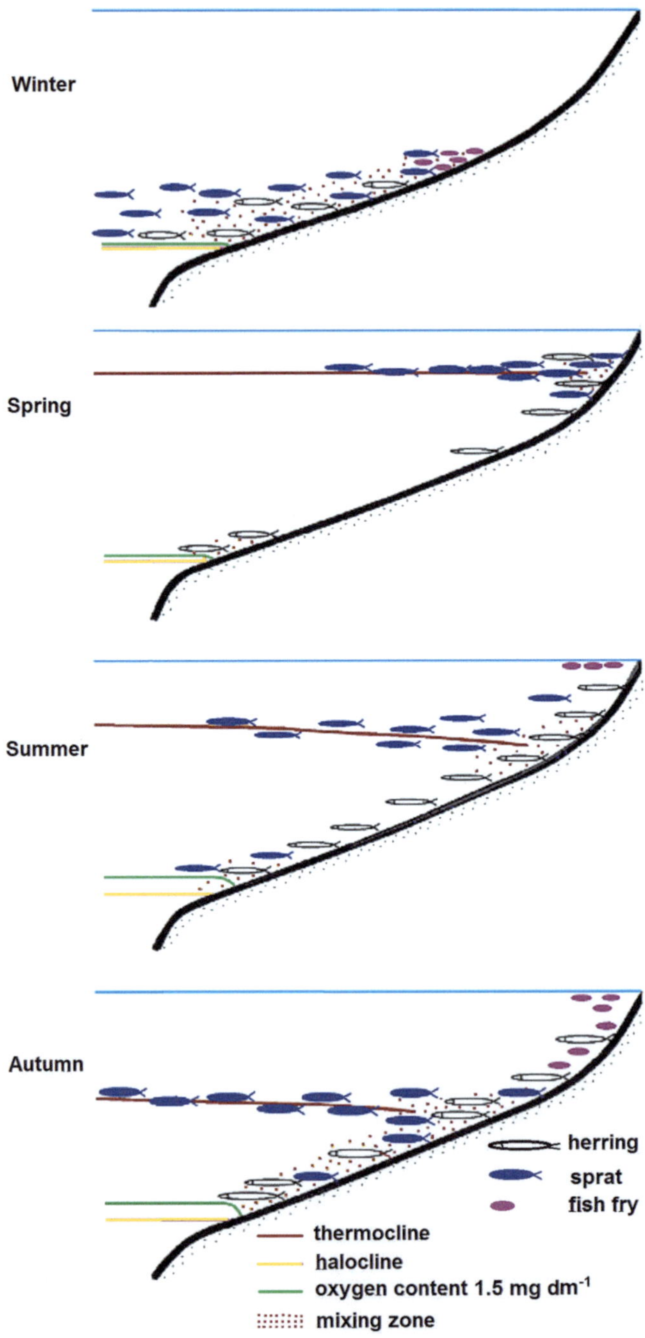

Fig. 3.20 Seasonal differences in distribution of herring shoals in the Baltic Sea (Ojaveer 2014)

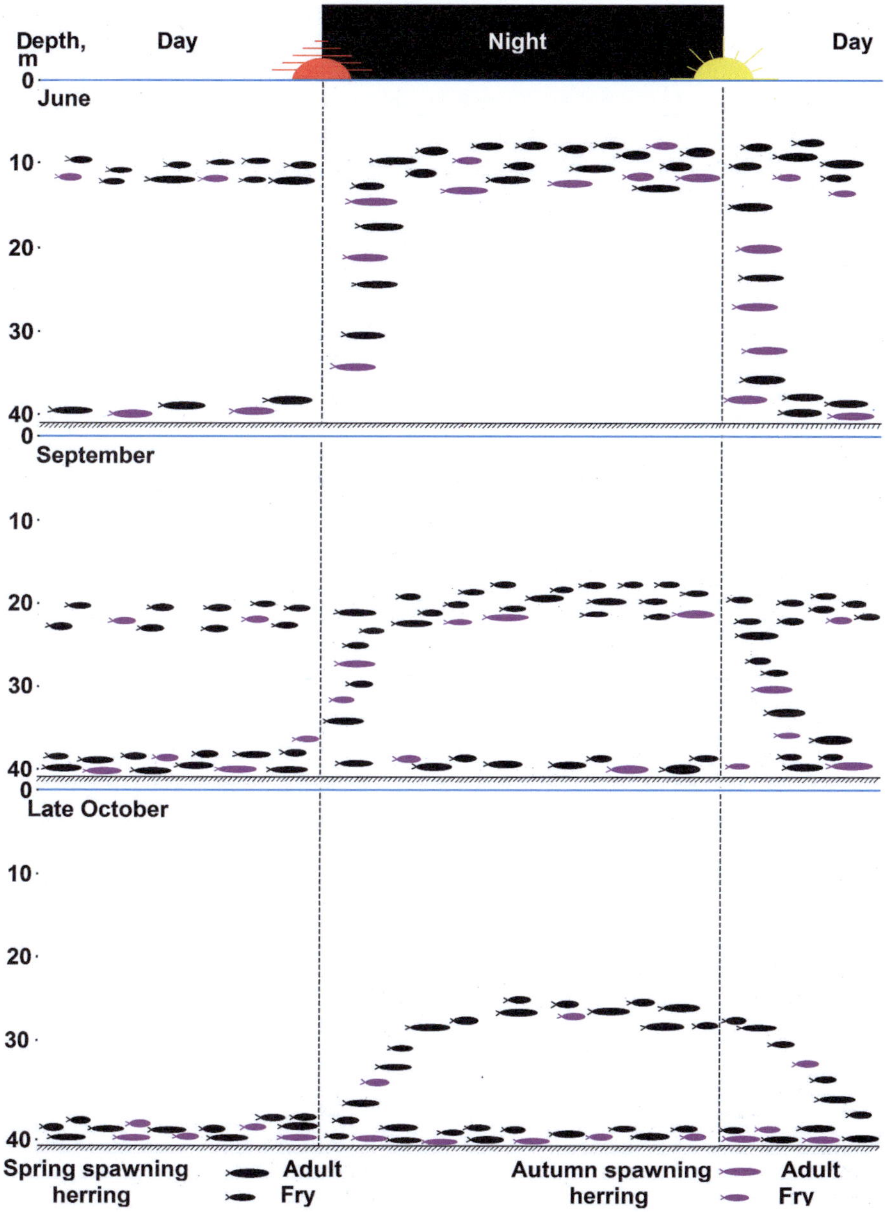

Fig. 3.21 Intensity of the diurnal vertical migration of herring and the composition of shoals in the Gulf of Riga in June, September and October (Ojaveer 2014)

fish to their daily horizons starts with sunrise and lasts, depending on the intensity of the sunlight, 1–3 h. The intensity of the daily vertical migration changes over the year. The migration depends on the temperature of water layers, oxygen content and other environmental parameters, as well as on the condition of the fish organism. Herring shoals start diurnal vertical migration at the end of wintering, with the migration being most active during the spring feeding period and at early summer. Spring spawning herring and sprat participate most actively in the vertical diurnal migration, as their prey organisms also actively perform this migration. Autumn spawning herring, as a fish more connected with the bottom layers, ascends to the surface layers in spring, but in summer and autumn, commonly keeps to layers deeper than spring spawning herring, rarely leaving the bottom layers in autumn. During the summer, the intensity of the migration gradually weakens. However, in tenous form, the diurnal vertical migration lasts until autumn. The diurnal vertical migration has been considered to be an adaptation that has allowed herring to be situated in the high density of food organisms for a long time, under favourable temperature and light conditions, and simultaneously be comparatively inconspicuous to enemies.

Migrations Between Various Sea Areas The most important information on herring migrations has been collected through taggings. However, indirect sources (investigations of otolith types and infection with parasites, examination of changes in age composition and in the average weight and length in certain sea areas) have also been applied to conclusions on migrations.

The most important general regularity clarified by tagging is a different migration strategy in sea and gulf herrings. Gulf herring's migrations are rather short, whereas sea herring undertakes much longer relocations. Otterlind (1962) found that sea herrings generally do not migrate into the Bothnian Sea and Bothnian Bay, and gulf herrings do not undertake mass migrations into the Baltic Proper. Also, the exchange of specimens between the eastern and western coasts of Bothnian Bay is limited – this takes place chiefly via the Norra Kvarken.

Finnish herring markings (Parmanne 1988) allow us to conclude that the herring of the Gulf of Finland does not usually undertake long migrations outside of the gulf. Characteristic changes in the average length and weight of herring in the various parts of the gulf and the NE Baltic show that after spawning, older herring migrates from the gulf westward and returns to the gulf at the beginning of winter.

The population of the Gulf of Riga spawns and feeds mainly in their home gulf. After spawning, some part of herrings undertakes (especially in years when the abundance of food is scanty) a feeding migration into the open sea. In autumn or early spring, the main portion of them returns to the home gulf.

Compared to the gulf herring, migrations of sea herring are much longer. Long-term studies by Otterlind (1962) showed that spring spawning herring from the Swedish east coast spawning grounds migrate to the Southern Baltic after spawning, mainly to the area of Bornholm Deep and the Stolpe channel, where they feed in the area of upwelling on the rich biological production from June–July to

October–November. The intensity of this regular migration varies by year. In the years rich in zooplankton, the abundance of herrings from the Swedish east coast spawning grounds in the feeding places of the Southern Baltic is less than in the years that are poor in zooplankton. The mass migration of herring from these spawning grounds to the highly productive feeding grounds indicates that the spawning grounds west of the Gotland Basin produce more herring than can feed the waters near the spawning places. Therefore, after spawning, herring shoals move in search of nourishment along the coastal slope southwards. Taggings have indicated that only a small fraction of herrings reproducing at the Swedish east coast migrate eastwards of the deep area in the center of the Gotland Deep. The probable cause of this is that the density of herring food organisms is rather low and the oxygen conditions and temperature comparatively unfavourable in higher water layers in the central part of the deep. Therefore, mixing of the Swedish east coast herring with the sea herring of the Northeastern Baltic is probably rather limited. Owing to different living conditions, the Swedish east coast herring is also comparatively different from the Hanö Bay herring. The Hanö Bay herring does not undertake long migrations – it does not migrate north of the southern Kalmarsund nor west of Bornholm (Otterlind 1962).

The main number of the herring spawning in Vistula Bay and the Gulf of Gdansk feeds in the gulf and its neighbouring areas. Some part of the herrings reproducing in the Gulfs of Pomorsk and of Gdansk (but not in Vistula Bay) has spent some time in the water of higher salinity outside the Baltic Sea and are infected with the larvae of *Anisakis simplex* (also dangerous to humans). This parasite lives at higher salinity than is found in the Baltic.

German herring scientists (e.g., Biester 1979) have clarified that the main part of the younger Rügen herrings live in the Baltic between 16°E (east of Bornholm), Kühlungsborn and the Gulf of Mecklenburg. The bulk of older Rügen herring and a part of the younger migrate to Kattegat and Skagerrak to feed. Their most common route passes through the Sound, where the shoals winter on their way back from the western feeding areas to the Baltic. The Rügen herring is closely connected with the spawning grounds situated in the area of Rügen Island. The herring population reproducing in this area spends a part of its ontogenesis in the Baltic Sea and another part in much higher salinity. Therefore, this population can be considered to be the transitional population between the Atlantic and Baltic herrings. The spring spawning herring of the southwestern Baltic (the Rügen herring) lives in a comparatively well delimited area, where mixing with other populations is limited, and therefore its population status is comparatively clear.

Reproduction The sex ratio in herring shoals is generally 1:1. The males attain sexual maturity at a somewhat younger age and smaller size than the females. Even among adult herrings, a small per cent of specimens with their gonads in the juvenile stage can be found, i.e., the sex of these fishes has not yet been determined.

In the Gulf of Riga, a specimen belonging to spring spawning herrings has been caught with both eggs and sperm in the prespawning stage in its sexual glands.

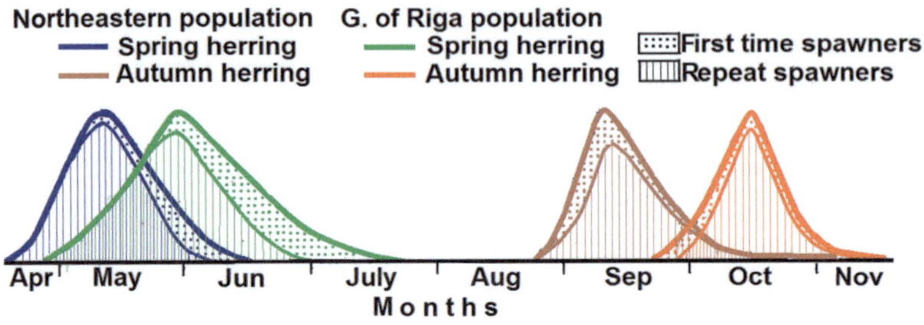

Fig. 3.22 Spawning periods of herring groups in the NE Baltic Sea (Ojaveer 2014)

In general, the sexual maturation of Baltic herrings takes place at the age of 2–3 years. In the Gulf of Riga, a small part of the spring spawners have already attained sexual maturity at the age of 1 year. Usually, herring spawning concentrations in the Baltic Sea consist of 2–6-year old fish (Table 3.2). Spawning is generally started by the age groups having wintered in the most favourable conditions, which stimulate development of the sexual glands. The order of arrival of herring age groups to spawning grounds is an adaptive process. This allows for reproduction of medium age groups under the most favourable conditions for egg fertilization and embryonic development (Fig. 3.22). Therefore, the reproduction potential of medium age groups having the highest relative fecundity and the sexual products of the highest quality is largest.

Fecundity Local populations of Baltic spring spawning herring differ from one another both in their average fecundity and the relationship between their fecundity, weight and length (Fig. 3.23). The correlation coefficients between these relationships are significant. The closest is the dependence between the fecundity and weight. The fecundity is lowest in the first-time spawners (recruitment). The specimens having the best growth produce more eggs than slow-growing specimens. The relative fecundity is highest in medium age-groups.

Spawning and Embryonic Development Investigations into herring reproduction were started in the Baltic Sea about one and a half centuries ago. The first studies were performed in bights and inlets in the Southwestern Baltic (Kupffer 1877).

In general, in the Baltic Sea, spawning of the spring spawning herring lasts from February to August. In the SW Baltic, spawning starts in February and finishes in June, and correspondingly in the northernmost part of the Bothnian Bay in June and August. The start of spawning of the spring spawning herring is adapted to the time of the maximum primary production in the corresponding sea area. In this case, the peak of abundance of copepod nauplii (the most important larval food of herring) falls to the period when the maximum herring larvae transfer to the external food. This adaptation is extremely important, as in this case, the transfer of the larvae from endogenous to exogenous feeding coincides best with the maximum abundance of their food organisms. In this case, the survival of the larvae is the largest and the abundance of the developing year class the highest.

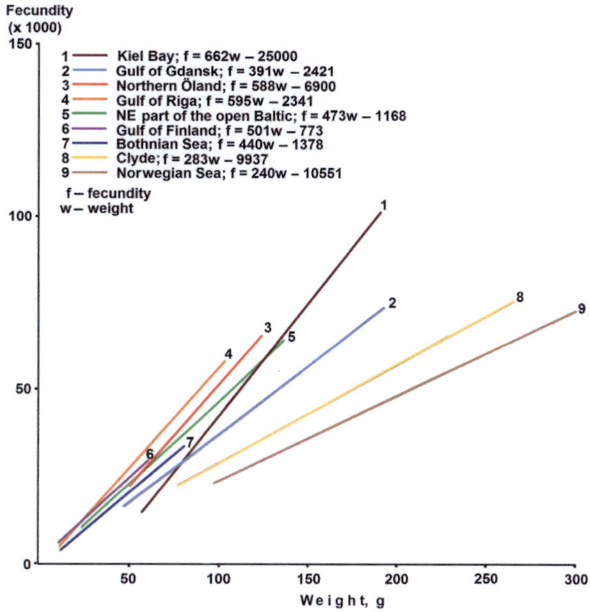

Fig. 3.23 Fecundity of spring herring populations in the Baltic Sea and in the neighbouring seas (Ojaveer 1988)

Spawning of the spring spawning herring starts just after the breaking up of ice, after the temperature of the water rises to 2–3 °C. The temperature becomes too high for spawning when it has risen to 18–20 °C. The temperatures measured at herring spawning in different sea areas are variable (Table 3.1). Later in the reproduction period, spawning shifts to deeper grounds where the temperature remains at an acceptable level. Spawning grounds are situated in the vicinity of coasts, mainly at the localities of turbulent water movements. The optimum temperature for the embryonic development of the spring spawning herring (which results in the highest average percentage of embryos hatched) is 7 °C for both the NE Baltic and the Rügen herring. The temperatures deviating too much from the optimum cause mass malformations in embryos. Developing at 3 °C, only 1.3% marine herrings of the NE Baltic were normal and the gulf herring embryos were all abnormal (Ojaveer 1981).

Salinity in spawning grounds varies according to the parts of the sea. Spring spawning herring spawns along the whole coast of the Baltic Sea, excluding comparatively limited (most freshened) areas in the easternmost part of the Gulf of Finland and in Vyborg Bay.

In Rügen herring, the lowest critical salinity of embryological development is about 4 psu (Klinkhardt, 1984). In the NE Baltic Sea, normal development of the species is possible beginning at the salinity of 2.5 psu, but some embryos start their development at 1 psu. The optimum salinity of the spring herring embryonic development is 5–20 psu (Ojaveer 1981). At higher salinities, deviations from normal development increase rather rapidly, and at ocean salinities, embryos are predominantly abnormal. Consequently, the herring populations in question have already differentiated from the populations spawning at oceanic salinities.

Table 3.1 Vital parameters of reproduction of spring spawning herring populations

Area	Spawning			
	Time (months)	Depth (m)	Salinity (psu)	Temperature (°C)
SW Baltic	(II) III–VI	<5–6	(3) 5–6 (15)	6–20
Hanö Bay	IV–VI	0.5–6	7–7.5	5.5–16.5
Gdansk – Klaipeda	IV–VI	0.8–20	4–6	3–18
Vistula Lagoon	IV–VI	0.8–3	3–4	3–18
East coast of Sweden	V–VII	0–19.5	6–7	3.8–18.5
NE Baltic	IV–VI	2–8	3–7	2–13
Gulf of Riga	V–VI	4–15	3–5	(6)9–16
Gulf of Finland	V–VII	0.2–17	>2–5	6–15
Bothnian Sea	IV–VII	0–8	5–6	3.5–18.5
Bothnian Bay	V–VII	0–10	3–4	8–15

Data from: Altukhov (1959), Aneer (1989), Ehnholm (1951), Hessle (1925), Kääriä et al. (1988), Ojaveer et al. (2003), Parmanne et al. (1997), Rajasilta et al. (1989), Strzyżewska (1969) and Weber (1971)

Herrings arrive at spawning grounds in shoals. As a rule, herring does not lay eggs on developing embryos. After fertilization, eggs stick on the substrate: bottom vegetation, stones, shells, gravel, etc. The spawning grounds are commonly open littoral areas of stony bottom, covered with vegetation. The main plant species important to the spawning substrate are *Pilayella littoralis, Ectocarpus confervoides, Furcellaria lumbricalis, Fucus vesiculosus*, etc. Herring avoids spawning on soft bottoms. Sparse bottom vegetation or its absence may be a cause of absence for herring spawning. The type of spawning substrate may vary by sea area and time period, depending also on the natural sequence of the composition of plant assemblages at the depth of the favourable spawning temperature.

The spawning grounds used by herring for reproduction differ by year, depending on the variation in temperature, possible spawning substrate and other factors, including human activity.

A limited percentage of herring eggs can start their development parthenogenetically, i. e., without being fertilized. Their development ceases commonly before cleavage or leads to malformations and elimination of the embryo afterwards.

Mortality during the embryonic period and the percent of abnormal embryos is lower in the embryos developing on bottom vegetation and in sparse spawn than on bottom sediments and in multilayered aggregations where oxygen conditions are worse. Egg mortality on spawning grounds depends on the environmental conditions: herring eggs are eaten by eelpout, whitefish, etc., can die due to low oxygen concentration or unsuitable temperature regime, can be eliminated by fungi or cyanobacteria (Ojaveer 2014).

Hatching of spring herring embryos starts at 1540–1970 degree-hours and the yolk-sac resorbes 2840–3500 degree-hours arter fertilization. At hatching, the sea herring embryo is clearly longer than the gulf herring one. In the NE Baltic, the average length of the sea herring embryo after embryonic development at 7 °C was 6.5 mm, and of the gulf herring embryo having developed at 17 °C, 5.8 mm. The

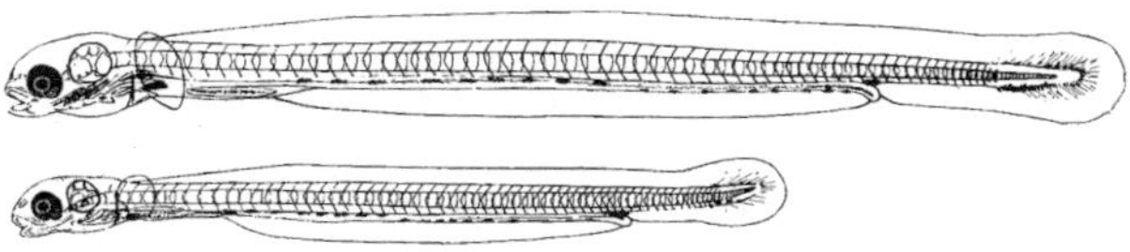

Fig. 3.24 Larvae of the spring spawning herring developed on the spawning grounds of the Gulf of Riga after resorption of yolk sac: *top* – sea herring developed in the embryonic period at 12 °C, *bottom* – gulf herring developed in the embryonic period at 17 °C (Ojaveer 1988)

difference in length between the sea herring larva having spent its embryonic period in colder water and the gulf herring larva having spent its embryonic period in warmer water (this difference is normal under natural conditions) can clearly be noticed (Fig. 3.24). This can be the basis for the development of different individual length and weight in the sea and gulf herring populations.

The bulk of the spring herring larvae dwell in the coastal zone, where the bottom depth varies between 10 and 25 m and the temperature between 12 and 19 °C.

During the first year of life, the mortality of spring spawning herring is highest in the embryonic period and at the beginning of larval development. The year class abundance is determined during a rather short period – in the time of transition of the larvae to external feeding. During this so-called critical period, at a length from 8 to 12 mm, the mortality of larvae is highest, the cause being food deficiency. The level of mortality depends both on the abundance of their main food, copepod nauplii, and on the percentage of the larvae with morphological and physiological abnormalities which do not allow them to catch or use the food.

Metamorphosis of the spring herring larvae takes place 2–2.5 months after hatching, when its total length is about 3 cm (Ojaveer 1988). In summer, the whitebaits live in the 5–10 m thick surface layer, where the temperature increases up to 18–19 °C. They leave the coastal zone only after a large decrease in temperature in autumn. Young herrings (age groups 0 and 1) remain in the energetically active areas where the biological productivity and the amount of food is largest. This means that young herring gathers in places of intense mixing of water layers, including frontal zones.

Feeding Generally, the period of mixed endogenous and exogenous feeding is absent in Baltic herring. Its first exogenous food is monotonous, consisting only of copepod nauplii and eggs of planktonic organisms. Large larvae also feed on copepodites. 8–16 mm larvae feed twice a day, in the morning and in the evening, while larger larvae have an additional feeding at about 2 p.m. The daily ration of herring larvae in the Gulf of Riga varies between 6 and 9% of the weight of larva. The main food competitors of herring larvae are larval gobies.

In summer, the whitebaits feed on copepods. In autumn, when their length exceeds 6 cm and migration to greater depth starts, they begin to prey on mysids, amphipods and some other invertebrates. The chief food competitors of young herring are sprat, sticklebacks and gobies.

The composition of the food of various ecological groups of adult herring differs clearly. The differences increase with age. A part of gulf herring feed during their whole life on small plankton. Therefore, their growth rate decreases considerably with age. On the other hand, in the northern and central Baltic Proper and in the adjacent areas, some spring herrings transfer rather early to predation on sprat, sticklebacks, large invertebrates, etc. These so-called giant herrings have much higher growth rates than the average (Ojaveer 1988).

The most stable component in the Baltic herring food is the copepods (*Pseudocalanus minutus elongatus*, *Limnocalanus grimaldii*, *Eurytemora* spp., *Temora* spp., *Centropages hamatus*, *Acartia* spp.). Of the cladocerans, *Evadne* spp. and *Podon* spp. in summer and *Bosmina* spp.in autumn are particularly important. Zooplankton is the most substantial food item during the most intense feeding period in spring and summer. In autumn, the abundance and biomass of zooplankton decrease, the feeding activity drops, and mysids and amphipods obtain important positions in herring food.

In the open Baltic, spatial and seasonal variations in feeding intensity are much smaller than in the gulfs. Because of sharp seasonal fluctuations in temperature and food abundance, in summer, the feeding of herring is more intense in gulfs than in the open sea, but weaker in autumn and winter. In gulfs, the importance of mysids and amphipods in herring food is higher and that of cladocerans lower than in the open sea.

In general, at temperatures <2 °C, herring does not feed. The longest fasting periods occur in gulfs, where winter temperatures are lower than in the open Baltic. In the III maturity stage of gonads, the feeding intensity of herring decreases clearly, and during the spawning time, its stomach is empty. The feeding activity of adult herring is highest after spawning, and in immature herring, in spring. Due to the temporal differences in spawning periods, intense feeding periods of the populations in the southern and NE parts of the Baltic Proper somewhat differ.

Herring feeds both on densely distributed smaller plankters and on the larger organisms that it selects by sight. It has 2–3 maximum stomach fullness periods a day. The main food competitors of adult herring are sprat, smelt, young cod and flatfish.

Herring is the most important consumer in the pelagic system of the Baltic Sea. About 50% of the food of herring populations is utilized by 1- and 2-year-old individuals. During its life, herring can transform from a consumer of the first order to one of the second or even third order, transforming gradually, preying on the energetically most valuable larger crustaceans and even fish. Therefore, during its life, herring's relative ration decreases from an average of 12 body weights in the 2-year-old herrings to 7 body weights in the 7-year-old fish (Ojaveer 2014).

Growth In the period of the transfer of the larvae to exogenous feeding, their length increment and the condition factor are the lowest. In the larvae living on exogenous food, the condition factor and the growth rate increase substantially. The growth in length drops again at the beginning of metamorphosis.

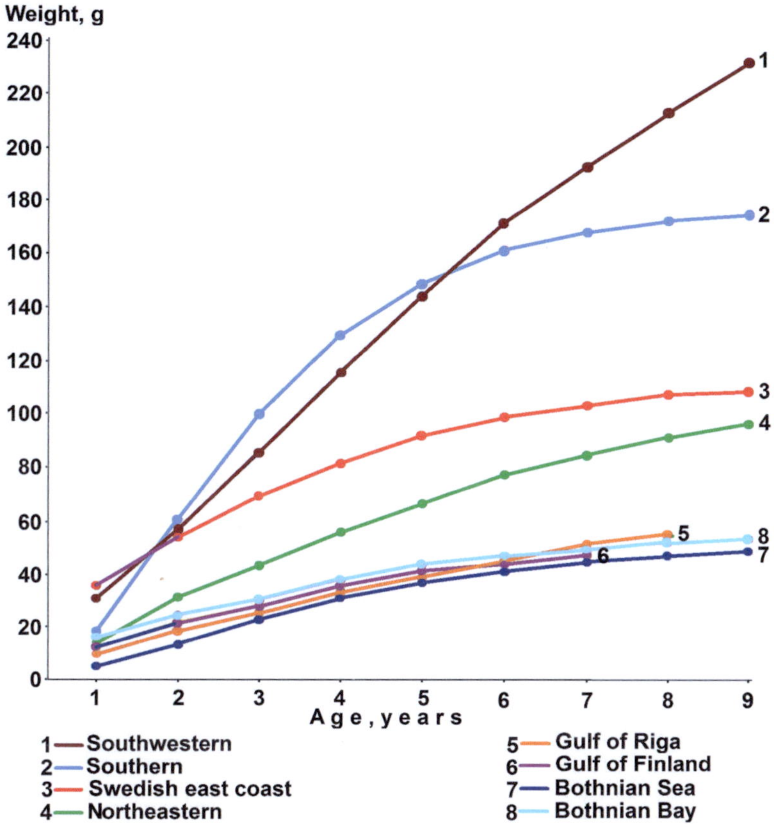

Fig. 3.25 Average growth in weight of populations of the spring spawning herring in the Baltic Sea during 1970–1986 (Ojaveer 2014)

The growth of individual herrings differs significantly in the limits of every population and also between populations. The asymptotic length and weight are larger in the open sea populations than in the gulf herring, and higher in sea herring in southern populations (Fig. 3.25). In winter, under the conditions of decreasing temperature and available food animals, the feeding activity and growth diminish. In this time, herring uses the energetic reserves accumulated in its organism earlier. Therefore, the weight of individual herrings can drop during winter. The increase of the body mass begins after the start of feeding, in younger herring earlier than in older specimens. In 1-year-old herring, the most rapid increase in body weight takes place from June to August (Fig. 3.26).

In adult herring, the dynamics of body mass depend substantially on the sexual cycle. The maximum body mass occurs just before and the minimum after spawning.

Fig. 3.26 Seasonal dynamics in average weight of the spring spawning herring in the open part of the northeastern Baltic (**a**) and the Gulf of Finland (**b**) (Ojaveer 1988)

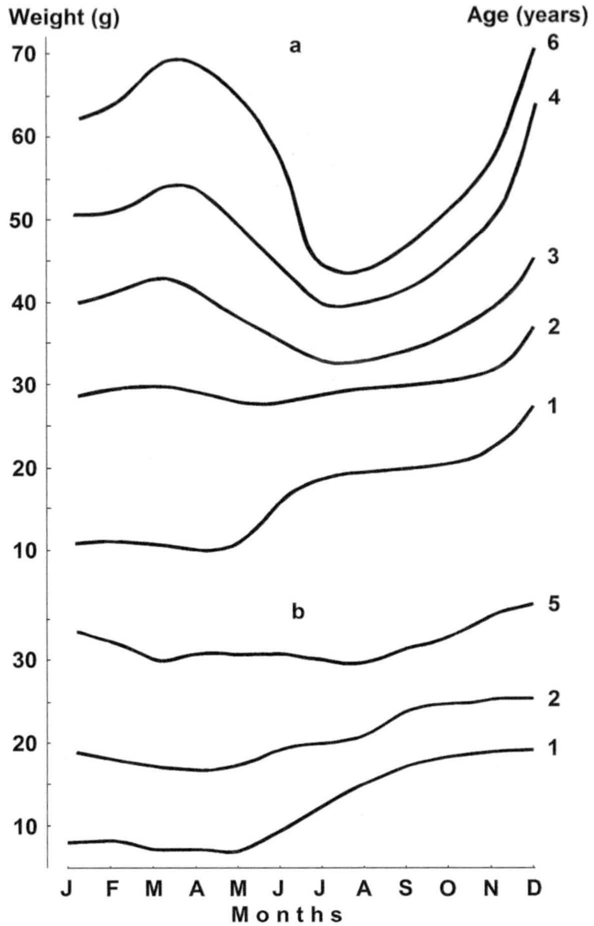

The growth rate varies due to the abundance and quality (species composition) of food and the conditions of food assimilation (salinity and temperature). During the 1960s–1980s, in the period of increase of biological productivity in the Baltic due to eutrophication, herring growth rate in several areas of the Baltic Sea markedly accelerated. The increase in growth rate coincided with the periods of intense mixing of water layers, which is usual after the intrusion of saline North Sea water into the Baltic Sea, resulting in an increase in biological productivity and herring food. In the times of deepwater stagnation, when the food abundance and habitation volume of herring decrease, its growth rate reduces. Especially substantial was the drop during an extraordinarily long stagnation from the early 1980s to the 1990s, when after a large decrease in salinity, important food organisms of herring (*Pseudocalanus*, etc.) retreated from large areas towards the southwest and the oxygen deficiency in deep layers contributed to the aggravation of the situation in herring food resources in many Baltic Sea areas (Fig. 3.27).

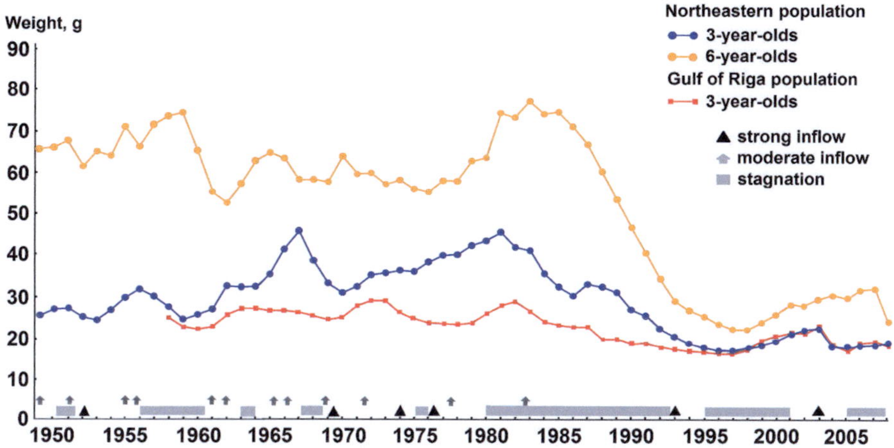

Fig. 3.27 Dynamics of the average weight in the spring spawning herring of the northeastern Baltic and the Gulf of Riga in relation to the inflows of saline water from the North Sea (Ojaveer 2014)

The maximum total length of the Baltic spring spawning herring has been measured on Saaremaa Island (51 cm) and the maximum weight (1050 g) has been found in Finland (Ojaveer 2014). The oldest Baltic herring caught in the Gotland basin was 20 years old, in the Gulf of Finland, 16, and in the Gulf of Riga, 10 years old. In the commercial fishery, the oldest herring age groups are in the Gulf of Riga and the Southern Baltic, 5–7 years, and in the Gulf of Finland and Northern and Central Baltic, 8–11 years old.

Enemies and Parasites Herring embryos on spawning grounds are preyed upon chiefly by eelpout, whitefish and sticklebacks. In the larval stage, its main enemies are three-spined stickleback, adult herring, smelt, etc. Adolescent and adult herring occupy an important position in the food of cod, salmon, sea trout, pikeperch, perch, turbot, river lamprey, sea lamprey, seals, seagulls, cormorants, etc. The most important consumer is cod. In the periods of high abundance of cod, the natural mortality of herring considerably increases, just to drop again after the decline in cod biomass.

Also, herring has many parasites and diseases. Dangerous fungal disease is caused by *Ichthyophonus hoferi*. To the many species living in the internal organs of herring as parasites belong *Costia necatrix, Hexamita truttae, Eimeria sardiniae, Epistylis lwoffi*, etc. Of trematodes on herring parasitize *Diplostomum spathaceum* and *D. baeri*, of nematods, *Contracaecum osculatum*, the species of the genus *Anisakis*, etc., of acanthocephalans, *Pomphorhynchus laevis*. The most abundant parasites in herring are those belonging to nematodes.

In the years 1978–1985, the composition of the parasite fauna was comparatively stable. Beginning in the year 1990, it has been found that herring has been infected with the parasites usually connected with freshwater environments (*Epistylis* spp., *Hexamita* spp., etc). It is possible that during recent decades, the pressure of para-

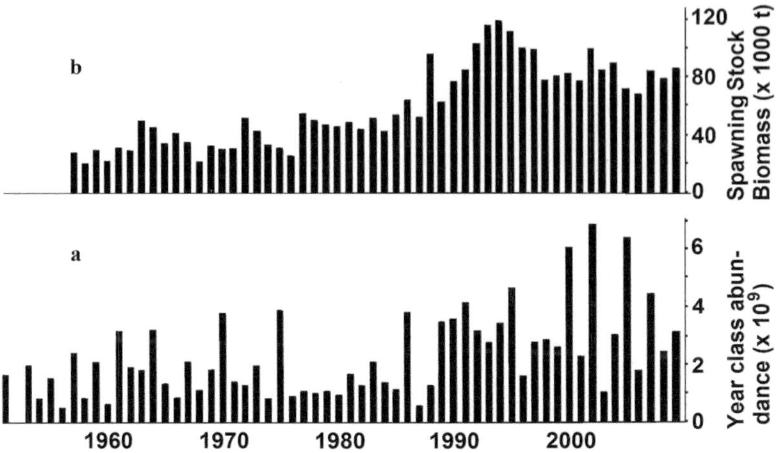

Fig. 3.28 Relative abundance of the spring spawning herring year classes (**a**) and the spawning stock biomass (**b**) in the Gulf of Riga in 1951–2009 (Ojaveer 2014)

sites upon herring has increased, as anthropogenic pollutants cause physiological, biochemical and other rearrangements in the host organism, weakening its immune defence.

Abundance Dynamics The abundance of year classes in spring herring is formed during their embryonic and larval life. The year class abundance depends on the level of embryonic and larval mortality. Important causes of elimination of the embryos are the quality of sexual products and the conditions of embryonic development on spawning grounds. It has been found, however, that the highest mortality during the formation of spring herring year classes occurs at the transfer of the herring larvae to exogenous feeding and depends on the abundance level of their main food during this period – copepod nauplii. In spring spawning herring rich year classes have formed only under the conditions of high numbers of nauplii (and correspondingly, comparatively low larval mortality) at the transfer of the bulk of herring larvae to exogenous feeding, especially when the development of copepod nauplii has started early. The time and intensity of reproduction of both herring and copepods depend on temperature and the mixing of water layers to create favorable conditions for rich primary production (Ojaveer 1988).

Formation of abundant year classes in sea herring populations spawning at open sea coasts, mainly in the vicinity of deeps, is facilitated by the advection of saline water stimulating mixing of water layers around the halocline and the transport of nutrient salts from deeps to surface layers on the spawning grounds.

Gulf herring spawns comparatively far from deeps. There, the level of primary production depends on the nutrients from rivers or shallow areas in the vicinity. Abundant gulf herring year classes have been formed in the conditions of rather mild temperature in the preceding winter and active westerly winds, both during the winter and the spawning period (which was generally the case in the 1990s and the first half of the 2000s, Ojaveer et al. 2011, Fig. 3.28).

Table 3.2 Vital parameters of autumn spawning herring reproduction in various parts of the Baltic

Area	Spawning period (months)	Depth of spawning (m)	Salinity (psu)	Temperature (°C)
SW Baltic	IX–XI	10–15	16–18 (20)	13–15
Southern Baltic	VIII–XI	10–25	12–16	
NE Baltic	VIII–XI	10–25	5–7	8–16.5
Gulf of Riga	IX–X	3–15	5–7	7–12
Gulf of Finland	VIII–IX	10–20	5–7	
Bothnian Sea	IX–X	10–15	4–5	8–14
Bothnian Bay	IX–X	5–20	4–5	6–14

Data from: Cieglewicz and Posadzki (1947), Ehnholm (1951), Halling (1978), Hessle (1925), Kosior and Strzyzewska (1979), Ojaveer (1988), Ojaveer et al. (2003), Parmanne et al. (1997), Rechlin (1964) and Weber (1971)

In addition to the natural conditions, during recent decades, the success of herring reproduction and the abundance of its year classes have increasingly been influenced by human activity. The main anthropogenic limitations are: (1) reduction of possible spawning places through changing of the qualities of bottom substrates; (2) reduction in the success of fertilization and the survival of embryos on spawning grounds through worsening of the water quality (e.g., through oxygen depletion, pollution, etc.); (3) excessive exploitation of the herring spawning stock.

The conditions regulating the abundance of year classes in spring spawning herring depend on the environmental parameters in various parts of the Baltic Sea – climate, mixing conditions and anthropogenic influence. As these parameters differ for the herring reproduction areas in each region of the sea (Table 3.2), then it can be expected that the abundance of the same year class developed in various regions of the Baltic Sea is different.

Assessment and Management Spring spawning herring is economically the most stable of the important fish species in the Baltic Sea. During recent centuries, it has played a very important role both in economy and policy. For centuries, herring was fished mainly with nets and beach seines from small sailing and rowing boats. Considerable intensification of resource exploitation was related to the mechanization of fisheries. Development of trapnet fishery on the spawning grounds started in 1939 in Tallinn Bay, but rapidly developed in the 1950s–1960s, causing considerable increase in catches in several parts of the Baltic. Trawl fishery was started in the mid-1920s in the southwestern parts of the sea. This gear was widely developed in the 1950s. Several trawl types were developed that allow for the catching of fish in any part of the water column. Fishing for herring is most important in the northern and eastern parts of the sea (Lithuania, Latvia, Estonia and Finland). The catches of the pelagic species are used for human consumption, reduction to oil and meal and to animal fodder. Stock assessments for management purposes have been performed in some areas (e.g., the Gulf of Riga and the NE Baltic) since the 1950s. Beginning in 1974, the stock condition of the Baltic herring populations has been assessed by a special ICES working group and the management of the resources has been based on the ICES recommendations.

3.9.1.3 Autumn Spawning Herring *Clupea harengus membras* L.

Intraspecific Units As a result of thorough investigations (Hessle 1925; Ehnholm 1951; Kändler 1951; Rechlin 1964; Strzyżewska 1969; Ojaveer 1988; etc.), it has been clarified that the autumn spawning herring of the Baltic Sea cannot be considered as a continuously intermingling population.

Distribution and Habitat The following local units can be differentiated: (1) the autumn herring of the southeastern Baltic (according to the opinion of a number of authors, this unit should be divided into four subunits: (a) the Belt herring spawning in the Great Belt; (b) the Sound herring spawning in the southern part of the Sound, eastward of Møn and Falster; (c) the autumn herring spawning in Mecklenburg and Kiel Bays and the Fehmarn Sound, and (d) the autumn herring of the Bornholm Deep and Hanö Bay), (2) the autumn herring of the southern part of the Baltic; (3) the autumn herring of the Swedish east coast; (4) the autumn herring of the Northeastern Baltic; (5) the autumn herring of the Bothnian Sea and Bothnian Bay (Fig. 3.29). These units differ from each other by morphological and biological parameters, and also in abundance dynamics. As compared with the differences between the spring herring populations, the differences between the autumn herring groups are substantially smaller. They can clearly be seen comparing certain body proportions, morphological characters, fecundity, the time of mass spawning, growth rate and behavior (Table 3.2).

Larvae of autumn spawning herring are more related to deeper water layers than spring herring larvae. Larval autumn herring winter in comparatively deep water (they have been found down to a depth of 62 m) at the coastal slope. The adult

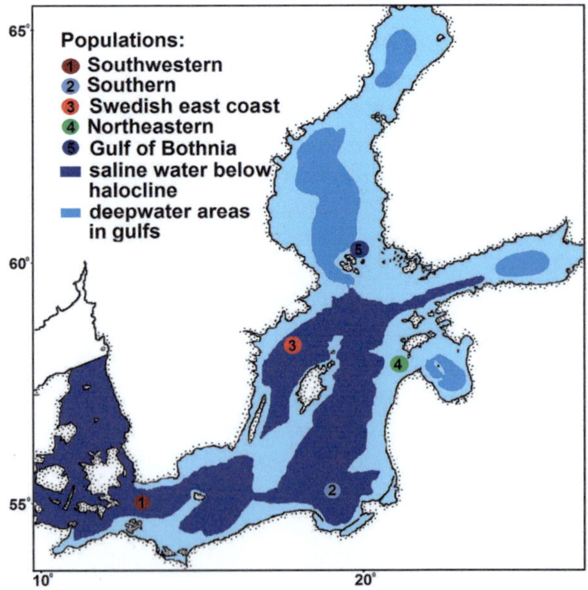

Fig. 3.29 (Sub) populations of the autumn spawning herring in the Baltic Sea after Hessle (1925), Rechlin (1967), and Ojaveer (1988)

autumn herrings are distributed in the deepest part of the herring shoals both in summer and winter. No large-scale marking experiments have been performed for investigation of autumn herring migrations. However, indirect evidence show that the migrations of the autumn herring are less intense than those of the spring spawning herring.

Reproduction Systematic studies on spawning places and spawning intensity of autumn spawning herring in the Baltic were started in the SW part of the sea, in the Danish, Swedish and German waters in the second half of the 1920s (Jensen 1950). The bulk of the autumn herring attains sexual maturity at the age of 3–4 years. A small part matures in their second year of life or only spawns for the first time when 5 years old. Males mature at a somewhat earlier age than females. The ratio between sexes is roughly 1:1. In older fish, rapid development of sexual glands starts in June-July, in males earlier than in females. Spawning migration usually takes place in July–August. Before spawning, the fat content of the autumn herring is comparatively high, and in prespawning concentrations, the fish does not feed. As younger herring feed higher in the water layer, at higher temperature, they mature earlier than older specimens and also migrate to the spawning places earlier (Fig. 3.22). In the Baltic Sea, spawning of the autumn herring starts in August and lasts up to October–November. As with spring herring, the medium age groups of autumn herring have the highest relative fecundity and the greatest percentage of viable hatch. These age groups spawn in the middle of the spawning season. Therefore, on the average, their offspring has the best possibilities for survival and gives the largest contribution to the year class abundance.

The individual fecundity of the autumn spawning herring is markedly higher than in the spring spawning herring of the same age. Similarly to the spring spawning herring, the fecundity is lowest in recruitment, gains the maximum values in medium age groups, and decreases again in older herring. It depends on weight, length and age of the fish and varies widely throughout the age groups, especially in recruitment. Variation of the fecundity in the autumn spawning herring populations and its dependence on weight, length and age of the fish is indicated in Fig. 3.30. Similar data on the North Sea herring spawning on Doggerbank are shown for comparison. Differences in fecundity between the subpopulations of the autumn spawning herring are clearly smaller than in the spring herring populations.

The northern part of the Baltic Sea represents the border of its area for the Baltic autumn herring. The pressure exerted by low salinity, low winter temperature and other external conditions on its reproductive system is obvious. The possible period of intense development of its gametes is short, lasting only 3–4 months. Compared with spring herring, the percentage of fertilization of autumn herring eggs and the proportion of normal embryos at hatch are markedly lower (Ojaveer 1981). Also, in the northern areas, a far greater number of deviations in the sexual glands (anomalous external appearance, accretion of gonads, their backwardness, the presence of clearly smaller and less developed oocytes among the main mass of uniformly and farther developed eggs in the gonad, etc.) are found in autumn herring than in spring herring. From time to time, female autumn herrings with hardened gonads and died,

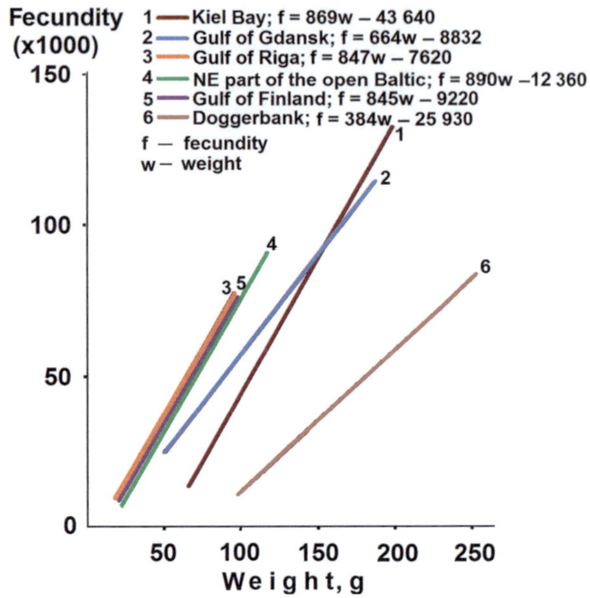

Fig. 3.30 Dependence of fecundity on weight of the autumn herring populations of the Baltic Sea and the North Sea autumn herring spawning on Doggerbank (Ojaveer 1988)

resorbing eggs can be found. Eggs die in the gonads mainly in the years of low fatness of spawners, after sharp drops in temperature during the spawning period. It is not clear to what extent this phenomenon can affect the mortality of herring, but in such gonads, the number of developing eggs is considerably smaller than in normal ones (Ojaveer 2014).

As in spring spawning herring, a small proportion of autumn spawning herrings can develop parthenogenetically. Commonly, development stops before cleavage. The only parthenogenetic embryo hatched in our experiments had abnormally thick pigment, but survived up to yolk resorption. In the period 1958–1978, the proportion of visibly deviating herrings (hybrids of herring and sprat, of spring and autumn spawning herring, individuals hatched outside the normal hatching period, etc.) in the Gulf of Riga constituted about 0.05%.

The spawning of autumn spawning herring is confined to the areas of steep coastal slopes or banks in the open sea with intense vertical mixing of water layers which create the best conditions both for the embryonic development (good oxygen supply, removal of metabolites, etc.) and the survival of larvae (high productivity of larval food). The bottom substrate is mainly sand or gravel at a depth of (5)–10–20 (25) m. The spawning begins in August in the vicinity of the thermocline, in a limited depth interval at about 16 °C. Mass spawning starts after the first autumn gale that mixes and aerates the water on the spawning grounds and broadens the near-bottom area with the optimum temperature for spawning. Spawning lasts until October–November, in the last half of the spawning period on the tops of offshore banks at (6)-12–16 °C. The salinity on the spawning grounds of the Baltic Proper chiefly varies around 5–7 psu, in the SE Baltic, 16–18 psu, and in the large gulfs, 4–5 psu (Table 3.2).

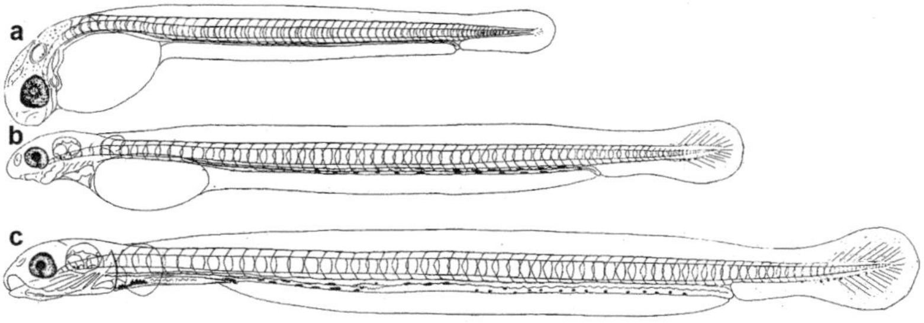

Fig. 3.31 Autumn herring larvae at the beginning (**a**), and at the end (**b**) of hatching period and after yolk sac resorption (**c**) (Ojaveer 1988)

The optimum salinity (5–20 psu) and temperature (7–12 °C) for embryonic development are higher in autumn herring than in spring spawning herring. The embryonic development lasts 1370–3600 degree-hours, and duration of hatching takes 900–1700 degree-hours. The embryos hatched at the beginning and at the end of the hatching period clearly differ in their level of development (Fig. 3.31). The average length at hatching is nearly equal after development at 7 and 12 °C (7.0 and 7.1 mm, respectively), but only 6.6 mm at 17 °C. The autumn herring larvae are longer than the spring herring larvae developed at the same temperature. In autumn herring larvae, the length after the yolk sac resorption is biggest (8.9 mm) after embryonic development at 12 °C. This temperature could be taken to be the optimum temperature of the Baltic autumn herring's embryonic development.

The positive reaction of the autumn herring larvae to light is mainly expressed as a general increase in activity, rather than moving in the direction of the light source like spring herring larvae do. This is probably due to the adaptation of autumn herring larvae to their development in deeper water and at weaker solar radiation.

Larval development of autumn herring takes place at a temperature from 12–17 °C at the beginning to about 7–10 °C at the end of the reproduction period. In winter, the temperatures decrease, in the northern Baltic more than in the southwest. Therefore, autumn herring larvae in the southwestern and southern Baltic clearly achieve a bigger size in the year of hatching than those in the northern parts of the sea. In the southern part of Kattegat and Great Belt, the length of the autumn herring fry reaches 25–45 mm in April and 30–44 mm in February–March in the Kiel Bight and Bornholm Deep, respectively. However, in the NE Baltic, their length in March–April of their first spring is 20–30 mm, and in the Bothnian Sea, a couple of months earlier in November, 14–20 mm (Jensen 1950; Halling 1978; etc.). In winter, the autumn herring larvae move into deeper layers. In the NE Baltic, they have been found down to a depth of 62 m, and in the Gulf of Riga, to 40 m. Water temperature at their wintering depth has been 0–3 °C. In the following spring, after the warming of surface layers to 2.5–3 °C, the largest concentrations of the autumn herring larvae have been found in the places of higher temperature where the density of their food organisms is highest. Metamorphosis of the autumn herring larvae takes place at a length of about 45 mm, in the NE Baltic, 8–9 months after hatching, in the southern areas earlier (Ojaveer 1988).

Feeding Fry of the autumn spawning herring are able to feed at a comparatively low temperature. Generally, the composition of food is similar as that for the spring spawning herring. However, the first food of the autumn herring fry is the winter generation of copepod nauplii, which are, by their size, larger than the individuals belonging to the summer generation. Seasonal feeding dynamics of the adult autumn herring differs by sea areas. In the southwestern and southern Baltic, it begins in March–April, reaches the maximum in the southwest Baltic in April and in the southern Baltic in May, and breaks for the spawning time to increase again in the Southern Baltic in October–November and in the Southwestern Baltic in November–December. In the northern parts of the sea, the feeding intensity of the autumn herring increases from April to June, declines in the time of rapid gonad maturation in July, and remains on the low level up to October. Intense feeding resumes in November and remains at that level up to the feeding breakage in winter (Ojaveer 2014).

Growth and Age In autumn, herring growth differences between the subpopulations are smaller than in spring herring. In the zero age group, remarkable increase in weight occurs from May to August and in November–December. In adult fish, the average weight clearly decreases after spawning. The decrease stops in April–May the following year. Weight increase starts in June and reaches the maximum in August.

The maximum age found in autumn herrings in the Baltic is less than in spring herrings. In the 1960s–1970s, the maximum age found in the autumn spawning herrings was in the SW part of the Baltic at 8–9 years, in the southern Baltic, 12 years, in the central Baltic, 11 years, in the Northern Baltic, 15 years, and in the Gulfs of Riga and Finland, 13 and 12 years, correspondingly.

Abundance Dynamics Larvae of the autumn herring develop over a comparatively long time (from hatching to the following spring) at rather low temperature and under mostly unfavorable feeding conditions. Therefore, the period of the possibly large mortality of the larvae lasts potentially much longer than in spring spawning herring – from hatching to the mass reproduction of copepods the following spring. Consequently, year class abundance of the autumn spawning herring depends on the food availability during the entire larval development. Therefore, it is possible to state that the year class abundance of both spring spawning and autumn spawning herring are determined by meteorological conditions influencing primarily biological productivity. In autumn herring, the year class abundance depends mainly on air temperature, but also on the wind direction and force during the first winter. Western winds and comparatively high temperature in January–March during the first winter and in April are extremely important in the development of abundant autumn herring year class.

Natural mortality of the autumn herring greatly depends on the number of predators in the area. The most important predator of herring in the Baltic is cod. Dependence of the natural mortality of autumn herring on cod abundance (cod invasions into the NE Baltic) is indicated in Fig. 3.32.

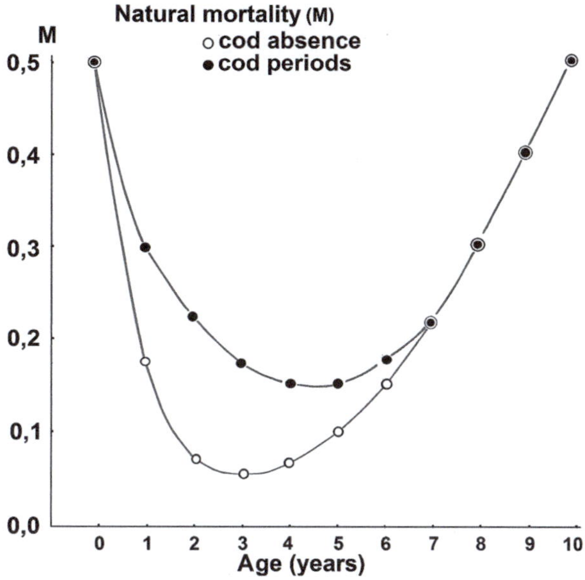

Fig. 3.32 Natural mortality of autumn spawning herring in the northeastern Baltic in 1963–1969 at the low abundance of cod and during the cod invasion (Ojaveer 1988)

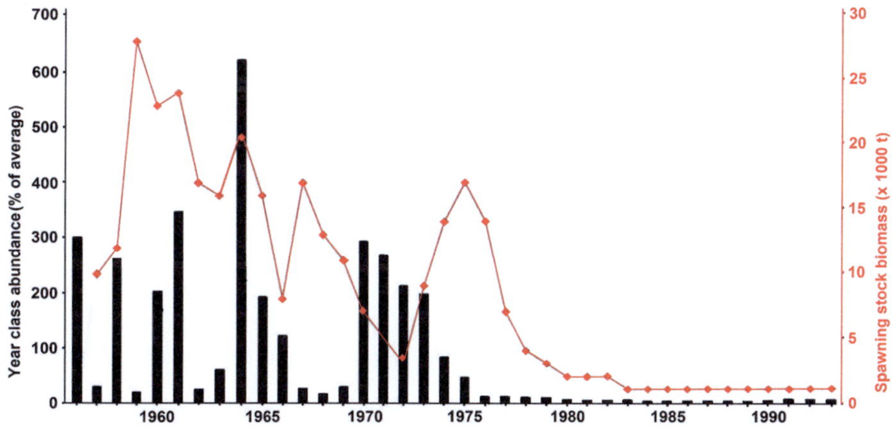

Fig. 3.33 Abundance of year classes and the spawning stock biomass of autumn spawning herring in the Gulf of Riga in 1956–1992 (Ojaveer 1988)

Generally, abundant autumn herring year classes have formed in the conditions of strong saltwater inflows into the Baltic Sea and mild winters, facilitating high biological productivity and the survival of autumn herring larvae. Investigations by Hessle (1925), Rechlin (1967, 1969) and Ojaveer (1988) have shown that all around the Baltic Sea or in several of its regions, abundant autumn herring year classes developed in 1903, 1909, 1913, 1917, 1918, 1922, 1924, 1929, 1936, 1937 (especially strong year class), 1938, 1942, 1948, 1952, 1954, 1956, 1959, 1960, 1961, 1964, and 1970–1972 (Fig. 3.33). Autumn spawning herring dominated in herring catches in the Open Baltic at the beginning of the last century up to the 1920s. From the 1940s to the 1970s, the percentage of autumn spawning herring in the herring concentrations diminished and its importance in catches decreased. However, in 1950–1960, it was estimated that the share of autumn spawning herring in the total herring resource was about 8% in the Northern Baltic Proper, 47% in the Gulf of Riga, and 14% in the Gulf of Finland.

In the Southern Baltic, the last abundant autumn herring year class formed in 1964, and in the NE Baltic and the Gulf of Riga in 1970–1972. Since then, no abundant year classes have developed in the Baltic Sea, despite several strong saltwater inflows and mild winters. Therefore, the conditions found to be important in the past have possibly lost their role. It is probable that in the mass anthropogenic eutrophication, which started in the Baltic in the 1950s, the oxygen deficiency has hit the reproduction system of the autumn herring (especially the mortality of embryos on the spawning grounds, which are situated deeper than those of spring herring) much more seriously than that of the spring herring (Ojaveer 2014).

Assessment and Management Presently, the biomass of the autumn herring is rather small in the Baltic Sea. Therefore, they have not been discriminated from spring spawning herring for assessment purposes and are currently added to the spring spawning herring for overall herring assessment. However, keeping in mind possible abundance/biomass fluctuations, it would be reasonable to start special autumn herring assessment and management in the Baltic Sea.

3.9.1.4 Baltic Sprat *Sprattus sprattus balticus* (Schn.)

The marine boreal sprat is one of the most abundant fish in the Baltic Sea food chain and constitutes an important part of fish catches. The species is distributed in almost all areas of the sea. But the limited salinity allows for its reproduction mainly in the open part of the sea, in the western part of the Gulf of Riga and in the western and central parts of the Gulf of Finland (Fig. 3.34).

Studies on the Baltic sprat had already started in the nineteenth century and works on sprat are constantly growing.

Intraspecific Groups The species *Sprattus sprattus* (L.) consists of three subspecies that are rather similar in their features. *S. sprattus sprattus* (L.) distributes in the vicinity of the European coast of the Atlantic Ocean from Tromsö and Lofoten Islands, including the North Sea. The area of the *S. sprattus phalericus* (Risso) involves the northern regions of the Mediterranean Sea and the Black Sea. *S. sprattus balticus* (L.) occupies the Baltic Sea, the Sound and the Belts up to the southern part of the Kattegat.

Sprat does not form a homogenous interbreeding stock in the Baltic Sea. However, no clear-cut sprat populations have been discerned in the Baltic either. A number of authors have studied regional differences in sprat morphology, growth rate, abundance dynamics and other parameters. The results can be summarized as follows:

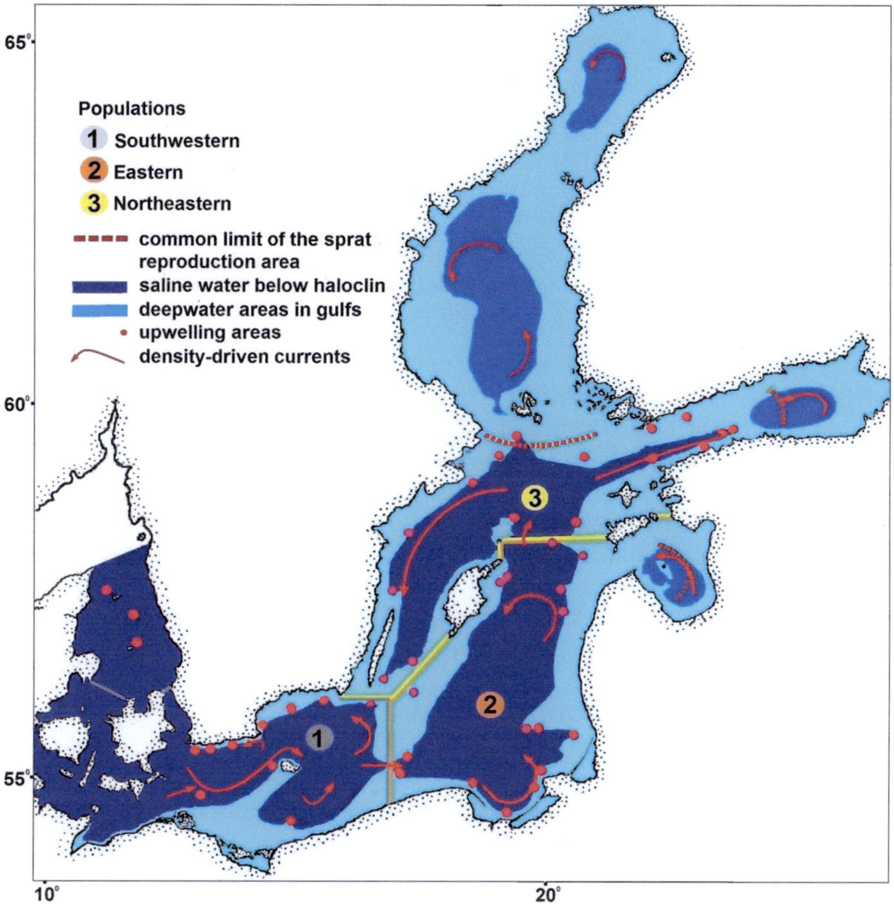

Fig. 3.34 Areas of sprat populations in the Baltic Sea (Ojaveer 2014)

1. Based on the comparison of morphological features, Seletskaja (1970) differentiated three stocks in Baltic sprat: the sprat of the Gulf of Gdansk, of the Bornholm Deep area and of the Gotland Deep. But like other sprat specialists, she stated that the borderlines between the stocks, their migrations and mixing possibilities are not clear.
2. In Baltic sprat eggs, temporal and regional differences have been found. The number of eggs and their diameter in the Bornholm and Gdansk areas are clearly different. At the start of spawning in the SW Baltic, the special gravity of sprat egg is the highest. With the shift of the spawning center of sprat northwards where the special gravity of sea water decreases, the specific gravity of sprat eggs also decreases (Nissling et al. 2003).
3. Lindquist (1971) was the first to show that the width of the first and second growth zones in sprat otolith is different in various areas of the Baltic Sea. This

regional difference does not change with the age of fish. In principle, this feature could be applied for discrimination of sprats of various regions.

4. In different areas of the Baltic Sea (Bornholm, Gdansk and Gotland basins), the average length and weight of sprat differs significantly (ICES 1987, 1989, etc.).

Spatial differences in characters of sprat could be related to the existence of large circulation systems (Fig. 2.12) in the Baltic Sea, influencing feeding, reproduction and wintering conditions, and the corresponding migrations of sprat. However, the stocks of sprat spawning in pelagic layers are tied to particular regions of the sea much more loosely than herring populations having demersal eggs (Ojaveer and Kalejs 2010). Therefore, differentiation of local sprat populations is more problematic than in the case of herring.

Distribution, Habitats, Migrations Sprat distribution in the Baltic Sea is mainly limited by reproduction conditions and winter temperature. For year-round habitation, sprat needs a very large volume of water with certain salinity, oxygen and temperature conditions. Sprat feeding shoals are mostly largest in the warm surface water in the zone of coastal slope and upwellings (the high-energy zone). Sprat reproduction is basically confined to the same area. Young sprat keeps in the productive near-coast domain, too.

Wintering areas of sprat depend on both the temperature and oxygen content of water. In mild winter, the possible wintering area of sprat is comparatively large. In severe winters, sprat shoals aggregate from the colder coastal water into warmer water layers at a depth from 70 to 100–130 m in the area of deeps (Fig. 3.35). In the Baltic Sea, sprat wintering depth varies by region. In the Gotland Basin area, sprat winters below the 3.5–4 °C, and in the northern areas, below the 3 °C isotherm. Its sparse assemblages, sometimes lacking shoaling structure, can occur at 2–2.5 °C. In large gulfs, sprat can be found in winter, even at temperatures below 0.5 °C (Ojaveer 1988). In wintering aggregations, older sprats remain in deeper layers of water and younger ones in higher layers (colder but richer in oxygen).

The diurnal vertical migration is most intense in spring, when the temperature gradient in the water column above sprat concentrations is least. The vertical migra-

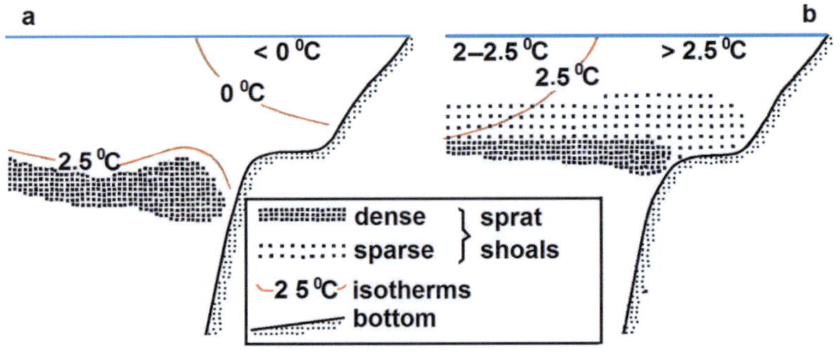

Fig. 3.35 Wintering concentrations of sprat in the Baltic Sea: (**a**) in severe winter; (**b**) in mild winter (Ojaveer 2014)

tion is most limited in severe winters, when the temperature of surface layers is close to 0 °C, and also in the spawning period.

Active migrations of sprat shoals between sea areas are mainly connected with the formation of wintering concentrations or feeding migration towards the coast in autumn. Sprat shoals can also shift passively, with movements of warm water masses. Under normal conditions (at temperatures above 2–3 °C), sprat remains above the isoline of oxygen concentration of 2 cm^3 dm^{-3}. Sprat shoals avoid the regions of vernal diatom bloom and of the blue-green algal bloom in summer.

Reproduction The ratio of sexes in sprat has generally been close to 1:1 up to the oldest age groups. Therefore, it can be supposed that the length of life in males and females is equal. Baltic sprat attains sexual maturity at the age of 1–4 (mainly 2–3) years. It spawns in batches. The average number of eggs in a batch varies depending on the size of the female. In the Bornholm Basin, the mean number of eggs in a batch is larger than in the Gdansk and Gotland areas. As an adaptation facilitating better buoyancy at low salinity, in the Baltic, the diameter of sprat eggs is bigger (1.10–1.70 mm) than in the North Sea (0.82–1.23 mm) (Nissling et al. 2003). The eggs are spawned at salinities of at least 5–6 psu. Spawning individuals and spawned eggs have been found at temperatures from 4 to 19 °C, but the optimum temperature range for embryonic development is 6–12 °C (Elwertowski 1957). The main spawning areas of sprat in the Baltic Sea are the open part of the sea, the western and central parts of the Gulf of Finland, the western part of the Gulf of Riga, the Arkona Basin and the Kiel and Mecklenburg Bights (Ojaveer 2014). The spawning season starts in the bottom layer of the SW part of the sea and in the Bornholm and Gdansk area in February-March, lasting until August. In northward areas, spawning begins progressively later. In the Gotland area, spawning extends from April to August. After mild winters, the spawning intensity is higher and the mass spawning occurs earlier in spring than after cold winters. The spawning intensity also depends on temperature (which is the chief factor limiting spawning) and the fatness of spawners. Depending on the thermic regime, the spawning period can be divided into two parts: (1) in deep layers (90–110 m) of the Bornholm and Gdansk deeps in February and in the 70–110 m layer in the Southern Baltic in March–April; (2) beginning from late May, with warming of the surface layers, sprat spawns in upper water layers, usually at a depth of about 40 m. This regional difference in the conditions of spawning and embryonic development may have created the basis for spatial differentiation of the Baltic sprat populations.

Sprat mass spawning takes place in a wide area above the coastal slope of the depths from 20–30 m to 140–160 m. The bulk of eggs are shed above the 80–100 m depths, in the area of the steep coastal slope. It has been found that the number of sprat embryos in the Baltic Sea declines towards the north. Grauman (1980) concludes from her decades-long studies that regardless of the large variations in hydrological conditions and spawning efficiency, sprat spawning places in the Baltic Sea are very stable. The character of cyclonic currents in the Baltic Sea (Figs. 2.12 and 2.14) limits the possible transport of sprat in their young stages from the spawning places to remote sea areas. These circumstances support the possibility of the development of regional sprat populations in the Baltic Sea.

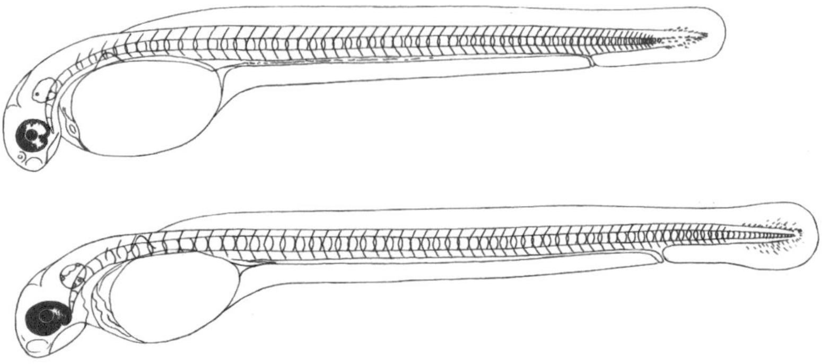

Fig. 3.36 The hybrid larva (spring spawning herring of the Gulf of Riga (♂) × sprat (♀), *top*) and the larva of the spring spawning herring of the Gulf of Riga (*bottom*) developed during embryonic period in uniform salinity and temperature conditions and hatched 1620 degree-hours after fertilization (Ojaveer 2014)

Sprat eggs are found up to the southern part of the Bothnian Sea, but in the Aland Sea and Bothnian Bay, no larvae have been caught (Sjöblom and Parmanne 1979), probably due to low salinity. The length of sprat larva at hatching is 2–4 mm. Its embryonic development lasts 3–7 days. The metamorphosis occurs at a length of 25–40 mm. Up to the end of metamorphosis, sprat keeps close to the coast. Whitebaits move to deeper water. After the first wintering, young sprats remain in higher water layers than adults, to depths of 40–50 m in the southern part and to 75 m in the northern part of the sea.

In the Gulf of Riga, some specimens with intermediate characters between sprat and herring have been caught. A special experiment in June 1971 confirmed that one of spring herring eggs fertilized with sprat sperm (salinity 4.95 psu, temperature of fertilization 13.4 °C, temperature of embryonic development 17 °C, oxygen content of water 9–9.7 cm^3 dm^{-3}) hatched after 1620 degree-hours long embryonic development. The hybrid larvae were clearly shorter than the herring embryos developed under corresponding conditions and the amount of pigment cells in their eyes was remarkably less (Fig. 3.36).

Feeding In the Baltic Sea food chains, sprat transfers energy from zooplankton (*Temora* spp., *Acartia* spp. etc.) to predatory fish, marine mammals, man, etc.

Sprat transfers to exogenous feeding at a length of 6–7 mm. The larvae feed on diatoms, flagellates, eggs and young stages of copepods. Sprat larvae prefer *Temora* spp., the lack of which may result in the formation of weak year classes. Larger larvae prey only on zooplankton. 20–30 mm long larvae consume the nauplii and copepodite stages of copepods, larval molluscs and the eggs of invertebrates. Beginning at a length of about 30 mm, the sprat's food is characteristically seasonal. The most common food items of 30 mm long and longer (including adult) sprat are calanoids – especially *Temora longicornis* throughout the year, *Pontoporeia* spp. and mysids at the beginning of winter, Pseudocalanus mainly in winter, and Eurytemora and Acartia mainly in summer and autumn. In summer, cladocerans

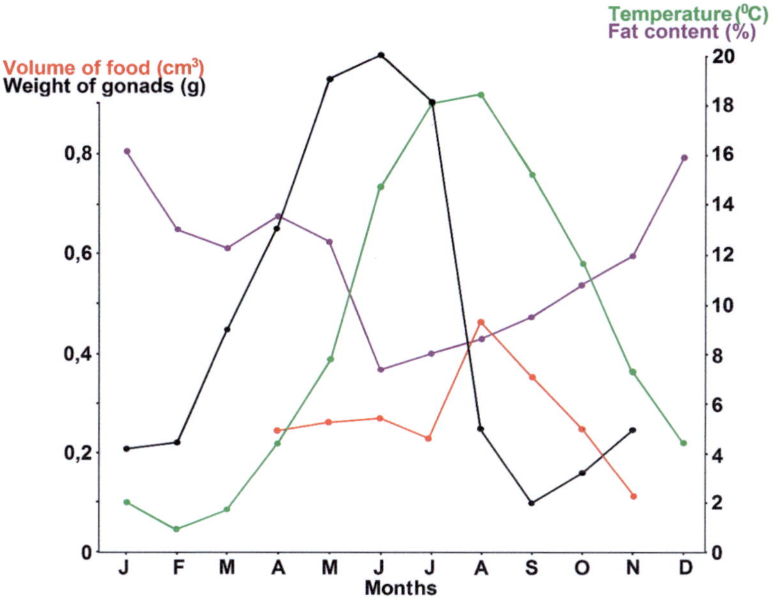

Fig. 3.37 Dependence of sprat feeding intensity, fat content and gonads weight on water temperature, after Elwertowski (1960)

(*Bosmina* spp, *Evadne* spp. and *Podon* spp.) also occur in the food (Miller 1969; Lankov 1988; etc.).

The most intense feeding period of sprat is August-September after spawning. Then, its gut is full of a reddish mass of copepods. The temperature of the surface layers is highest in this time. The relations between the sea surface temperature, the sprat's feeding intensity, the stage of its gonad development and fatness in the Southern Baltic is shown in Fig. 3.37. Sprat may feed at 0.2–0.3 °C, but the largest sprat concentrations occur in winter at temperatures between 2.5 and 4 °C. In cold winters, sprat's feeding intensity markedly decreases.

In the yearly cycle, another period of intense feeding for sprat falls in the April–May period. Sprat also feeds during its prolonged spawning time. The daily rhythm of feeding involves 2–3 maxima, varying by sea areas and seasons.

Growth and Age On average, sprat females are longer and heavier than the males of the same age. In the NE Baltic, the difference in the total length ranges from 0.1 cm in the 0-group sprat to 0.4–0.5 cm in the 4–8-year-old individuals. In the NE Baltic, intense feeding of sprat begins in May. However, in May-June, assimilated food is obviously used for gonad development. In June, a new growth zone is noticed only in the otoliths of a few individuals. The length of growth of adult specimens starts at the end of June or in July. The main growth and formation of the growth zone in the otolith takes place in the main feeding period in autumn. Growth depends on temperature and continues up to October–November, in mild winters, even to February–March. The dynamics of the weight growth of sprat is similar to

Fig. 3.38 Average weight of sprat in 1970–1986 (Ojaveer 2014)

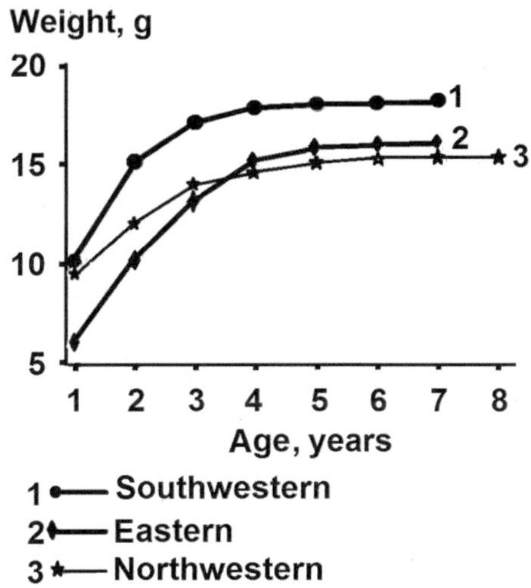

its growth in length. During winter and spring (mainly from November to May), sprat's weight decreases.

Clear periodic and regional differences occur in sprat growth in the Baltic Sea. During the period 1960–1990, growth gradually accelerated. The 4–7-year old sprats of the 1975 year class were, on average, 1.4 cm longer than the individuals of the 1959 year class at the same age. The phenomenon was probably due to the increase in the biological productivity in the Baltic Sea caused by growing eutrophication and also to an obvious decrease in sprat abundance from the late 1950s to the early 1980s. After a period of comparatively stable growth in the 1980s, the mean weight at age rapidly diminished beginning in the 1990s when the abundance of the species markedly increased and the abundance of valuable food organisms for pelagic fish decreased.

In the Baltic Sea, the growth rate of sprat decreases from the western areas towards the east and north (Fig. 3.38). This phenomenon has been related to the decrease in salinity and shortening of the growth period in the same direction (Elwertowski 1960; Lindquist 1971). The rather stable decrease in length of the 1-year-old sprat towards the north and east is also reflected in its otoliths (Lindquist 1971). Therefore, the width of the first growth zone can be applied as a "natural tag" in the differentiation of individuals of various sea areas. However, this feature cannot be used as a strictly individual tag, because of its comparatively wide variation due to the long spawning period of sprat.

Age composition of sprat stock depends on the abundance of its year classes and variation in its mortality rate. In catches, the share of the 8-year-old sprat rose from 0% in 1961 to 50–60% in 1966–1967, only to drop to 5–6% in 1977–1978. The oldest sprat caught in the Baltic Sea SW of the Åland Islands had 21 winter rings on its otolith.

Enemies and Parasites The main predator of sprat in the Baltic Sea is cod. It chiefly consumes sprat of the length group from 6 to 12 cm (Uzars 1989). The impact of cod on sprat stock is clearly different in various parts of the sea. According to the calculations of the ICES Working Group on the Assessment of the Pelagic Stocks in the Baltic, in the Southern and Southwestern Baltic with constantly abundant cod stock, the natural mortality coefficient of sprat reaches 0.4–0.5 or even higher. In the northern areas of the Baltic Sea where cod stock is usually much more limited, the natural mortality rate of sprat is estimated to be 0.2. During the period of high abundance of cod stock in 1978–1984, dense cod shoals were usual everywhere in the open part of the sea and also in the large gulfs. In this period, the natural mortality of 1–7-year-old sprat in the Baltic Sea rose to 0.60–0.90, and the instantaneous predation mortality to 0.6–0.7 (Hoziosky et al. 1989, etc.). After the cod stock decrease, the mortality of sprat in the northern areas of the sea dropped. Sprat is also preyed upon by salmon, sea trout, garfish, giant herring, turbot, seals and other predators.

As with other fish, the composition of parasitofauna of sprat probably depends on water salinity and other environmental factors. Therefore, during recent decades, several changes in sprat parasites have occurred. One of the most important parasites is *Ichthyophonus hoferi* (Fungi). *Eimeria sardinae* (Coccidiomorpha), *Glugea hertwigi* (Microsporidia), *Trichodina nigra* (Ciliata), *Proteocephalus exiguus* (Cestodea), *Diplostomum spathaceum, Hemiurus ocreata* (Trematoda), *Hysteriothylacium aduncum, Contracaecum osculatum, Anisakis* spp. (Nematoda), *Pomphorhynchus laevis* (Acanthocephala) have also been found in sprat (Ojaveer et al. 2003).

Abundance Dynamics Variation in sprat year class abundance is wide and the number of strong generations comparatively limited. Compared with herring, the sprat stock is much less stable in the Baltic Sea. Abundance dynamics of sprat in the Baltic Sea substantially depend on the thermal regime of the sea. Sprat distribution in winter before spawning is much conditioned on water temperature. In the case of low temperature, sprat does not feed during this time. In some years, in northern areas, the fast can last up to a half of the year. The wintering conditions can be estimated as normal if the water temperature is 3 °C or higher and the oxygen concentration at least 2 cm^3 dm^{-3}. In warm winters, especially when oxygen conditions are acceptable, the volume of water for possible sprat distribution increases. This favors sprat wintering and normal development of its sexual glands. Such conditions also support reproduction and growth of sprat's food organisms, creating preconditions for sprat's normal feeding before and during spawning.

In the northern areas of the Baltic Sea, the abundance of sprat year classes depends on the temperature of the upper 40 m thick water layer in spring and summer. Grauman (1969) estimates that 50–96% of sprat embryos perish in spawning places. Strong sprat year classes are formed at comparatively high temperature and a high number of copepod nauplii in spawning areas. The abundance of sprat year class can substantially decrease during the larval and the subsequent 0-group phase. The period of high mortality in forming sprat year class lasts from spawning up to the following spring. Therefore, mild temperature conditions during the first winter are important for the formation of abundant sprat generation.

Fluctuations in sprat stocks in the Baltic Sea have always been comparatively large. The changes in sprat resources have been sudden and usually unexpected. Therefore, they have been negatively reflected in economy. It has been clarified that up to now, the sudden variations in sprat resources have been caused by (1) the large variation in the abundance of sprat generations and (2) the changes in the sprat stock induced by the main consumer of sprat – cod. Grauman and Yula (1989) and Uzars (1989) are of the opinion that the predation mortality of sprat induced by cod feeding can be much higher than the fishing mortality and can cause depression of sprat stock. On the background of periodic variations of the Baltic Sea ecosystems, in comparison with other fishes, in the period 1890–1936, sprat abundance was comparatively high, from 1937 to 1980, moderate, and beginning with the mid-1980s to the present time, has dominated in the Baltic Sea.

Assessment and Management Since the end of the 1970s, assessment of sprat resources in the Baltic have been carried out using the VPA technics. In the period 1977–1988, the ICES Working Group on Assessment of the Pelagic Stocks in the Baltic assessed the sprat resources in three units which correspond to the natural regions in the open part of the Baltic and the adjacent sea areas: (1) ICES Subdivisions 24 + 25 (the Arkona and SW regions); (2) ICES Subdivisions 26 + 28 (the Eastern region, the area of the Gotland and Gdansk Deeps); (3) ICES Subdivisions 27–32 (the Northwest region and adjacent areas of the Gulfs of Finland and Bothnia). Such a scheme of areas represented a compromise of the biological and practical considerations which enabled us to get datasets and compose biologically well-founded assessments (Fig. 3.39). Sprat abundance fluctuations in these three units are different, both due to the notably higher predation mortality in the SW and southern regions and lower natural mortality in the eastern, northwestern and adjacent regions caused by higher cod abundance in the southern areas, as well as the different relative abundance of sprat year classes in the southern and northern parts of the sea. Beginning in1989, sprat resources of the Baltic Sea have been assessed as one unit.

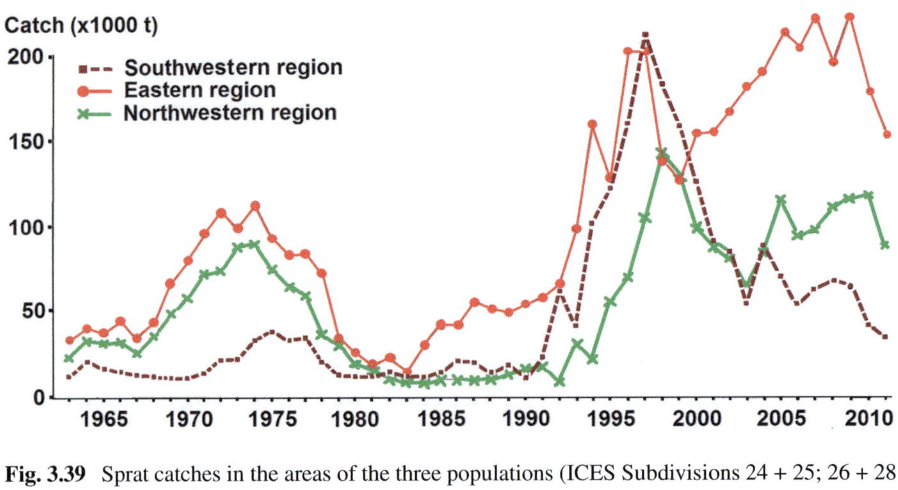

Fig. 3.39 Sprat catches in the areas of the three populations (ICES Subdivisions 24 + 25; 26 + 28 and 27, 29 and 32) (Ojaveer 2014)

This methodologically very important decision was made without relevant background investigations, which are still absent up to the present (ICES 1995, 2015a).

Sprat exploitation is mainly performed with pelagic trawls catching a mixture of herring and sprat. The proportion of the two species in the catches varies according to area and season.

In the management of sprat stock, it is important that a minimum mesh size for sprat fishery (16 mm) be established. The catches are limited by the quotas for the whole Baltic Sea. Discarding of sprat is prohibited in the Baltic Sea.

3.9.2 Marine Demersal Fish

Compared to the pelagic species, Baltic cod, flounder, plaice, dab, and turbot have somewhat more limited distribution. Their reproduction areas depend on water salinity, and therefore are situated in the southwestern, southern and central areas of the sea. But their economical value is comparatively higher. As in the case of pelagic species of marine background, the resources of the demersal marine species adapted to the conditions in the Baltic Sea are assessed and managed in an international collaboration based on scientifically sound methods.

3.9.2.1 Baltic Cod *Gadus morhua callarias* (L.)

Cod is distributed in the northern part of the Atlantic and Pacific Oceans and in the adjacent seas. In the Baltic Sea, the species is represented by its subspecies *Gadus morhua callarias* L. A number of investigations into meristic characters and body proportions, otolith types (Berner 1968; Berner and Vaske 1985; Bagge and Steffensen 1989), taggings for investigations of the migrations (Otterlind 1976, 1985a; Aro 1989, etc.), analyses of allele frequencies of the loci, coding LDH, IDH and PGI, as well as studies on the haemoglobin and serum transferrin types (Sick 1965; Jamieson and Otterlind 1971, etc.) have shown that two different cod stocks exist in the Baltic. The western unit populates the area west of Bornholm Island, the eastern, the area east of the island. The western population should be treated as a transitional group between the North Sea cod and the Baltic Sea eastern population. This was also the conclusion of the study carried out by Nielsen et al. (2003). The area of the population reaches from the southernmost part of the Kattegat up to 14°30′E. Eastwards of this line to 63°N extends the area of the eastern population (Bagge et al. 1994). The borderline between these populations can somewhat vary depending on the relative abundance of these populations. The transition between the populations is gradual and some mixing occurs between them. The location of the border between the populations also depends on the intensity of the drift of young cod of the western population into the area of the eastern cod. This drift is intense in mild winters when large water masses may move under the pressure of strong winds through the Transition Area to the east. Presence of the ice cover can drastically reduce both the influence of winds and the transport of young cod. Westward migrations of the cod of the eastern stock are rare. The activity of the spermatozoa and the buoyancy of eggs of the eastern population also enable their reproduction in the area of the western population. But the eggs of the western population sink in the area of the eastern population, due to their larger specific gravity, even in the Bornholm Deep to such a depth where their survival is problematic (Hüssy 2011).

Sharp changes in climate since the 1980s and their impact on cod migrations have induced new studies with applications of genetic methods on the status of the western and eastern cod populations in the Baltic. Stroganov et al. (2013) found that the investigation of cod samples from the eastern and western parts of the Baltic Sea demonstrated a high degree of their identity. The authors explain their result with the increased migration of cod at the beginning of the present millennium from the eastern Baltic to the waters of the Bornholm Basin and the distribution of a part of the migrants from the Bornholm Basin to the Arkona Basin. It can be expected that in the periods of drastic environmental variations, intermingling between the members of forming local populations certainly increases. Contrasting with Stroganov et al. (2013), Pocwierz-Kotusa et al. (2014) state, based on genetic comparison of cod samples from the Kiel Bight and the eastern Baltic, that the samples from the eastern and western Baltic well differed from one another. They conclude that their study provides further justification for separate management of western and eastern cod populations.

Distribution and Habitat The area of both the western but especially of the eastern population is variable. During the periods of small stock size of the eastern stock, the population distributes chiefly in the southern part of the sea. When the stock biomass increases, both young and adult specimens migrate to the feeding grounds situated in the eastern and later also in the northern parts of the sea. Embryos and larvae can move with currents, young cod both passively and actively to the coastal areas rich of food. Young cod can endure environments that are practically freshwater for long periods of time, and adult cod can also live at salinities below 5 psu (Aro 2000). During the last period of a large increase of abundance of the eastern cod population in the Baltic Sea, large shoals consisting mainly of 16–28-cm-long cods invaded the NE Baltic and the Gulfs of Riga, Finland and Bothnia.

It has been found that cod aggregations populating various parts of their area can differ in meristic characteristics and body proportions (Berner and Vaske 1985; etc.). Cod of the eastern population migrate widely and are not connected with any specific spawning ground. Spawning migration of this population is directed towards deeps. The emigration rate of cod from the northern part of its area to the southern and southwestern spawning grounds is highest during periods of stagnation. Adult cods from the Gulfs of Riga, Finland and Bothnia begin their spawning migration in December–March. It has been noticed, however, that a certain part of mature cod does not migrate from its feeding places in the Bothnian Sea towards the south. During periods of high salinity and large biomass of cod population, some part of the population can even spawn in the western Gulf of Finland (Ojaveer et al. 2003).

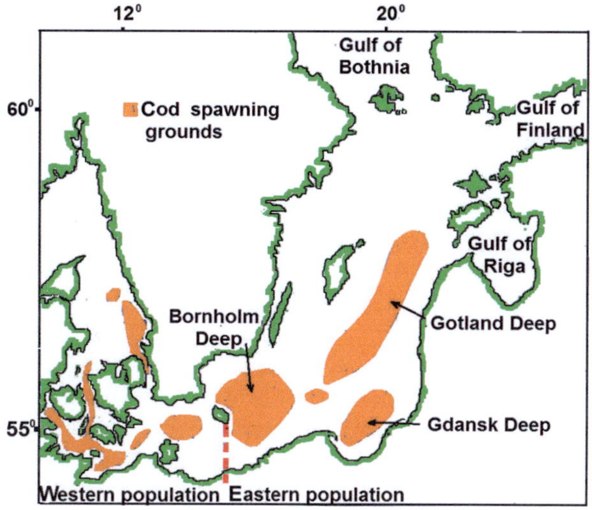

Fig. 3.40 Chief spawning places of cod in the Baltic Sea (Ojaveer 2014)

Reproduction Sexual maturation of cod in the Baltic Sea differs by region. In the western population, cod matures beginning at the age of 3 years. Some part of the cod of the eastern population matures at the age of 2 years, but the majority do so a year later. In both populations, all specimens are mature at the age of 7 years. In the Gulf of Finland, cod gains sexual maturity at the age of 4 or even 5 years when its total length is 36–45 cm. The first time spawners of the western population are about 5 cm longer than those in the eastern population. In the Bornholm Basin, the male cod attains sexual maturity beginning when they are 17 cm and females when they are 22 cm.

In the area of the western population, the most important spawning grounds are situated in the Kiel and Mecklenburg Bays, in the Great and Little Belt, in the southern part of Kattegat and also in the Arkona Basin. In the latter region, cod both of the western and eastern populations (Bagge et al. 1994; Hüssy 2011) can supposedly spawn.

The main spawning grounds of the eastern population are situated in the Bornholm, Gotland and Gdansk deeps (Fig. 3.40). The eastern population also spawns in Slupsk Furrow and in the periods of high salinity of the sea northwards of the main spawning places. During recent decades, the most important spawning area of the eastern population has been the Bornholm Basin.

In the western population, spawning begins in January. The peak occurs in February–April and reproduction stops in May. The eastern population starts reproduction in March and reaches the highest intensity from the beginning of May to the middle of June. In recent years, the spawning period has had the tendency to stretch up to July–August, supposedly due to lower water temperatures and the increase in importance of the first-time spawners.

Older and larger females spawn at the beginning of the spawning period, while towards the end of the period, the importance of younger fish increases. In the female cod, the spawning period lasts 7–11 weeks. No size difference in various periods of spawning has been noticed in males of the western population. This phenomenon exists in the eastern population, but unclearly.

The fecundity of cod depends mainly on the weight and length of the fish. The fecundity of cod with a length of 32–87 cm in the Gdansk area varied between 150,000 and 3,500,000 eggs (Kosior and Strzyzewska 1979). The western and eastern populations differ in their fecundity. However, the fecundity of cod spawning in three different spawning grounds in the area of the eastern population has shown no real differences. According to Bagge et al. (1994), the fecundity of the Baltic cod is twice as high as that of the North Sea or Icelandic cod and can be calculated as $AF = 3.87Tl^{3.1862}$.

The eastern population spawns at 3–6 °C. At 5 °C, the larvae hatch in about 18 days. The number of eggs below 1 m^2 is highest in the Bornholm Deep, while in the Gotland Basin and the Gdansk Deep, egg densities are less (Bagge et al. 1994; Hüssy 2011). The vertical distribution of eggs depends on salinity. The minimum salinity needed for the floating of cod eggs is 10.5–11.0 psu. Eggs of the western population develop at a depth of more than 20 m from the sea surface. The eggs of the eastern population develop in the Bornholm area at depths over 55 m (the largest number occurs at a depth of 60–70 m), in the Gdansk Deep, below the 80 m isobath (the maximum egg number occurs at a depth of 80–110 m). In the southern part of the Gotland Deep, the maximum density of the cod eggs is found below the 90–130 m isobaths. During adaptation to the conditions in the brackish Baltic Sea, the diameter of the eastern Baltic cod eggs have increased to 1.38–1.99 mm against 1.16–1.60 mm in the North Sea cod (Bagge et al. 1994).

Environmental conditions regulate both fertilization and embryonic development. For the activation of the spermatozoa of the eastern population, salinity ≥11–12 psu, and in the western population, >15–16 psu is needed. The salinity of the neutral buoyancy of eggs is, in the eastern population, 14.5 psu, and in the western population, 20–22 psu. Optimum temperature for hatching in the eastern population (Bornholm Basin) is 2–10 °C (the temperature under 2 °C or higher than 10 °C has a negative impact on embryonic development and also on the length of larvae at hatching), in the western population, 5.5–8.5 °C (Bagge et al. 1994; Nissling and Westin 1997; Hüssy 2011).

Embryonic mortality depends mainly on the oxygen concentration in the layer of egg distribution. The minimum oxygen concentration necessary for embryonic development is 2 mL L^{-1} O_2 (Aro 2000). The average mortality of embryos in the western population is 96–99% (Hüssy 2011). Mean embryonic mortality in the Bornholm, Gotland and Gdansk areas has been different, varying from 91 to 99%. The percent of survivors has varied mainly between 0.1 and 20. Plikshs et al. (1993) are of the opinion that development of an abundant cod year class can be hoped for if the embryonic survival exceeds 11%. Survival of embryos is usually higher in the areas where the volume of water suitable for cod reproduction is large (Fig. 3.41), *i. e.*, in the Bornholm and other westward deeps.

The area of distribution of cod embryos and larvae is nearly the same. They distribute both above and below the pycnocline. Yolk-sac larvae perform a feeding

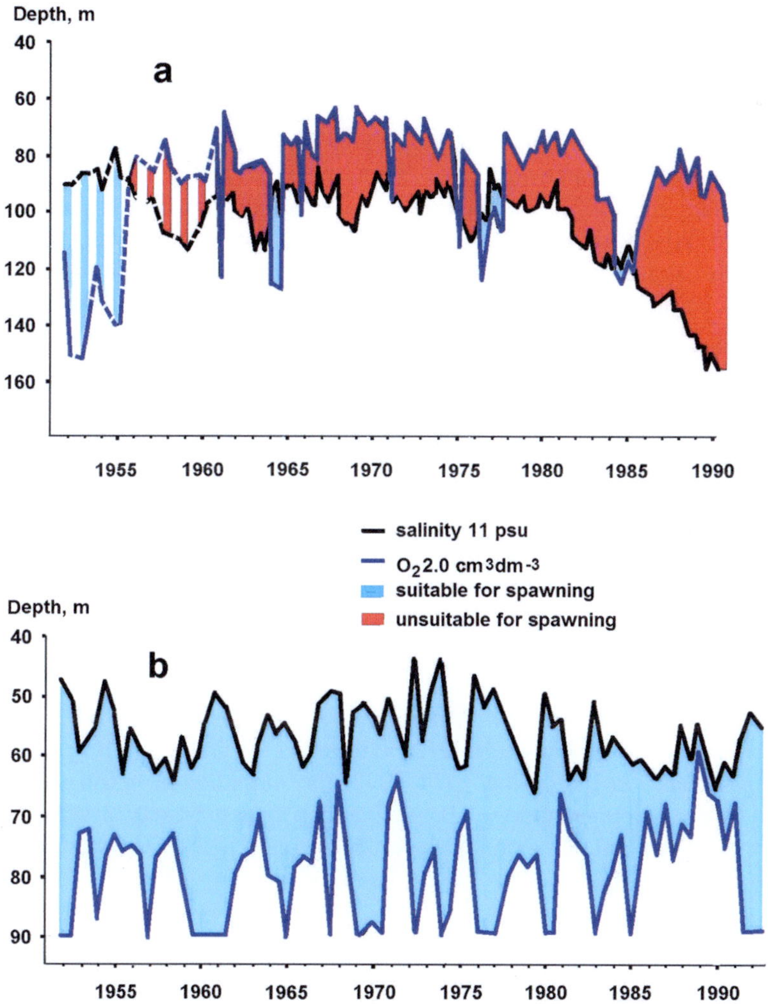

Fig. 3.41 Depth interval of possible normal development of cod embryos in Gotland (**a**) and Bornholm (**b**) spawning grounds (Ojaveer 2014)

migration – they ascend upwards, into the photic layer. Cod larvae can be carried by currents into shallower areas. In the productive coastal zone, their survival is supposedly much higher than above the deeps. Young cods distribute in the pelagic layers up to May–June of the following year. Then, they transfer to the bottom layers down to a depth of 60 m and live there until they are about 2 years old (Bagge et al. 1994).

Feeding Cod fry transfer to exogenous feeding when they reach a length of 4–5 mm. Their food consists mainly of copepod eggs, nauplies and copepodites. *Podon* spp., *Synchaeta* sp. and phytoplankton are also used. Larger fry (>6 mm) consuming *Pseudocalanus elongatus minutus* in the upper part of the halocline can feed mainly during the time when there is enough light to catch the prey (Zuzarte et al. 1996).

In the central Baltic, the 0-group cod of the eastern population consume mysids which dominate in their food during the whole year. Also important are *Harmothoë* spp. and other invertebrates.

At the length of 18–20-cm, the food basis of cod diversifies. The food of cod of the length 20–30 cm is mainly *Saduria* spp., *Harmothoë* spp., *Mysis* spp., *Pontoporeia* spp., and other invertebrates. The importance of fish (sprat, herring, gobies) in its diet has increased (Uzars 1975, 1989; Uzars and Plikshs 2000, etc.). Cods over 30 cm in length mature and live mainly on fish food. Common food objects are herring, sprat, young cod, gobies, smelt, four-bearded rockling, flounder, eelpout, common sand eel and also other fish. Cannibalism is widespread. The number of young cods in the stomachs of adults increases after the development of a rich young generation. A number of specialists are of the opinion that the importance of cannibalism on the abundance of cod year classes is negligible (Uzars 1989; Uzars and Plikshs 2000; Hüssy 2011, etc.). The importance of fish in the diet varies both by year and area. This is mainly caused by fluctuations in the biomass of prey species, as well as variations in the distribution of cod and its prey due to environmental impacts (e.g., oxygen content of water layers). The feeding intensity of the adult cod markedly declines during spawning time.

In the open part of the Baltic, adult cod feeds chiefly on herring, sprat and invertebrates – the most important of them is *Saduria entomon* (Chrzan 1962, Fig. 3.42). If the abundance of herring decreases, cod transfers to sprat, preferring the 6–12 cm long specimens. In the period 1978–1992, the mortality in herring age groups caused by cod amounted to M = 0.03–0.57 and in sprat to M = 0.22–0.74 (the highest mortality was that of 1-year-old fishes, Bagge et al. 1994). The mortality caused

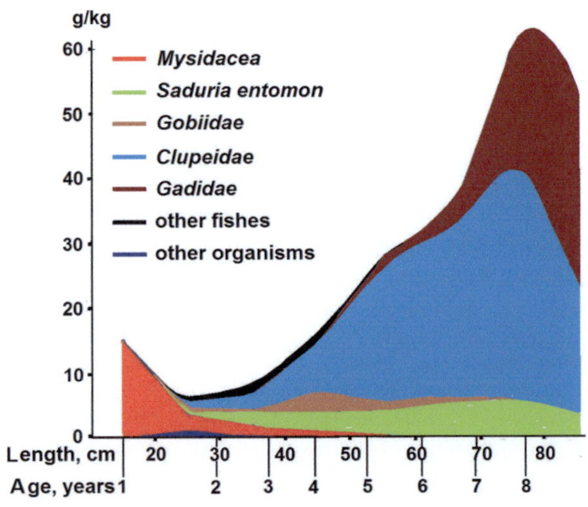

Fig. 3.42 Cod feeding during the life in grams per kilogram of cod weight after Chrzan (1962)

by the feeding of adult cod on 0- and 1-year-old cod was comparatively large, M = 0.1–1.0.

The western population of cod mostly consumes crustaceans, polychaetes and fish. The food of 11–15 cm long cod constitutes by weight about 62% of crustaceans, more than 30% of polychaetes and about 7% of fish. 11–35 cm long cod in Kiel Bay consumes crustaceans (*Diastylis rathkei, Crangon crangon, Gastrosaccus spinifer*) at about 36% (by weight), molluscs (mainly *Cyprina islandica*) at about 27%, polychaetes (*Harmothoë sarsi, Nephthys* spp., *Pherusa plumose* etc.) at about 27% and fish (*Pomatoschistus minutus*). The cod of a length over 35 cm preys more and more on fish and takes ever longer specimens, whereas the importance of polychaetes and crustaceans continuously decreases. In the feeding of the western cod population, sprat has higher importance than herring. Data collected in 1994 allowed us to state that the predation mortality caused by cod was in herring M = 0.01–0.21 and in sprat M = 0.12–0.30. Of invertebrates, the cod of the western population preys mostly on *Arctica islandica*. In Kiel Bay, cod feeds on very abundant coenosis of polychaetes. *Cyprina islandica* is also important in cod food, especially in the case in which the shell of the mollusc has been damaged by trawling.

In the yearly cycle, cod feeds more intensely in the second half-year, as during the spawning time, its feeding is weaker. The feeding of young cod is influenced by the low temperature of deep water after winter. In Kiel Bay, cod of a length over 20 cm usually catches its prey at dawn or by twilight, while the fish under 20 cm in length feed more or less continuously (Arntz 1974).

The yearly ration in cod changes with age. In 1977–1981, in the eastern population, the ration constituted from 628 g in 1-year-olds to 9846 g in 7-year-olds (Zalachowski 1985). Cod is the most important consumer of herring and sprat in the Baltic Sea. From the late 1970s to the mid-1980s, cod stock consumed more pelagic fish in the Baltic than was taken with commercial catches (Aro, 2000).

Growth and Age Variation in length and weight of the 1-year-old cod in the eastern population is well-correlated with the changes of temperature in the area of this population. Investigation of growth of the older cod has been complicated (especially in the eastern population) because of problems in age determination by otoliths. The materials collected on research vessels have shown that in the western population, the average length of age groups are clearly higher than in the eastern population.

Enemies and Parasites In the Baltic Sea, both the young and adult cod is preyed upon by seals, with the young also preyed upon by the adult part of the population. Therefore, in various parts of the population area, the losses are different (Bagge et al. 1994). Moreover, cod suffers from a number of diseases and parasites. Infestation with parasites varies in the Baltic Sea both in time and space. On cod, protozoans *Loma morhua* parasitize in the gills and *Goussia gadi* in the airbladder. Microorganisms of the genera *Vibrio,* Aeromonas and Pseudomonas cause skin ulcers and hemophilia accompanied by hyperaemia, a decrease in vitamin A, the lipid content in the liver, and other symptoms. Cysts from a dangerous fungus disease *Ichthyophonus hoferi* have been found in the liver, spleen and kidneys.

Important parasites are also *Cryptocotyle lingua* (black cysts in the skin), *Diplostomum spathaceum* (in the eyes), *Bothriocephalus scorpii* (in the intestines), *Hysterothylacium aduncum* (in the intestines and stomach), *Contracaecum osculatum* (in the liver and body cavity*)*, *Cucullanus cirratus* (in the intestines), *Anisakis simplex* (in the body cavity), *Echinorhynchus gadi, Pomphorhynchus laevis, Corynossoma semerme, C. strumosum*, *Lernaeocera branchialis*, etc. (Lang 1988; Mellergaard and Lang 1999; Bagge et al. 1994; etc.)

Abundance Dynamics Abundance of both the western and eastern populations has varied in dependence on the strength of year classes and also on the intensity of exploitation. The abundance of year classes is formed under the impact of environmental conditions. The majority of specialists (Plikshs et al. 1993; Bagge et al. 1994; etc.) are of the opinion that the decisive factors are water salinity, oxygen conditions and temperature in the reproduction places favoring hatching of viable larvae and their survival. However, optimum values of abiotic conditions alone cannot determine the abundance of forming year class. The impact of the spawning stock in the formation of the abundance of year classes has varied in different periods. Also, the presence of a sufficient food base for larvae and fry is extremely important. If the production of planktonic organisms suitable for larval food is rich, then preconditions for the transfer of cod larvae to exogenous food and the formation of abundant generation are favorable. The numbers of the forming year class can be negatively influenced by the cannibalism of older cod and the feeding of pelagic fish on cod embryos. Substantial predation on cod eggs by sprat and herring was found by Köster and Möllmann (2000). However, the authors evaluate that cod larvae are not seriously affected by these pelagic species, as the spatial overlap between the prey and predator species is limited.

Formation of rich generation increases cod biomass and induces its migration to the coastal waters in the southern and southwestern Baltic and to the eastern and northern parts of the sea.

Although cod is confined mainly to the western and southern parts of the Baltic Sea, it presents, among demersal fish, the highest economic interest throughout the whole Baltic. In the period of comparatively low salinity from the 1890s to the second half of the 1930s, cod catches were moderate. After the salinity increase in the mid-1930s, the reproduction conditions for cod improved. Beginning in the second half of the 1930s, cod catches increased. In the 1940s and the subsequent decades, a number of comparatively good/rich year classes (1941, 1942, 1945, 1947–1950, 1953, 1954, 1957, 1963, 1964, 1965 – Ojaveer 2014) formed in the eastern population. At the same time, the biomass of cod food animals (mainly herring and sprat) also increased. For this increase, the obvious cause was the eutrophication of the sea that evidently started in the 1950s and intensified in the 1960s–1970s. Together with the increase in the biomass of potential food, the period of the strong saltwater intrusions into the Baltic Sea continued. The situation for the formation of strong cod year classes and the increase in cod biomass was unique. The spawning stock of the eastern cod population and the catches from this stock as well increased to a very high level (Fig. 3.43).

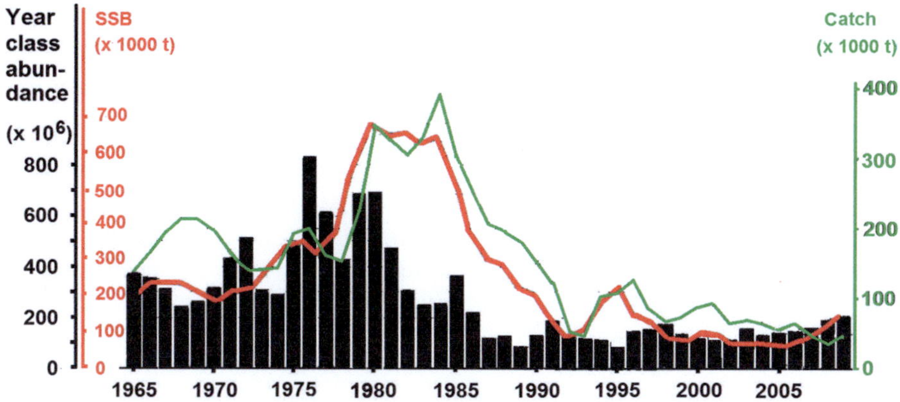

Fig. 3.43 Dynamics of year class abundance (*black bars*), spawning stock biomass (*red*), and catches (*green*) of cod in 1965–2009 (Ojaveer 2014)

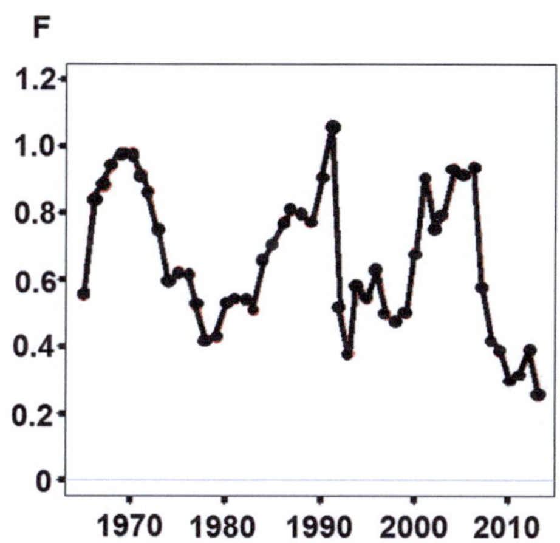

Fig. 3.44 Fishing mortality (F) of the eastern population of Baltic cod during 1965–2013 (ICES 2015b)

After a rapid decrease in salinity beginning at the end of the 1970s, the abundance of cod year classes dropped to a low level. Fishing mortality, which had increased during the years of cod high biomass, and the steep decrease in recruitment caused a very low level/depression in the eastern population of cod. Vallin et al. (1999) summarize that the major reasons for the low abundance of poor cod recruitment were many years of unfavourable hydrographic conditions, accompanied by an extensive mortality in adults due to commercial fishing. Severe methods of management (including decreasing of fishing mortality, ban of commercial fishing in some years, Fig. 3.44) applied to increasing the biomass of both cod populations in the Baltic have contributed to the stabilization of their spawning stock size.

Assessment and Management Assessment of stock size of the Baltic cod was started in 1957, when the ICES special session for the improvement of the condition of demersal fish stocks in the Baltic Sea took place. But the results remained meagre. Already then (like later on), an insurmountable obstacle – mistakes in age reading (especially in the eastern population) – seriously tangled the work. The age-composition-based VPA was used for stock assessment for the first time in1978. Beginning in 1981 estimates of the abundance of the 1-year-old year class (which were, however, rather different from later estimates, like the applied coefficients of fishing mortality) were used for the assessments. The VPA-based analytical assessment is also in use at present. Estimates of the total catch, changes in the annual average catch per fishing gear, data collected on the cruises arranged for stock assessment, data on discards, etc., compose the basis for the assesssments. Reporting on catches and discards has not always been correct.

Presently, cod is fished chiefly in the southern and southwestern areas of the Baltic Sea. The main gears are pelagic and demersal trawls, gillnets, hooks, etc. A considerable part of the catch is taken by amateur fishermen. Beginning in 2004, exploitation of cod stock has been regulated with the catch quotas agreed to by ICES, calculated separately for the western and eastern population, as well as with seasonal and other fishing restrictions (mainly on the most important reproduction areas, etc.), minimum mesh size and other technical conditions. Beginning in 2007, the basis for management of cod stocks is a special plan for management of the Baltic cod.

3.9.2.2 Flounder *Platichthys flesus trachurus* (Duncker)

Flounder is distributed from the White and Barents Seas to the Mediterranean, Black and Azov Seas. It has formed a number of subspecies. The Baltic Sea is populated by the subspecies *Platichthys flesus trachurus* (Duncker). Throughout the Baltic Sea, flounder exhibits considerable plasticity in the formation of geographical and biological groups. These are connected to environmental differences in their areas, notably in their spawning places, which are mostly situated in deep basins, but also in some areas off coasts (Strodtman 1918; Mikelsaar 1957; Otterlind 1967; etc.). Mikelsaar (1957) has expressed the opinion that the deep-spawning flounder (the pelagic-egg flounder) could have immigrated into the Baltic Sea from the west, while the coastal-spawning flounder (the demersal-egg flounder) after the Ice Age came from the northeast.

Intraspecific Groups By reproduction biology, genetic and other features, the flounder in the Baltic Sea can be differentiated into two large groups – the deep-spawning flounders developing during embryonic life in the pelagic water layers and the coastal-spawning flounders which spend their embryonic stage on the bottom.

The deep-spawning flounders spawn in the deeps where the eggs of comparatively low specific gravity float in the water layer. The eggs are significantly larger than in coastal-spawning flounder. The diameter of the eggs in the Baltic deep-spawning flounder is clearly adaptive and has formed to enable the embryos to float

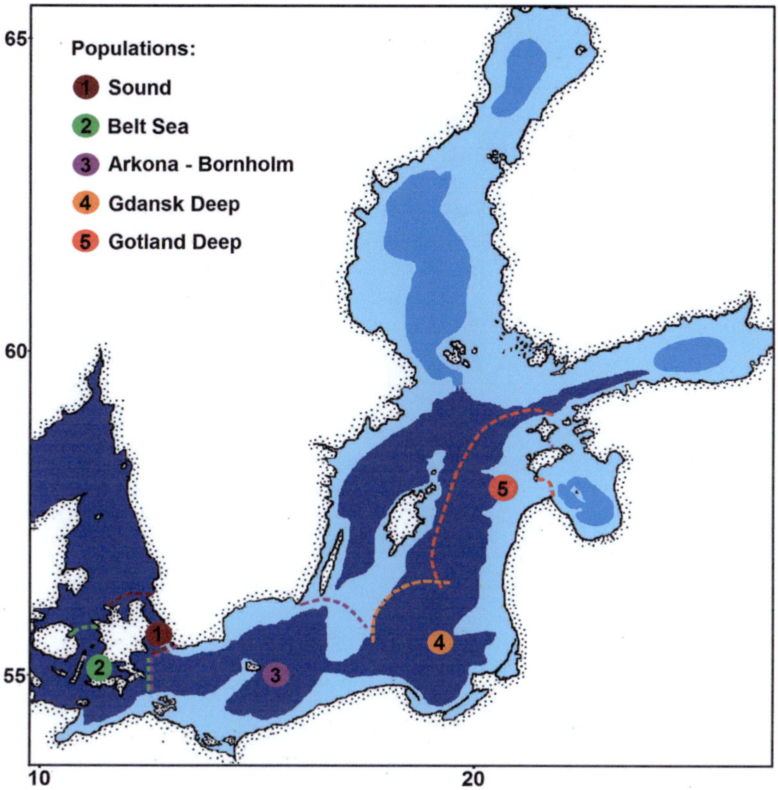

Fig. 3.45 Identified populations of pelagic-egg flounder in the Baltic Sea (ICES 2011)

in such water layers where the (oxygen) conditions for their development are acceptable. In the deep-spawning flounder reproducing in various spawning places, the diameter of the eggs is different: in the Sound, 1.12 mm, in the Arkona Deep, 1.34 mm, and in the Gotland Deep area, 1.43 mm (Mikelsaar 1957; Nissling et al. 2002; etc.). Spermatozoa of the deep-spawning flounder activate at a salinity of 9–12 psu, while they activate in the Bornholm flounder population at significantly lower salinity than in the Sound and Arkona populations. Results of genetic studies, comparison of the salinities of the neutral buoyancy of eggs, tagging data, morphometric and biological information collected for a long period have enabled us to distinguish five deep-spawning flounder populations in the Baltic Sea, along with their main areas: (1) the Sound; (2) the Belt Sea; (3) the area of the Arkona and Bornholm deeps; (4) the Gdansk deep and adjacent areas; (5) the Gotland Deep and the open sea areas situated northwards (Strodtman 1918; Mikelsaar 1957; ICES 2011, etc., Fig. 3.45).

Coastal-spawning populations live mainly in the northern part of the sea and the Gulfs of Finland, Riga and Bothnia, at lower salinity. Their eggs are smaller (diameter 0.83–1.20 mm, the mean 1.025 mm) than the eggs of the deep-spawning

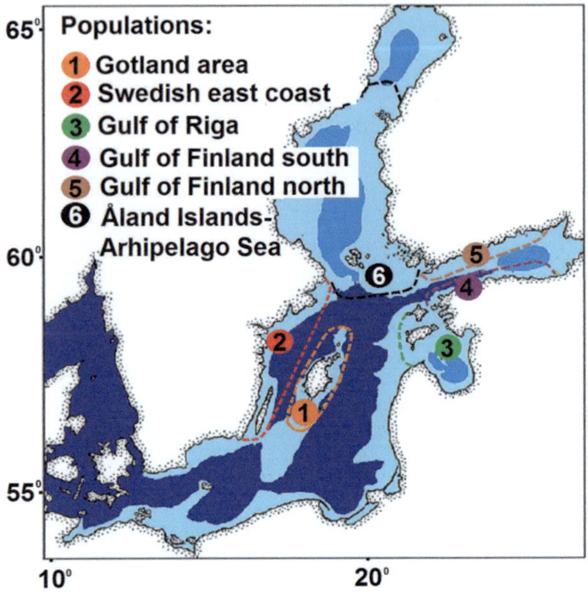

Fig. 3.46 Identified populations of demersal-egg flounder in the Baltic Sea (ICES 2011)

populations and have higher specific gravity. Spermatozoa of the coastal-spawning flounder activate at a salinity of 3–4 psu, but fertilization of eggs can be possible only at salinities greater than 6 psu. Fertilized eggs fall to the bottom and develop at comparatively limited depth (4–27 m), at low (>6 psu) salinity, and at temperatures of about 5–7 °C (Mikelsaar 1957). In addition to the characteristic distribution and reproduction environment, the coastal-spawning flounder has a lower average number of vertebrae and also a smaller mean figure of rays in the anal and dorsal fins. The following populations of coastal-spawning flounder have been distinguished in the Baltic with their areas (ICES 2011; Fig. 3.46): (1) around Gotland Island; (2) at the Swedish east coast; (3) in the Gulf of Riga; (4) at the south coast of the Gulf of Finland; (5) at the north coast of the Gulf of Finland; (6) in the southern part of the Bothnian Sea. In addition to these populations stated by ICES (2011), Strodtman (1918) has mentioned the coastal-spawning population spawning its eggs of an average diameter of 1005 mm on the bottom at a salinity of 7.5 psu on the Odra Bank, which has not been referred to of late. The exchange of specimens between the populations of the northern and southern coasts of the Gulf of Finland exists, with the main migration direction moving from north to south (rather unimportant in numbers).

The centers and reproduction places of flounder populations are situated in areas of higher biologial productivity, developed under the impact of bottom relief and currents. The basic parts of the populations are comparatively well-differentiated from each other. Nevertheless, exchange of specimens occurs even between the remote populations.

The decisions on the population structure of flatfish and the possible immigration routes of this species into the Baltic Sea cannot be considered to be conclusive, and corresponding investigations are ongoing.

Distribution and Habitat Flounder is a common fish species throughout the Baltic Sea, excluding the northernmost part of Bothnian Bay and the easternmost Gulf of Finland, where the species is rare. Flounder can occasionally, particularly in river estuaries, enter into freshwater environments. The species prefers to live on sandy and clayey bottoms, where younger specimens keep to shallower and older to deeper ground. 0- and 1-year-old specimens distribute in summer on 0–3-meter-deep sandy bottoms. The densest shoals of older flounder gather for intense feeding in summer on 30–40 m deep bottoms. Well-known feeding areas are the Irbe Sound and the Hiiu madal bank (north of the Hiiumaa Isle) in the Northeast Baltic. In autumn and winter, adult flounder concentrates at depths of 80–100 m, their distribution depending on the oxygen concentration of water layers. Shoals of prespawning adult coastal-spawning flounder gather in the vicinity of coastal slopes, while those of deep-spawning flounder, to the spawning places in deeps. Flounder is a species of clearly nocturnal activity.

Migrations of flounder are mostly up to 80–95 km long. However, flounder is also capable of longer travel, including travel over deeps (Otterlind 1967). The frequency of longer migrations increases with age. Flounders undertake long migrations in cases in which the oxygen concentration or other environmental conditions in the wintering or feeding grounds are unfavorable. Migrations can be influenced by water salinity. The longest known migration of flounder is about 700 km–the flounder tagged at Toila (in the eastern part of the Gulf of Finland) was caught at southeastern Gotland. However, in flounder, the fraction of long migrations is comparatively small.

Reproduction Males of the deep-spawning flounder mature at the age of 2–5, females at 3–6 years. Eggs are spawned in deeps, embryos develop floating in water. The deep-spawning population of the Gotland area spawns in April–May.

The fecundity of flounder depends mainly on its growth rate, and this can vary both by population and time period. The fecundity of the deep-spawning flounder of the North Sea, Kattegat, Kiel Bay, and Arkona-Bornholm population and that of the Gulf of Gdansk population differ significantly from each other and depend statistically on the length and weight of the fish. In the Transition Area and in the Baltic Sea, the fecundity of flounder is higher than in the North Sea. Kändler and Pirwitz (1957) suppose that this fact is due to adaptation to the more complicated reproduction conditions of the species in the Baltic Sea. In the 1970s, fecundity of the deep-spawning flounder of the eastern population (Gotland Deep area) reached from 180,000 eggs in the 3-year-old flounder (average weight 91 g, average length 20.6 cm) to 1,350,000 in the 8-year-old flounder (mean weight 377 g, mean length 31.6 cm) (Ojaveer 2014).

The coastal-spawning flounder reproduces in May–June on stony bottoms at a depth of 4–27 m (the density of eggs was highest at 10 m). Males of the coastal-spawning flounders first spawn at the age of 3, females at 4. After the data collected in the 1940s–1950s, the fecundity of the flounders with lengths from 18 to 34 cm varied between 220,320 and 2,041,000 eggs (Mikelsaar, 1957). Embryonic development lasts 5–10 days. The length at hatching is, on average, 3–4 mm. At the length of 9–10 mm, flounders are already asymmetric and one eye moves to the other side of the body.

Feeding 4–6 mm long larvae of the eastern population of deep-spawning flounder have quite a wide food spectrum: phytoplankton, copepod eggs, nauplii and other copepod young stages, rotifers, adult *Acartia bifilosa* and cladocerans (Grauman et al. 1989). The larvae 6–10 mm in length eat eggs, nauplii and copepodites of copepods, *Bosmina* spp. and *Podon* spp. The larvae of 10–15 mm transfer to a demersal way of life and to consumption of zooplankton, especially cladocerans, which constitute 90% of their food by weight.

Flounder 2–3 cm in length feed on nektobenthic organisms – Chironomidae, Harpacticoidae, mysids and demersal diatoms. Whitebaits of 3–10 cm consume Chironomidae, Gammaridae, Harpacticoidae, Oligochaeta, mysids, etc. Their food differs markedly by area.

At the length of 10–19 cm, juveniles gradually transfer to the food of adult flounder, consuming Polychaeta, Mollusca, etc. In the Gulf of Gdansk area, the role of plants, Oligochaeta, Copepoda, Ostracoda, Bathyporeia and larval Chironomidae decreases with the growth of flounder and the importance of *Macoma baltica* and *Saduria entomon* increases in its diet (Mulicki 1947).

Similar changes in the flounder diet also occur in the NE Baltic. In 1961–1963, north of Hiiumaa, the share of crustaceans in flounder feeding decreased from about 47% in the 10–19 cm length group to about 14% in the 25–30 cm length group (the proportion of *Monoporeia affinis* dropped from about 36% to 1%). Simultaneously, the percentage of Mollusca increased from 41% in the 10–19 cm length group to 84% in the 25–30 cm length group. With increasing fish *length* from 10–19 cm to 25–30 cm, the proportion of *S. entomon* in its diet rose from about 7% to 11%, and that of *Mya arenaria* from 3% to 5%, but the share of *Gammarus* spp., *Mytilus edulis* and *Pygospio elegans* declined. In addition to the above food items, in the Hiiumaa area, *Pontoporeia femorata, Idothea baltica, Cardium lamarcki, Hydrobia ulvae, Nereis diversicolor* and larval Chironomidae occurred in the stomachs of mainly smaller flounder, whereas in the nourishment of chiefly larger flounder, *Bathyporeia pilosa, Halicryptus spinulosus* and plants were present. In the Hiiumaa area, fish and *Crangon crangon* were also found in flounder food. The food composition depends on both the structure of zoobenthos and the length composition of flounder stock (Štšukina 1970).

In the Gulf of Riga, three quarters of the food of smaller flounder is composed of *Monoporeia affinis*, while *S. entomon* and *M. baltica* are of much smaller importance. Large flounder feed chiefly on *S. entomon* and *M. baltica*, with *M. affinis* having low importance.

Starting with the 22–24 cm length, flounder prey on large molluscs. At Vilsandi Island west of Saaremaa, large flounder swallow three-spined and nine-spined sticklebacks in addition to molluscs (Ojaveer et al. 2003).

In the Central Baltic, flounder feeds most actively in June. Its feeding activity decreases rapidly from July to August and further slows up until November. In the Hiiumaa area, stomach fullness achieves its maximum in larger specimens in July, but in smaller and medium-sized fish in August.

Growth and Age The growth and aging processes of males and females are different. The average length and weight also differ by population. In the eastern Gotland population, $L\infty$ for females equals 36.5–37.8 cm, for males, 32.3–33.4 cm, and for the whole population, 36.5–37.8 cm. $W\infty$ for females amounts to 765 g, for males, to 466 g, and for the whole population, to 629 g.

In the Central Baltic, the proportion of specimens older than 7 years is rather low in commercial catches. In the Gulf of Finland, flounder may be abundantly represented in catches up to the age group of 10. In the Baltic Sea, the age of flounder can reach 16 years.

Enemies and Parasites During the early, pelagic stages of development, flounder is exposed to potential predators. After settlement on bottom, it lies hidden there. Therefore, the loss due to predation is probably comparatively low. However, west of Saaremaa, flounder with scars from lamprey teeth have been found. On the other hand, steady contact with possibly polluted bottom sediments exerts influence on its health, because infection with both diseases and parasites depends on environmental factors – salinity, temperature, as well as pollutants (Bylund and Lönnström 1994; etc.)

In regard to externally visible diseases, lymphocytosis, acute skin ulceration and bacterial fin rot have been found in flounder in the Baltic. Regarding ulcers, *Vibrio anguillarum*, *Aeromonas* spp., *Pseudomonas* spp., Achromobacter Moraxella, Flavobacterium, Planococcus, *Aeromonas hydrophila*, and *Pseudomonas fluorescens* have been isolated. Ulcers are mainly located on the blind side. Males are more often affected than females by both ulcers and lymphocystis. The prevalence is highest in polluted waters (Bylund and Lönnström 1994; etc.).

By prevalence of infestation, the main parasites found in flounder are *Trichodina jadranica* (in the gills, infestation up to 100%), *Diplostomum spathaceum, D. baeri* (m., in the eyes), *Cryptocotyle concave* (m., in the gills), Rhapidascaris acus and *C. osculatum* (larvae, in the mesentery and liver), *Pomphorhynchus laevis* (larvae and imagoes). The most stable parasites in flounder are *Glugea stepheny* in the intestine wall and Myxosporidiae in the urinary bladder. In the flounder of the Gulf of Riga, *Nicolla skrjabini*, *Scolex pleuronectis*, and in the Eastern Baltic Proper, *Hysteriothylacium aduncum* and *Bothriocephalus scorpii* have also been found (Ojaveer 2014).

Assessment and Management Flounder resources of the Baltic Sea have been under rather effective exploitation for at least a century. The catches were comparatively good during the 1920s–1930s. From the mid-1970s to the mid-1990s, the annual landings frequently reached 10,000–15,000 tonnes. From the mid-1990s up to the present, they have fluctuated mainly between 15,000 and 20,000 tonnes. In recent years, the landings have been rather good, especially in the southwestern part of the sea (ICES 2015a). It is probable that the deep-spawning populations have contributed considerably to the catches.

In the Baltic Sea, fishery intensified earlier than in the oceans and large open seas. Also, the problem of regulation of fisheries rose in the Baltic earlier than in other seas. Therefore, during the first years of the existence of the ICES, the protection of flounder and plaice resources (primarily exploitation of young fish) was repeatedly under discussion. Flounder fishery was treated in the First Baltic Sea Fishery Congress in Riga in 1910. In 1913, an agreement on the protection of resources of flounder and plaice in the western part of the Baltic Sea was harmonized between Denmark and Germany. In 1934, the minimum legal size of flounder was increased to 21 cm in the straits between Gedser, Arenshop and Utlängan – the borderline between Germany and Poland. Already earlier (in 1925), a convention on the protection of fish resources in the Gulf of Riga had been concluded between Estonia and Latvia which included provisions on the regulation of flounder fishery.

Today, flounder is caught both as bycatch in cod bottom trawl fishery and in special flounder fishery. Out of the flounder fished as bycatch, much more is probably discarded than landed. Flounder resources are protected according to the size of total catch recommended by the ICES and a number of measures varying by sea area: minimum mesh size, minimum legal fish size, closed seasons, etc. Utilization of the Baltic Sea flounder for reduction is prohibited. Flounder fishery in the central and southern Baltic Sea has intensified in recent years.

The quality of the material collected on the flounder stock has not been high enough to enable representative analytical assessment.

3.9.2.3 Plaice *Pleuronectes platessa* L.

Distribution, Intraspecific Groups Plaice is distributed near the European coast of the Atlantic from the western coast of the Novaya Zemlya to Portugal and in the western part of the Mediterranean Sea. It occurs in the areas surrounding Iceland, at the southern coast of Greenland and in the western and southern parts of the Baltic Sea. The plaice of the Baltic Sea differs from the species elsewhere in its smaller size and certain other features (Valle 1934). In the Baltic Sea, its main area reaches up to the Gulf of Gdansk. Sporadically, it is found as far up as Gotland, in the Stockholm Archipelago, the Gulf of Finland, and in the vicinity of Saaremaa and the Aland Islands (Mikelsaar 1984, etc.).

Based on differences in morphometric characters, migrations and the buoyancy of eggs, the plaice of the Baltic Sea has been divided into the following populations (ICES 2011): (1) the Sound population (ICES Subdivision 23); (2) the Belt Sea population, including the ICES Subdivision 22 and the western part of the Subdivision 24; (3) the population in the Arkona, Bornholm and Gdansk Deeps and in the eastern part of the Gotland Deep (Fig. 3.47).

Biology Plaice is a shoaling demersal fish that prefers sandy bottoms. Young plaice distributes mainly at depths up to 10 m. The species feeds by day. It performs regular feeding migrations from spawning grounds to the feeding places in late spring, and from there to the deep wintering-spawning grounds in autumn-winter.

In the Baltic Sea, plaice matures chiefly at the age of 2–3 years, males earlier than females. As with flounder, the fecundity of the species and the increase in fecundity with growth in body weight are higher in the Baltic than in the North Sea but lower than in the Transition Area. In the Kiel Bight, the fecundity of a 35 cm-

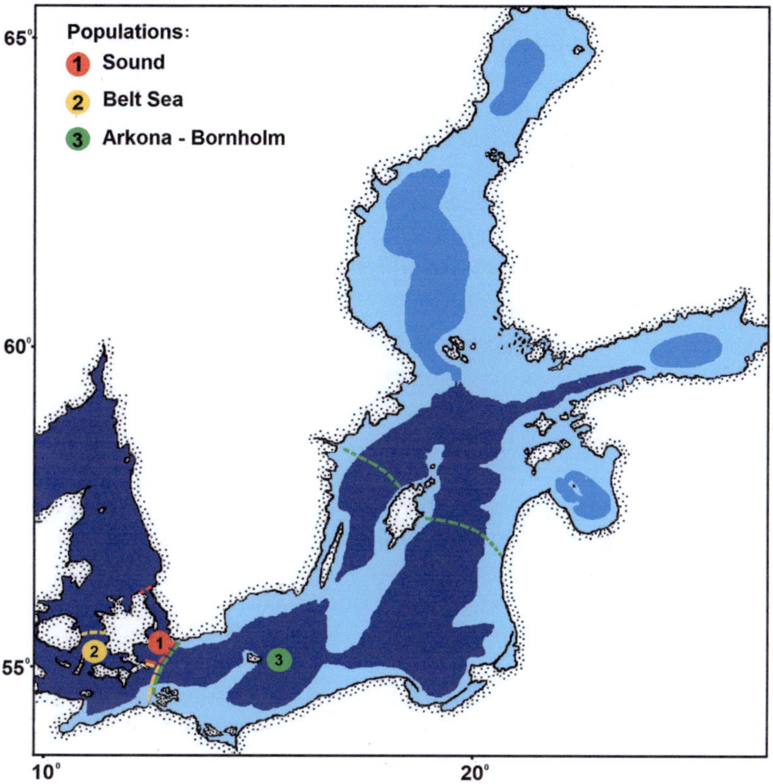

Fig. 3.47 Suggested populations of plaice in the Baltic Sea (ICES 2011)

long plaice is 308,000, in the Arkona–Bornholm area, 250,000, and in the North Sea, 64,000 (Kändler and Pirwitz 1957).

Plaice spawning grounds are situated in the Arkona and Bornholm Deeps, and also probably in the Slupsk Channel (Voipio 1981).

Plaice spermatozoa activate in the Southwestern Baltic (Arkona and Bornholm deeps) at a minimum salinity of 10 psu, in the Gotland deep, at 9 psu. Neutral buoyancy of plaice eggs occurs in the ICES Subdivision 24 at a salinity of 15.0–15.7 psu, in the ICES Subdivision 25, at 14.0–17.7 psu, and in the ICES Subdivision 28, at 16.3–18.2 psu. The lower salinity limit for successful spawning was estimated at 12.6–13.6 psu. Swimming spermatozoa were recorded at 9–12 psu (the percent of fertilization is low at these salinities) and the duration of their mobility increased up to 15 psu. Successful spawning of dab, plaice and flounder in the Baltic Sea is not restricted by spermatozoa activation, but rather by the ability of the eggs to develop at low salinities. It can be concluded (Nissling et al., 2002) that with respect to salinity requirements, opportunities for successful spawning of plaice exist regularly in the Arkona basin (SD 24) and the Bornholm basin (SD 25), but also occasionally in the Gdansk and Gotland basins (SD 26 and 28). In the southern North Sea and the Baltic Sea, plaice can form hybrids with flounder and dab.

The plaice spawning period lasts from November to May. Embryos need a salinity of at least 12–13 psu for normal development. At 6–7 °C, the embryonic development lasts 20–22 days. The length of the larvae at hatching is about 6 mm. The metamorphosis starts when the length of larva reaches 10 mm. The young plaice transfers to the demersal life when it reaches 12–14 mm long.

Plaice is bentophagous. Young fish live mainly on crustaceans, until later when their basic food is constituted by molluscs. Polychaeta and small demersal fish are also consumed.

The lifespan reaches 10 years. Growth rate decreases from the west to the east. Females grow faster and live longer than males.

Stock Assessment and Management Plaice is a very important fishing object in the North Sea. The stock is also economically significant in Skagerrak, Kattegat and in the western part of the Baltic. The species has been investigated for quite some time. After the foundation of the ICES, the organization gave their first recommendations on fish stocks in 1905, prohibiting the catching of undersized plaice in Kattegat. Later on, in the 1920s–1930s, several international agreements were concluded concerning the protection of plaice stock in Skagerrak, Kattegat, the Danish Sounds and the Baltic Sea (Ojaveer 2002).

In the Baltic Sea, about 95% of plaice catches are fished by Danish and Polish fishermen in the ICES Subdivisions 21, 22, 24 and 25. The size of catches has been clearly periodic. In the 1920s–1930s, the catches were comparatively good, the best taking place in 1928 (5201 tonnes). In this period, plaice was quite often also caught in the central part of the sea. In 1928, flounder catches in the Liepaja area constituted about 1% of plaice and dab. An increase in plaice catches in the Baltic Sea was also recorded from the late 1950s to the 1960s, and during the period 1972–1980 when the yearly landings amounted to 14,000–19,000 tonnes. The bulk of the landings was fished in the Transition Area. From the late 1980s to the mid-1990s, the catches did not reach 1000 tonnes but increased thereafter. Over the last decade, annual plaice catches have been rather stable, at a level of 2000–3000 tonnes (ICES 2015a). In the Baltic Sea, plaice is protected according to minimum commercial length and mesh size, and closed periods have also existed in the past. The data collected up to the present do not allow for regular analytical assessment of stock size and presentation of management recommendations on plaice populations.

3.9.2.4 Dab *Limanda limanda* (L.)

The area of dab is situated along the European coast of the Atlantic Ocean. It reaches from the Gulf of Tsosh (east of the White Sea) to the Bay of Biscay, in the coastal areas of Spitzbergen and Iceland. In the Baltic Sea, the species lives in the western and southern regions up to the Stockholm Archipelago. Dab is less tolerant of low salinity and high temperatures than plaice and flounder.

Activation of dab spermatozoa occurs at 11–14 psu, with an increase in the duration of mobility with salinity. Dabs of the Sound and the Arkona basin significantly

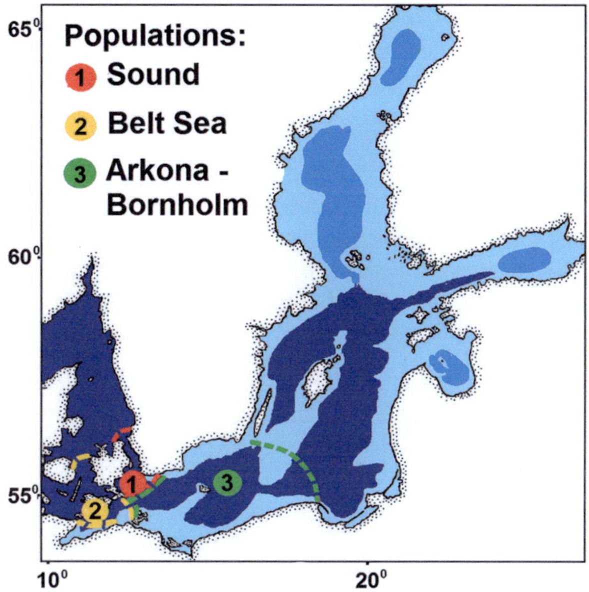

Fig. 3.48 Approximate location of populations of dab in the Baltic Sea (ICES 2011)

differ in this feature. The salinity of neutral egg buoyancy is significantly higher in dabs of the Sound and the western part of the Arkona basin than in dabs of the central and eastern areas of the named basin and those of the Bornholm deep (Nissling et al. 2002). The neutral buoyancy of 1% of developing embryos is guaranteed with a salinity of 17.8 psu. Salinity of this level occurs rather seldomly in the Arkona and Bornholm deeps, e.g., in the early 1950s and the late 1970s. Low salinity excludes the possibility of dab reproduction in the Gdansk and Gotland deeps. It is assumed that dab constitutes three populations in the Baltic Sea (ICES 2011; Fig. 3.48): (1) The Sound population (ICES Subdivision 23); (2) the Belt Sea population (ICES Subdivision 22 and the western part of the Subdivision 24 south of Møn Island); (3) the population of the Arkona and Bornholm deeps (eastern part of the ICES Subdivision 24 and Subdivision 25).

In the northern parts of the Baltic Sea, dab is rarely encountered. Some decades ago, the species was found in Finnish waters and at Saaremaa and Hiiumaa Islands.

Biology Dab prefers sandy bottoms. The species is more abundantly found at a depth of 20–80 m. Its migrations are rather short. Dab's fecundity in the Baltic is higher than in the North Sea but lower than in the Transition Area. The fecundity of the 35 cm-long dab in the North Sea is 420,000 eggs, in Kiel Bay, 765,000, and in the Arkona – Bornholm area, 675,000 (Kändler and Pirwitz 1957). The salinity of the neutral buoyancy of eggs is 25.8–27.1 psu in ICES subdivision 23 (the Sound) and 19.2–22.6 psu in ICES subdivision 24 (the Arkona basin). Spermatozoa activate in Subdivision 23 at 12.9 psu, in Subdivision 24, at 11.9 psu, and in subdivision 25, at 11.7 psu. Consequently, these populations show good adaptation to the salinity conditions in their environment. However, compared to flounder and plaice, dab is less well-adapted to the low salinity of the Baltic and is therefore more influenced by the oxygen deficiency in the deep layers of the sea. The most eastward possible spawning ground of dab in the Baltic Sea is situated in the Bornholm basin. The salinity is too low in the Gdansk and Gotland basins for dab spawning (Nissling et al. 2002).

Dab spawns in the Baltic from April to August at a depth of 35–40 m. Embryonic development lasts 7–14 days. Larvae have been caught in the water layers under the halocline in June-July. Young dabs transfer to the demersal way of life, descending to a muddy bottom of about 10 m depth probably in July–August.

Dab lives on demersal invertebrates and fish (molluscs, Polychaeta, sand eels, etc.). Up to 13-year-old dabs have been found in the Baltic. In the Belts, the age groups up to 7 are important.

Management In the Baltic Sea, dab is fished mainly in ICES subdivision 22, less so in Subdivisions 24 and 25. Some specimens have also been caught in Subdivisions 26–30. The main dab catches have been landed by Danish and German fishermen, less so by their Swedish counterparts. Catches were best in the late 1920s through the 1930s, with annual landings over 2000 tonnes. Later, the stock diminished, probably owing to unfavourable spawning conditions (decrease in oxygen content in the water layers of embryonic development of dab in deeps). Consequently, the main cause of comparatively low abundance of dab in the Baltic is insufficient adaptation of the species to the conditions in this brackish sea. Therefore, dab has populated only the Transition Area and the Bornholm deep area, and an increase in its abundance occurs only after larger inflows of saline water. In Belts, dab catches improved again in the 1970s. In 1981–1990, the yearly catches fluctuated by around 2000 tonnes, and since 2003, mainly around 1500 tonnes (ICES, 2015a). Dab is fished mainly with a demersal trawl. The quality of data has not enabled analytical assessment of dab populations in the Baltic. In the Baltic, the species is protected according to the minimum commercial length and minimum mesh size.

3.9.2.5 Turbot *Psetta maxima* (L.)

The species is distributed in the eastern part of the Atlantic Ocean from eastern Finnmark to the NW coast of Africa, including the Baltic and Mediterranian seas and the area from the British Islands over to Orkney and Shetland Islands and on to Iceland.

In the Baltic Sea, turbot is one of the largest fish species and top predators. It occurs in the Transition Area, the open Baltic, the Gulf of Riga, the Gulf of Finland up to the Narva River, and the Gulf of Bothnia up to the Northern Quark. No genetically differing groups have been found in the turbot of the Baltic Sea. However, studies on spermatozoa morphometry allow for differentiation of two local populations (ICES 2011), one of which occurs in ICES Subdivisions 24 and 25 (the Arkona and Bornholm regions), the other in ICES Subdivision 26 (the Gdansk deep). As studies on population structure of turbot have not yet been concluded and the spawning ground fidelity of this species is notable (which favours formation of local populations), further investigations may change the present point of view.

Distribution Young turbots keep to rather shallow water (with a depth of 0–1.5 m in summer), with older specimens preferring the deeper water. In winter, turbots distribute in deeper water, usually up to 40 m, seldomly from 50 to 90 m. Turbot does not form large shoals. It is a solitarily occurring, stationary fish which likes sandy, stony or muddy bottoms. The importance of the species is highest in the southwestern and southern Baltic, but in some years, it has composed up to about a half of the flounder-turbot mixed catches in the Gulf of Finland (Mikelsaar 1984). Turbot occupy an important niche in the ecosystems of the Baltic Sea. Turbot as a consumer of small fish simultaneously regulates the abundance of the latter and competes for food with predatory fish. The life span of the species reaches 30 or even more years, and in the Baltic, 14–16-year-old turbots have been caught (Stankus 2003; Draganik et al. 2005).

Migrations Migrations of turbot in the Baltic Sea are rather short, especially at spawning grounds. The length of the greater part of its migrations remains under 10 km, with only 3% of them exceeding 20 km and longer ones than that being even more seldom. The longest known trip was made by a turbot marked at the NE coast of Gotland and recaptured at Læsø Island in Kattegat. Turbot usually travels in shallow water along the coast and avoids longer migrations across deeps. Its migrations from deeper wintering areas to shallower spawning and feeding grounds are of limited length. Turbot is a species of nocturnal activity.

Reproduction Turbot attains sexual maturity at the age of 3–4 years. The males mature generally in their third and females in the fourth year of life. In spawning shoals, in younger age groups, males are in the majority, but beginning with the age group of 9, females dominate. The spawning shoals of turbot are composed of fish of ages between 3 and 16 years. Dominating age groups of females are 4–7, with a body length of 24.6–39.0 cm, and of males of 4–6 with a body length of 23.1–29.0 cm (Stankus 2003). As with other fish, turbot fecundity depends on body weight, length and age. The fecundity in the SW part of the Baltic is at the level of the species in the North Sea but lower than that in the Transition Area. Kändler and Pirwitz (1957) estimated that the fecundity of a 35 cm-long turbot in the Kiel Bight was 1,200,000 eggs, in the North Sea, 570,000, and in the Arkona-Bornholm area, 600,000 eggs. The relative fecundity is the lowest in the first time spawners and the highest in the 6–7-year-old specimens. Therefore, the latter age groups perform the most significant role in the renewal of the population. Before spawning, turbots actively migrate towards the coastal zone, where gonad maturation and spawning take place. The timing and duration of turbot spawning are regulated according to hydrometeorological circumstances. In the northern Baltic, turbot spawn from the middle of May to late June, at the east coast of Gotland, from early May to the end of August. Spawning grounds are situated at a depth of 5–40 m on shallows and along the coast. The species' fidelity to their spawning place is notable (Florin and Franzen 2010). For successful reproduction, the salinity on spawning grounds should be at least 6–7 psu and the temperature 11–12 °C (Stankus 2003). The diameter of the pelagic eggs of turbot is 0.9–1.2 mm. Embryonic development lasts 7–9 days. The length of the newly hatched larva is about 2.5 mm. After metamorphosis occurs at a length of 25–27 mm, it transfers to the demersal way of life (Mikelsaar 1984).

Feeding In the early stages of development, turbot feeds on benthic and nektobenthic invertebrates, but shifts to a fish diet later on. In the Gulf of Finland, the turbots of a length up to 21 cm feed mainly in twilight on demersal invertebrates (mysids, amphipods, isopods). From this length onwards, turbot begins preying on fish (herring, sprat, three-spined stickleback, nine-spined stickleback, boltnose, common sand eel, gobiids, flounder, minnow, perch, etc.), but also consumes molluscs (Mikelsaar 1984).

In the open part of the central and southern Baltic, the food depends on the body length of the turbot and considerably varies with the years. Young turbot feed on planktonic crustaceans and demersal invertebrates. For the length group below 10 cm, mysids are extremely important, constituting about 30% of the ration by weight and occurring in stomachs of more than 2/3 of individuals. At the length of about 7–8 cm, turbot transfers to a predatory way of life. From this length onwards, the most important food items are sprat (about 34% of the ration by weight) and gobies (36%). Diet of the 10–20 cm length group (2–3 years old) consists of sprat (69%), goby (8%), shrimp (15%), mysids, etc. The most variable feeding is that of the 4–5-year-old turbot (at a length of 21–30 cm) which constitutes herring (60–70%), sprat, flounder, gobies and other fish, as well as demersal invertebrates. The bulk of food of the 7–8-year-old (31–40 cm long) turbots consists mainly of herring (about 81%), flounder, three-spined stickleback and gobies (Stankus 2003). In the central Baltic Sea, the food depends on the composition of the potential prey species in the area of the turbots' distribution. Also, the size of prey is of importance – pieces of food of a length of 4–11 cm are preferred. In the coastal zone, turbot preys on 9–16 cm long herrings, in the open sea, on 12–26 cm long ones. In the open central Baltic, the feeding of turbot is most intense in September–October. During this restoration period after spawning, turbot eats about two thirds of its annual ration. Turbot consumes the smaller part of its annual ration during its spawning period from May to July and after the restoration period in winter. In June–July, the increase in the feeding intensity runs parallel to the water temperature, the most favourable temperatures for feeding being from 15.5 to 19.3 °C. In August, when the females retreat to deeper water layers, young specimens and males feed in coastal areas mainly on invertebrates and fish fry. 17% of turbot's food is constituted of crustaceans, but the main food consists of herring (38% by weight), smelt (15.8%), cod (14.7%) and gobies (10.9%). In September–October, in the open sea, the main food of turbot is herring (68–81%), sprat (14–27%) and young flounder (2–5%). The food of female and male turbots overlaps by only about 79%. An important difference resides in the fact that beginning from a body length of 28 cm, the females of higher growth rate begin to feed on young flounders, which is not the case in males of slower growth. Instead, males consume smaller food animals – mysids, etc. The annual ration of turbot constitutes 327% of its body weight. Over 75% (1436 g) of its yearly ration is constituted by herring and 15% (287 g) by sprat. The share of crustaceans in turbot's annual ration amounts to only 1.7% (33 g).

Growth In the Baltic Sea, the growth rate of turbot decreases towards the north and east. The growth of the species is highest during the first 3 years, with the maximum occurring in the second year of life. In the third year, sexual difference becomes apparent – females exceed males in size. It is probable that growth rate is related to water temperature. The growth of turbot can well be described with the von Bertalanffi equation. Applying this equation, the theoretical maximum length of female turbots in the southern Baltic (Lithuanian, Polish and German waters) was found to be $L\infty = 51.9–55.0$ cm and the maximum weight, $W\infty = 2799–3539$ g, while for the males, correspondingly, $L\infty = 33.4–35.0$ cm and $W\infty = 677–778$ g (Stankus 2003; Draganik et al. 2005).

Enemies and Parasites Owing to its size, turbot does not have very many enemies. The parasites in turbot belong mainly to the species living in the eumarine environment. Almost all turbots are infected with *Bothriocephalus bipunctatus* (in the Gulf of Finland) or *B. scorpii*. From Nematoda, *Rhapidascaris acus* and *Hysterothylacium* spp. and *Contracaecum* spp. are found in turbot. *Echinorhynchus gadi, Corynosoma strumosum, C. semerme and Metechinorhynchus salmonis* are also among the parasites found in turbot (Ojaveer 2014).

Management Historical sources report that turbot was an important component of fish catches in the Southern Baltic from the Curonian Spit to Pomeranian Bay as early as the second half of the nineteenth century (Draganik et al. 2005). The species has been recorded in the statistics of Baltic fisheries since the early twentieth century, but the quality of the catch data was very variable. Frequently, turbot catches were reported together with flounder (constituting a part of the catches of flounder or "Baltic flatfish"). Turbot was not included in the earliest international agreements regulating fish catches (including minimum legal length) in the Baltic. After World War II, due to a lack of demand, in the east coast countries of the Baltic Sea, this species was mostly caught as bycatch in fisheries for demersal species. Larger individuals were valued only in coastal localities, where there was a longstanding tradition of marine fish consumption. Smaller turbot was considered less valuable than flounder. Generally, in the period 1965–1982, turbot catches did not exceed 200 tonnes (Stankus 2003; Draganik et al. 2005). Nowadays, turbot is fished mainly in ICES Subdivisions 22–29, i. e., in the Danish Sounds and the open Baltic. The interest in turbot fishery markedly increased in the early 1990s. The cause was an increase in prices paid for turbot in the western European countries (where turbot is much more valued than in the east coast countries, as the body of turbot is rather poor in fat). Because of their markedly higher price as compared with other species, turbots were picked out from other fish in mixed fisheries. In Subdivisions 26 and 28, net fishery for turbot started in the early 1990s. The reported annual landings rather rapidly increased from 1986 to 1993, exceeding the limit of 1000 tonnes in the period 1993–1996. Thereafter, annual landings diminished down to about 250 tonnes in 2014 (ICES 2015a). In the last decade, turbot fishery was closed in Latvia and Lithuania because of low abundance of the stock. The mortality of young turbots as a result of their mass bycatch in fisheries directed towards other species can be very troubling from the point of view of management of the species in the Baltic.

For protection of turbot stock in the Baltic Sea, measures backed by the EU have been applied: minimum legal landing size of fish (30 cm), minimum mesh size, closed areas and closed seasons varying by country. The quality of the data has not enabled composition of analytical assessments of the turbot stock in the Baltic.

3.9.3 Diadromous Fish

Valuable populations of fish migrating between salt and fresh waters – salmon, sea trout, European whitefish and vendace of sophisticated historical background – are distributed chiefly in the northern half of the sea. They fell under severe exploitation long ago, and presently, a number of original stocks of these species have vanished, have been overexploited and depressed, or suffer under unreasonable management of their environment. In this group of migratory fish, eel, garfish and river lamprey can also be considered.

3.9.3.1 Salmon *Salmo salar* L.

Atlantic salmon is an anadromous fish which spawns in the rivers discharging into the northern part of the Atlantic Ocean. In Europe, the species occurs from Portugal in the south to Novaya Zemlya in the north, in the rivers of Southwest Iceland and Greenland, and in the east coast rivers of North America from the Labrador Peninsula to the Connecticut River. Landlocked lake salmon *Salmo salar m. sebago* Girard is distributed in large deep lakes (e.g., Ladoga, Onega, Vänern, Saimaa, and Pielinen in Northern Europe) and does not leave for the sea. It spawns in the rivers flowing to lakes. In many areas, salmon populations are considerably reduced nowadays, mostly due to human impact.

Distribution, Intraspecies Groups Baltic salmon is an isolated group of the populations of the Atlantic salmon. This group reproduces in the rivers falling into the Baltic, east of the 13°E. Baltic salmon rarely mixes with the populations reproducing in the rivers discharging westwards of 13°E. Salmon colonized the Baltic Sea basin by at least three glacial lineages, today represented by salmon in the Gulf of Bothnia, southeastern Sweden, and the southern Baltic Sea, including the Gulf of Finland. The Baltic salmon is characterized by a distinct population structure which mirrors the postglacial colonization history. Today, the Baltic salmon reproduces naturally in about 30 rivers in Sweden, Finland, Latvia and Estonia. This number has historically been higher – probably reaching about 100. In a number of rivers discharging into the Baltic Sea, the salmon population has become extinct. Each river has a genetically unique population. On the species level, based on the IUCN criteria, salmon (like sea trout) have been categorized by HELCOM as vulnerable (ICES 2015b).

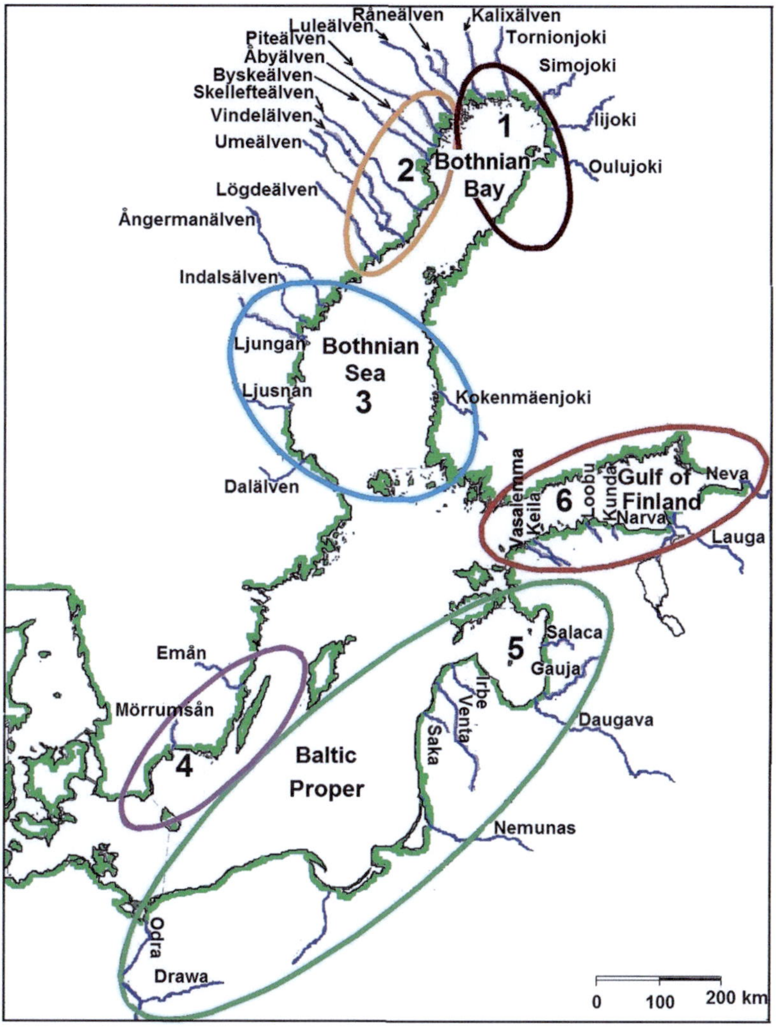

Fig. 3.49 Local populations and assessment units of salmon in the rivers falling into the Baltic Sea. Genetic difference between the populations of certain assessment units is smaller than between the asseessment units. Also, migration habits of populations of one assessment unit are similar (Ojaveer 2014)

Salmon populations are assessed by the units formed on the basis of biological and genetic features of the local populations reproducing in different rivers. Altogether, six asssessment units have been composed (ICES 2011; Fig. 3.49). The main part of the Baltic salmon stocks is related to the northern rivers of the catchment area. Salmon does not propagate in German and Danish rivers. The only salmon-inhabited river in Poland is the Drawa, a tributary of the Odra River. However, a number of rivers (Kemijoki, Kymijoki, Jägala, etc.) have lost their natural salmon populations due to the construction of dams, log driving, water pollution and other human activity.

Reproduction Salmons descending from various rivers intermingle on feeding grounds, but their receptors differentiating the composition of the water of the home river from others rather unerringly direct them to the river where they were hatched for spawning. The spawning migration can be very rapid (up to 100 km per day), but it often contains some resting breaks. Rapids and smaller waterfalls (up to 4–5 m) are crossed by jumps. In rivers, salmon do not feed, and their digestive glands and intestines degenerate. During the spawning migration, salmon live on, and the development of their genitals and nuptial plumage takes place as a result of reserves stored during the feeding period.

The goal of the spawning migration is the spawning area with a cool 0.5–3 m deep oxygen-rich water with sandy or gravelly bottom. The greater part of the rivers flowing into the Baltic Sea are comparatively short, and therefore, the migration to find suitable a spawning place is usually rather limited. The migration takes place in the conditions of high water level in autumn. Spawning begins at a temperature of 5–6 °C, but the main number of salmons spawn at 3–4 °C. The spawning procedure lasts up to several days. After spawning, both females and males are exhausted and descend back into the sea. Some of them perish but usually over half survives. After the feeding period and a new spawning migration, they can participate in propagation again (Rannak et al. 1983).

Salmon eggs are comparatively large, their diameter varying between 5.3 and 6.7 mm. Depending on weight and age, but also on population, the fecundity of salmon can significantly differ. The fecundity of the salmon of the Daugava River, having spent 1–4 years in the sea, is 7100–18,800 eggs, but that of the Kymijoki salmon is13,000–17,600 eggs. It is thought that a female salmon produces 1000–1200 eggs per kilogram of weight. The percent of fertilization and survival of embryos in nests is commonly high.

River Life Salmon populations play an important role in maintaining the balance in riverine food webs, both by harvesting invertebrates and also by providing important food resource for other predatory species. After the embryonic period lasting 5–6-months, 1.6–1.7 cm long larvae hatch in April–May. A rather large yolk-sac grants the larva a one and a half month-long period of endogenous nutrition. Before the yolk sac resorption, salmon larvae that have turned photophilic dig themselves out of the sand and sediments covering the nest and rise to the surface, where their swimming bladders are filled with air. At the beginning of the mixed endo- and exogenous feeding, the length of the larvae is 27.2–29 mm, and the period of mixed feeding lasts two-three weeks. The first exogenous food is copepod nauplii. After the resorption of the yolk sac, the food of the larvae is copepods, cladocerans, larvae of insects, etc. Formation of scales starts at the age of 41 days and lasts about 60 days. When the length of the young salmon reaches 50 mm and scales cover their whole body, it is called a parr. Parrs' habitats are rapids in the spawning area or nearby. They exhibit territorial behaviour. Mortality of parrs in rivers is caused mainly by predatory fish (pike, burbot, etc.), birds and water pollution. Parrs spend winter in rivers between stones and rocks. In winter, the mortality of parrs increases due to their limited energy resources and severe environmental conditions. Salmon juveniles are one of the few species that can utilize freshwater habitats in the rivers situated in the northern environment. The abundance of a year class is formed during the first year of river life. Good year class develops under certain environmental conditions: cool summer, high water level, etc. (Valle, 1934; Rannak et al. 1983). It has been found that a positive correlation exists between the abundance of parr in the Bothnian Bay salmon populations and the abundance of young herring in Bothnian Bay.

Young salmon spends 1–5 years in a river, and after this migrates to the sea. In the physiology and morphology of salmon parrs preparing for the descent to the sea, substantial changes occur in spring. Their probable background is changes in the photoperiod and temperature. Parr turns into slender smolt, with silvery sides and a greenish grey back. The definitive adaptation to marine life takes place in the estuary of the river, in water of salinity of 2–3 psu. The migration of smolts towards the sea starts in the southern Swedish rivers in the last week of April or the first week of May. In northern rivers, the migration starts later. Most young salmon descend in night, but in peak time, the migration continues by day.

Some part of the males of better growth attain sexual maturity in the river (in Latvian rivers, the males are practically all mature after 4 years of life) and participate in spawning as precocious male parrs. After spawning, the surviving dwarf males winter in the river and migrate into the sea the following spring.

Marine Life Young salmon, having migrated into the river estuary, adapt over a period of some weeks to brackish-water conditions in the 5–15 km wide coastal zone. In this time, young salmon feed on insects, isopods, fish fry, etc. During the first months of marine life, the majority of young salmon fall to the share of predatory fish (pike, perch, burbot, etc.), seals, cormorants, seagulls and other marine birds. It is probable that in avoidance of enemies, the reared young salmons are more defenceless than those hatched under natural conditions (Rannak et al. 1983). During the initial period of marine life, with the length of the young salmon being below 25 cm, its mortality achieves much higher values than later. This is probably one of the critical periods in the life of salmon. Later on, salmon transfers to a fish diet and starts feeding mainly on pelagic fish, mostly sprat, but also herring. In addition, salmon consumes sticklebacks, sand eels, cod, other salmon, garfish, perches, roaches, smelts, eelpouts, etc. In salmon food, invertebrates occupy only a moderate position.

The usual lifespan of the Baltic salmon is 4–6 years, the very oldest reaching the age of 10 years. In the sea, the young salmon follows the direction of currents, which are of cyclonic nature in the Baltic Sea. Most of the salmon gather in the southern part of the sea. However, the main number of the salmon having descended from the rivers of the Gulf of Finland, and some of those descended from the rivers of the Gulf of Riga, have been captured in the Gulf of Finland. Also, some part of the young salmon having descended from the rivers of the Gulf of Bothnia probably do not migrate into the southern Baltic, but remain in their home gulf where their growth rate is comparatively modest.

Diseases and Parasites Recently, one of the most important salmon diseases in the Baltic Sea has been the early mortality syndrome (M74/EMS) that first occurred in the Indalsälven hatchery. In the late 1980s, this disease was distributed very widely. During the last decade of the twentieth century, salmon populations in the rivers of the northern part of the Baltic Sea especially suffered under this disease (less so the populations propagating in the rivers falling into the Gulf of Riga). The main clinical symptoms were coordination disorders, lethargy, spiral swimming journey, etc. The disease often resulted in death before the first feeding. The studies allowed us to draw the conclusion that the cause of the M74/EMS was a lack of thiamine in the salmon eggs. It was supposed that the situation was related to the activity of thiamine in one of the main salmon food animals – herring. Later on, the disease was also found to be related to the toxic water pollution (HELCOM 1996, 2002; etc.). At the beginning of the twenty-first century, the outbreaks of this disease weakened and the mortality resulting from this syndrome in the salmon populations spawning in the rivers discharging into the Gulf of Bothnia decreased below 5% (ICES 2011). The newest explanation for the M74 syndrome suggests that it is related to a diet with a too low concentration of thiamine in relation to fat and energy content (especially the diet composed of young sprat – ICES 2015b).

The most important salmon parasites are from the flagellates *Costia necatrix*, the myxosporidians *Myxidium oviforme*, the infusorians *Trichodina acuta, T. nigra and T. meridionalis*, the cestodes *Eubothrium crassum* and *Bothriocphalus scorpii*, the trematodes *Hemiurus ocreatus*, *Diplostomum* spp., the nematodes *Rhapidascaris acus*, the acanthocephalans *Acanthocephalus anguillae, Echinorhynchus gadi* and *Pomphorhynchus laevis*, the hirudineans *Piscicola geometra*, and the crustaceans *Lepeoptheirus salmonis* and *Argulus coregoni*.

Assessment and Management Salmon muscles contain plenty of polyunsaturated fatty acids, which are beneficial for the human circulatory system. However, as the top predator, it accumulates harmful substances – environment toxicants. In the Baltic Sea, salmon landings have decreased from 5633 tonnes in 1990 to 881 tonnes in 2010. The landings increased to 1020 tonnes in 2014 (ICES 2015b). Since 2011, the harvest rate in the off-shore fishery has strongly declined. The harvest rate in the coastal fishery reached the lowest values in 2013–2014. In general, the exploitation rates in the sea fisheries have reduced to such a low level that most of the stocks are predicted to recover. Weak stocks need stock specific rebuilding measures, including fishery restrictions in estuaries and rivers, habitat restoration and the removal of potential migration obstacles. The decrease in landings was contributed to by more severe regulation of fishery, the impact of seals on catches and problems in the marketing of salmon related to its high dioxin content. Salmon stock management is complicated by incorrect declaration of catches.

Over two decades, up to 2005, the management of Baltic fish stocks was arranged by the International Baltic Sea Fishery Commission (IBSFC 1975). In its yearly recommendations, the IBSFC fixed the allowed salmon catches and technical conditions for the exploitation of stocks (closed seasons, minimum legal length of the fish in catches, the number and type of allowed fishing gear, etc.). This organization accepted the plan for management of salmon stocks in the Baltic Sea (Salmon Action Plan – SAP) for the years 1997–2010. This was the first plan for the multiannual management of the Baltic Sea fishery. The main goal of the plan was to achieve the 50% level of the estimated possible production of salmon young stages in corresponding rivers. During recent decades, the success in increasing the abundance of young stages of Baltic salmon has been achieved mainly on account of the smolts reared in hatcheries. However, stocking volumes of salmon have lately somewhat decreased.

But the abundance of the smolts hatched on the natural spawning grounds has gradually increased, notably in the Gulf of Bothnia. Also, wild salmon stocks in the Gulf of Finland show recovery. In other sea areas, only a very little increase, if any, is predicted. The current total production of all Baltic Sea rivers is around 2.84 million wild smolts, which corresponds to about 65% of the overall potential smolt production capacity of salmon stocks. About 4.7 million reared salmon smolts were released into the Baltic Sea in 2014. The situation in the Gulf of Finland is illustrated in Fig. 3.50.

Beginning in 2005, the management of fishery in the Baltic Sea has been a matter for the European Union and the Russian Federation. SAP has never been a part of the EU fishery policy. However, as salmon catches have significantly decreased during recent decades, improving the management of this species is very necessary. The EU has developed the multiannual salmon management project. The usage of drift nets has been prohibited. This substantially decreases exploitation of the mixture of populations in high seas. Also, a number of closed areas have been added in the estuaries of the salmon rivers. Restoration of the populations in the recent salmon rivers has begun. In this activity, success has been less than expected. The reintroduction of salmon populations has proved to be a time-consuming process, and one which has included surprises.

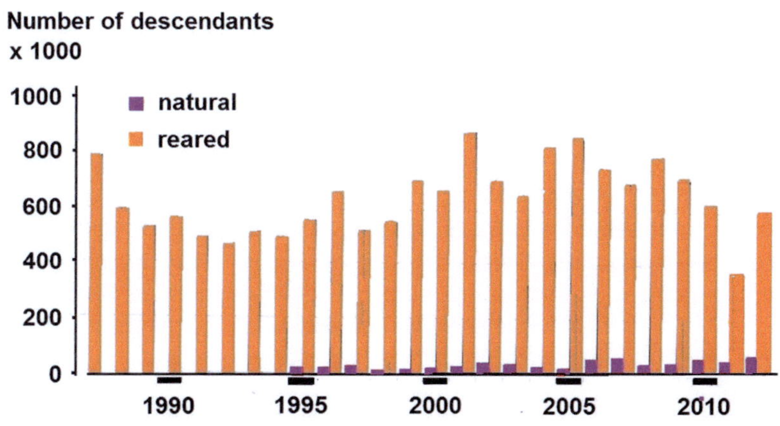

Fig. 3.50 Addition of natural salmon offspring and reared young stages into the Gulf of Finland in 1988–2012 (ICES 2013)

The salmon fishery has changed considerably since the beginning of the 1990s. Salmon was fished using driftnets before 2008, when driftnetting was banned and hook and line became more extensively used. Offshore exploitation using hook and line has decreased and is now even lower than in 2008. While feeding in the sea, salmon are caught by long lines, and during the spawning run, they are caught along the coast, mainly in trap nets and fixed gillnets. Where fisheries are allowed in the river mouths, set gill nets and trap nets are used. The status of the roughly 30 stocks of Baltic salmon is estimated by 6 assessment units using a Bayesian approach (Kuikka et al. 2014). The TAC is counted in numbers more often than in weight.

The Baltic Salmon and Trout Assessment Working Group (ICES 2015b) stated that the post-smolt survival has declined from the late 1980s up to the mid-2000s, but indications of improvement have been noticed since then. The current survival is estimated to be about 14% for wild and 4% for reared post-smolts. The positive turn in survival will probably lead many salmon stocks to recover closer to their target state. As a result of positive development in spawner abundances in 2012–2014, however, a gradual improvement in the stock status is expected for most of the northern stocks by 2021.

3.9.3.2 Sea Trout *Salmo trutta* L.

Sea trout is ecologically a very plastic anadromous species which can adapt in the coastal waters of oceans and also in small rivers and lakes. A number of endemic forms of the species live in the Mediterranean, Black and Caspian Sea basins. Anadromous sea trout of the Atlantic Ocean populates areas from North Spain to the White Sea, Iceland, the British Islands and the Baltic Sea. Between anadromous lake and river trouts of the same area, no substantial genetic variations have been found. Also, the populations can be only partly anadromous (i.e., the other part of the population may continuously live in the river). In this case, mainly females migrate to the sea.

In the rivers discharging into the Baltic Sea, about 1000 sea trout populations have been found. Sea trout reproduces both in salmon rivers and in smaller rivers and brooks throughout the NE coast of the Atlantic Ocean, including the Baltic Sea. Differently from salmon, sea trout spawns in the tributaries of larger rivers and smaller rivers/brooks. The majority of the populations live in the rivers falling into the open Baltic (ICES 2011). No data exist as to how many sea trout populations have existed in the Baltic Sea Basin. Many rivers have lost their sea trout populations because of migration obstacles (dams, water constructions), pollution or too intense exploitation. Nowadays, sea trout rivers can be found in all Baltic states, including Denmark and Germany (where no natural salmon populations remain). Presently, the Baltic Sea trout and salmon populations are assessed by the same methods (ICES 2011).

The number of artificially-reared sea trout smolts has increased much more than the number of salmon smolts, especially in the rivers discharging into the open Baltic.

Reproduction Sea trout spawning grounds are found in a much large number of rivers than those of salmon. Sea trout ascends into the spawning river in summer or autumn. The great majority finds its home river unerringly. In the Gulf of Bothnia, the first to enter into spawning rivers in May–June are larger specimens, with the smaller coming later, even in late autumn. The spawners having arrived later may not spawn in autumn. In the Vistula River, two sea trout populations spawn: one ascends into the river in summer and spawns in lower reaches, the other arrives later and spawns in the following year in the tributaries, far from the river estuary. In the case that salmon and sea trout spawn in the same river, salmon spawns nearer to the central part of the river, sea trout in the coastward places where the current is slower. The spawning temperature (mass spawning at 3–4 °C) and spawning behaviour are similar to those of salmon, only sea trout spawns somewhat earlier than salmon. Ripe sea trout eggs (diameter 4.5–5.5 mm) are smaller and paler than those of salmon. However, these differences are not always clear and do not always allow for differentiation of salmon eggs from those of sea trout. The fecundity of sea trout varies in large limits: the fecundity of 77 cm long sea trout of the Vääna River discharging into the Gulf of Finland was 12,800 eggs, the average fecundity of the fish over 70 cm in length, 8500 eggs, the fecundity of 60–69 cm long fish, 5800 eggs, and of the fishes below 50 cm, 2800 eggs (Rannak et al. 1983). The average relative fecundity (the number of eggs per 1 g weight without internal organs) varies between 2.3 and 3.7, i. e., it is higher than in salmon. The form of salmon and sea trout nests is rather similar, but those of salmon are larger than those of sea trout. Eggs in the nest are covered with about 20 cm thick sand. The development of the embryos lasts 4–5 months.

River Life Like salmon, young sea trouts spend 1–5 years in the river and then migrate to the sea. Newly hatched sea trout larva is shorter (13–15.7 mm) than that of salmon. The prelarvae of sea trout at the transference to mixed feeding are 21.5–25.6 mm long, i. e., a couple of millimeters shorter than the salmon prelarvae at the same stage of development. In sea trout, the formation of scales starts at about 30 mm in length. Sea trout parr are darker than those of salmon. During the first year of life, the sea trout parrs live in the vicinity of river banks in the 20–30 cm deep rapids with stony bottoms and a current of 10–50 cm sek^{-1}. Larger parr choose their living sites in deeper water. Sea trout parr, like those of salmon, defend their territory in river. Living together, sea trout parrs, being more aggressive, can limit the area of salmon parrs. The food of sea trout parr consists of gammarids, isopods, turbellarians, roe of sea trout, small fish, etc. Sea trout parrs grow faster and descend having become larger than the salmon parrs (Rannak et al. 1983). Survival of salmonids in winter may be low both because of energy deficiency and predators. During the river period, the most widespread sea trout parasites are *Metechinorhynchus salmonis*, *Acanthocephalus anguillae* (both in the intestines), *Diplostomum spathaceum* (in the eyes), *Costia necatrix* (in the gills), and *Trichodina nigra* (in the body). Some of them are also salmon parasites.

Marine Life The marine life of sea trout is basically comparable to that of salmon. The most important difference is that sea trout populations do not usually migrate as far as salmon. The migration of sea trouts originating from the rivers falling into the Gulf of Bothnia is limited to this gulf. Most sea trout populations move up to about 150 km from the coast. In the Danish archipelago, sea trouts only migrate very close to the coast. These populations are exploited by the local fishery. The sea trouts tagged in the Gulf of Finland have been recaptured in their home gulf, but also in the open Baltic and in the Gulfs of Bothnia and Riga. Like salmon, sea trout preys chiefly on herring and sprat. In the Baltic Proper, herring is preferred. In coastal areas, other important food items are garpike, smelt, gobies, sand eels, etc. (Valle 1934; Rannak et al. 1983, etc.). In salmon spawning time, sea trouts catch salmon eggs drifting downstream and also take parr. The growth of sea trout in the sea is slower than that of salmon. Sea trouts having spent 1 year in the sea and caught in September–October 1985–1991 in the Gulf of Finland were 42 cm long and weighed 780 g, whereas the specimens after 6 years in the sea were, on average, 71 cm long and weighed 4100 g.

The sea life lasts from half a year to 5–6 years, and after that, sea trout migrates to spawn in its home river.

Parasites In the sea, the main number of parasites are common to salmon, but some of them are different. The most widespread sea trout parasites are: *Echinorchynchus gadi* (in the intestines), *Hemiurus ocreatus* (in the digestive tract), *Myxidium oviforme* (in the gall bladder), *Bothriocephalus scorpii* (in the intestines), *Eubothrium crassum* (in the intestines), *Tinascaris aduncum* (in the liver), *Piscicola geometra* (on the body, in the oral cavity), and *Caligus lacustris* (on the body).

Assessment and Management ICES assesses the condition of the populations on the basis of the relation between the actual number and the potential possible abundance of parr. Additionally, data on catches, taggings and countings of descendants are used. No methods for analytical sea trout assessment exist. Therefore, it is not possible to compose quantitative fishing recommendations.

Sea trout is fished mainly near the coast and in rivers, less in high seas. Main catches originate from the Baltic Proper, but the share landed in the Gulfs of Finland and Bothnia is very important, too. Total landings have decreased from1563 tonnes in 1990 to 756 tonnes in 2009 and 219 tonnes in 2014. In 2014, 77% of the total catch was taken in the Main Basin. Recreational catches constitute a very important fraction of the total catches; in the Gulf of Bothnia, this is larger than the commercial catch. No correct catch statistics exist on free-time fishery and on trout bycatch in salmon fishery. In the Gulfs of Finland and Bothnia, most sea trouts are caught before their sexual maturation. Catch of sea trout poses a problem for the recovery of the threatened species. Therefore, well-founded minimum legal length, closed seasons and protected areas for the species, both in the rivers and sea, should be introduced (ICES 2015b).

The possibility of a free approach to spawning rivers and retention areas of young fish notably varies in the Baltic Sea. In the northern and eastern parts of the sea, the condition of populations is estimated to be bad in this respect, i. e., the abundance of the populations in the rivers is lower than the estimated potential abundance/

carrying capacity of the river. It is estimated that in the rivers of the southwestern Baltic, the abundance of sea trout populations is relatively higher (ICES 2015b).

3.9.3.3 European Whitefish *Coregonus lavaretus* (L.) s. l.

A number of whitefish groups exist in the Baltic Sea, but their taxonomic status is not very clear. Some of them spawn in the sea, but some are migratory, spawning in rivers (Ojaveer 2014). Sea-spawning whitefish groups travel widely, being most abundant in Bothnian Bay and the Quark, but they are also found in the Gulfs of Finland and Riga, in the Väinameri, etc. In sea-spawning populations, the number of gill rakers is small, as a rule 22–24, and their growth rate is lower than in the migratory type of whitefish. Migratory whitefish usually has more than 30 gill rakers, the population spawning in the Pärnu River falling into the Gulf of Riga being an exception with 24–26 gill rakers. Also, whitefish of an intermediate gill raker number are rather frequent, probably produced by man by means of uncontrolled crossbreeding of sea-spawning and migratory populations in hatcheries. Moreover, the two whitefish types are found to easily hybridize in the areas where both of them spawn (Lehtonen and Böhling 1988).

Females mostly attain sexual maturity at the age of 4–5 years, males usually a year earlier. Fecundity differs by population and increases with body size. In 5-year-old fish, the fecundity usually varies between 20,000 and 30,000 eggs, but in older and larger fish, can reach up to 90,000 eggs or even more (Ojaveer 2014). Spawning takes place on stony or gravelly bottoms. Most populations spawn in shallow water of 0.5–2 m in depth, but the spawning grounds of some populations (e.g., the population spawning at Ruhnu Isle in the Gulf of Riga) are situated down to about a 15 m depth. The spawning period usually lasts from October to December, with the larvae hatching in spring with the rise in water temperature. In the sea-spawning sparsely-rakered whitefish, the length of larvae at hatching is 11–14 mm, their yolk-sac is absorbed at a length of 14–16 mm, and metamorphosis takes place at a length of about 40 mm. This usually takes place in the southern Gulf of Bothnia at the end of June, and at the Estonian west coast, probably 1–2 weeks earlier.

Whitefish is an oxyphilic fish preferring cool water. The first food of larvae consists of zooplankters available in early spring, mainly calanoid nauplii and copepodites, and occasionally also rotifers. During summer, the proportion of nektobenthic and benthic organisms in the food gradually rises. The nourishment of adult whitefish varies by season and area. In the postspawning period, it undertakes rather wide feeding migrations in the sea. Its main food is benthic invertebrates, fish (mainly herring) eggs and small fish (chiefly gobies).

The presence of whitefish populations has been noted at the coasts of most countries around the Baltic Sea. The stock is most abundant in Bothnian Bay and the Quark. Whitefish is also exploited along the Swedish and Finnish coasts of the Bothnian Sea and the Baltic Proper, in the Gulfs of Finland and Riga, and in the Väinameri (Lehtonen and Böhling 1988; Thoresson 1996; Ojaveer 2014).

Southwards, the abundance of the species is rather low. The species is present in the Gulf of Gdansk, Pomeranian Bay and Szczecin Firth, but catches are small and only sporadic (Bartel 1993). In the Darss-Zingst area, findings of the species are very rare (Winkler 1996).

The abundance of whitefish populations has considerably fluctuated. In 1963–1985, the professional annual whitefish catch in the Gulf of Bothnia fluctuated between 910 and 2010 tonnes. The catches have been biggest in the Quark and Bothnian Bay. In 1975–1984, the non-professional whitefish catch in Finland varied between 352 and 769 tonnes a year (Lehtonen and Böhling 1988). In the NE Baltic, the catches were largest in the first half of the 1950s (the average annual catch of the sea-spawning whitefish in Estonia was 205 tonnes, the catches of other groups only a few tonnes). At that time, cotton nets were replaced by kapron ones; exploitation of the almost unprotected stocks increased considerably. By the mid-1970s, the annual catches decreased gradually up to the 1980s, when a number of small stocks became depleted. In recent decades, some signs of abundance recovery of some whitefish stocks have been noticed. Significant factors limiting the size of the whitefish populations are the muddying of their spawning grounds, eutrophication, and migration obstacles (dams) in their spawning rivers. Also, warm winters affecting the normal regime of embryonic development have unfavourably reflected on the year class abundance of whitefish. Regulation of whitefish exploitation and protection of its stocks have been based on national legislation. Usually, the minimum legal length for catches, regulation of mesh size in gears, closed areas and closed seasons are established. In the 1980s, a rather large number of whitefish young stages were stocked in the Gulf of Bothnia or the rivers falling into it (Lehtonen and Böhling 1988).

3.9.3.4 Vendace *Coregonus albula* (L.)

The area of vendace includes the basins of the Baltic, White and Barents Seas; it is also found in some lakes of the upper course of the Volga River and in the British Islands. The species is distributed mainly in lakes with favourable oxygen conditions in winter, but also in coastal brackish waters and in the northern part of its area in rivers. The vendace living in the sea migrate in autumn into the rivers for reproduction. Vendace is a shoaling pelagic fish. It requires waters of high oxygen content and is sensitive to eutrophication and pollution.

Vendace commonly attains sexual maturity at the age of 2 years. Rapid gonad development begins in July. The fecundity of vendace of 15 cm fork length spawning in the eastern Gulf of Finland and Lake Peipsi is about 4000–4500 eggs. Spawning begins in mid-November at a water temperature of 0.2–3 °C and lasts, on average, 23 days up to the middle of December. The slightly sticky eggs are laid on the hard sandy, gravelly or stony bottom at a depth of 1–6 m. Embryonic development lasts 165–176 days. The on average 8.6 mm long larvae hatch in April, and over about 10 days, they feed endogenously, using the resources of their yolk-sac.

Depending on the quantity of food and the water temperature, they transfer to exogenous feeding 8–23 days after hatching (Ojaveer 2014).

Vendace is a typical planktonophagous fish. It feeds year round in the pelagic layers and has the ability to select bigger zooplankters for food. Cladocerans and copepods usually make up to 80–90% of its food. Insects, larval smelt, perch and pikeperch are also used.

The size of vendace considerably varies by basin. In Finland, the maximum weight of vendace has been found to be 1.2 kg. Vendace grows rapidly before attaining sexual maturity, after which its growth rate considerably decreases. The growth rate depends on the food supply and the abundance of food competitors.

Vendace is a rather short-lived fish – its highest age in the Gulf of Finland is 5 years (Ojaveer 2014). The total mortality rate of vendace in Bothnian Bay is low compared with the rate in lakes, probably due to the fact that the natural mortality is lower in the sea (Lehtonen 1983).

In the Baltic Sea area, the vendace populations are largest in the northernmost parts, especially in the Gulf of Bothnia. In Sweden, trawling for vendace began in 1959, in Finland, in 1969. The increase in fishing effort gave high yields in the early 1970s, when old and little exploited stocks were harvested (Lehtonen 1983). Pelagic trawl fishery targeting vendace takes place mainly in the northern part of Bothnian Bay during the spawning period in October–November. Swedish catches varied in 1991–2004 between 200 and 1400 tonnes, and Finnish catches in 2001–2011, between 77 and 190 tonnes. In other parts of the Baltic Sea, the stock has been much smaller and of variable size. In many areas, the stock size is limited by eutrophication and the damming of spawning rivers. The main protective measures for vendace are the minimum legal size in catches, closed seasons and closed areas established by national legislation (Ojaveer 2014).

3.9.3.5 Eel *Anguilla anguilla* (L.)

Eel is widely distributed in the northern part of the Atlantic Ocean, from the Sargasso Sea in the west up to Europe and North Africa. Its natural area of distribution includes coastal waters and inland water bodies directly connected with the coastal regions of Europe and North Africa, starting from the Petchora River to the Black Sea.

European eel spawns in the Sargasso Sea (newest data show that other eel species also spawn in this sea), probably at a depth of 100–400 m. The spawners apparently perish after reproduction. The larvae are transported with the Gulf Stream to the European and North African coasts. Their journey lasts about 3 years. They undergo metamorphosis in coastal waters, their greater share entering rivers and distributing through freshwater bodies.

Eel is a thermophilous species of nocturnal activity. It feeds mainly on bottom invertebrates (larvae and pupae of Chironomidae, molluscs and aquatic insects, isopods, etc.) and small fish (gobies, sticklebacks and their eggs) (Mikelsaar 1984, etc.). The feeding intensity is highest in May–June and August–September. The highest temperatures in summer and low temperatures in winter inhibit its activity. Eel does not feed at temperatures below 8 °C. In winter, it can hibernate in a passive state. At favourable temperature and plentitude of food, eel grows rather rapidly.

The males live chiefly in brackish water, females in fresh water. Males of the age of 7–14 years with their length up to 51 cm and females of the age of 9–19 years with their length up to 150 cm undertake spawning migration. They descend in rivers to the sea. Their back grows black and their abdomen silvery, their eyes become larger, their intestines atrophy, and their anus closes. The migration starts in autumn/beginning of winter and is directed to the Sargasso Sea. The circumstances related to the spawning migration and reproduction of eels requires further investigation.

In many areas of the Baltic Sea, eel fishery is one of the most important sources of income for coastal fishermen. Eels are caught with nets, fish traps, hooks and other gear. Glass eels are caught on their appearance in the coastal zone of oceans/seas for introduction into lakes and other freshwater basins.

During recent decades, eel catches in the Baltic Sea have seriously decreased. The obvious cause is the continuous lessening of the recruitment of the European eel in the Baltic area that has taken place since the mid-1960s (Ask and Westerberg 2005). Due to misreporting, it is complicated to estimate the true amount of landings. As the stock has seriously decreased, in the countries of the EU, the management of eel resources has taken place based on national management projects, which include the minimum legal size of eel in catches and other measures.

3.9.3.6 Garfish *Belone belone* (L.)

The subspecies *Belone belone belone* is distributed in coastal areas of the Atlantic Ocean, from Finnmark, the Murman coast and Iceland to France, including the Baltic Sea. Its abundance diminishes towards the north.

Garfish migrates into the Baltic Sea through the Sound, chiefly in May. At the south coast of Skåne, the species appears in April–May, in southern Gotland, in May–June, in the southern Bothnian Sea, regularly but not numerously from June to August–September. It is probable that after spawning, garfish feeds in the open Baltic. The main garfish shoals leave the Baltic Sea through the Sound at the end of August and the first half of September. However, small numbers of garfish have been caught in the Baltic Proper up to the Aland Sea until October–November. At times, garfish can also be encountered in the Baltic Sea in December, but not in January. It has not been excluded that some garfish overwinter in the southern Baltic, but the main wintering area of the species is located west of the British Isles. Details of the garfish migration between its wintering areas and the spawning grounds in the Baltic Sea need clarification. In the Baltic Sea, the abundance of garfish is higher in the southern and central parts (Draganik and Kuczyński 1983; Otterlind 1985b; Winkler 1996). In the Gulf of Bothnia, the species has been found up to the Quark, in the Gulf of Finland, up to Viipuri. Garfish is found also in the Estonian Archipelago and the Gulf of Riga (Valle 1934; Ojaveer 2014).

Reproduction Upon arrival in the Baltic Sea, garfish ovaries contain eggs of various size. Draganik and Kuczyński (1983) found that in Puck Bay, the number of eggs in garfish ovaries varied from 8800 to 35,000, increasing with the size of the fish. However, the fecundity was not found to represent the number of eggs which would be laid in the current spawning season or could be capable of fertilization. Mikelsaar (1984) found that a female garfish has 2040 to 10,105 eggs at spawning and stated that, together with ripe eggs, small undeveloped ones are also present. The author assumes that the species may spawn in batches. The diameter of ripe eggs is 3.11–3.49 mm. Fertilized eggs are fixed to the substrate by their chorion filaments (Erm et al. 1970).

Garfish spawning is reported in Puck Bay (Draganik and Kuczyński 1983), in the Väinameri between Muhu and Hiiumaa Islands and Matsalu Bay, at the south coast of Saaremaa, and probably in small bays at the open sea coasts of Saaremaa and Hiiumaa. In Swedish waters, mature garfish are found regularly as far north as Öregrundsgrepen at the beginning of June and occasionally off Väddö in the Åland area at the beginning of July. However, findings of small fry have not reached farther north than the waters NE of Gotland. Garfish also spawns in the Wadden Sea (Otterlind 1985b; Erm et al. 1970).

In the Väinameri, garfish spawns in the areas rich in underwater vegetation, at a depth of 4–6 m and at temperatures 13.3–14.8 °C. Garfish embryos have been found in the second half of June and early July, mainly fastened to demersal algae (Erm et al. 1970). At hatching, garfish is rather well developed. Its length amounts to 12–13 mm. The larvae live chiefly in the algal belt. The yolk sac resorbs at 14–15 mm. The formation of the beaklike jaws starts with the elongation of the lower jaw. The upper jaw begins to grow fast in 9–10 cm long garfishes. The larvae and fry are distributed in upper water layers. In Estonian waters, they have been caught in the Väinameri and the northern part of the Gulf of Riga beginning in July. The young stages are distributed in the area of the coastal slope. Later, they leave for the open sea. Young garfish grow comparatively quickly and leave the Baltic Sea in autumn when they are "as big as pencils" (Otterlind 1985b).

Feeding A young garfish feeds on zooplankton and insects. Later on, in the Atlantic, it lives on crustaceans, euphausiids, isopods, amphipods, decapod larvae, fish (especially clupeids) and insects. In Estonian waters, garfish consume fish (sprat, herring, sand eel, sticklebacks, etc.) and invertebrates (*Idothea* spp., *Gammarus* spp., *Corophium* spp., insects) (Erm et al. 1970).

Growth and Age In the Baltic Sea, variation of length of 1–6-year-old garfish is 25–36, 45–59, 59–65, 66–70, 70–72, and 72–84 cm. respectively. The maximum age reported is 11–13 years (Draganik and Kuczyński 1983, etc.). The biggest specimen caught in Estonia was 77 cm long and weighed 985 g (Mikelsaar 1984).

Enemies and Mortality Draganik and Kuczyński (1983) estimated that the impact of fishery upon the survival rate of garfish is small, but the total mortality of the species is very high. Probably, the main mortality of garfish occurs outside the Baltic Sea, where tunas and killer whales feed on it (Otterlind 1985b, etc.).

Garfish can be infested with *Lecistorchynchus tenuis*, which does not occur in the Baltic. The parasite fauna of garfish seems to be limited in Baltic waters. Two species of helminthes have been found: *Ascarophis filiformis* and *Hysteriothylacium aduncum* (in the intestines). In addition, the coccidian parasite *Eimeria gadi* was detected in the intestine wall, and since the mid-1990s, larvae of the pathogenic nematode *Anisakis simplex* have been found in garfish, too (Draganik and Kuczyński 1983, etc.).

Assessment and Management Garfish stock size and catches fluctuate considerably. In the years of high temperature of water surface layers during the reproduction period, abundant year classes hatch. In the Southern Baltic, the 1959, 1975 and 1976 year classes of garfish were abundant (Draganik and Kuczyński 1983). In the NE Baltic, the alternation of rather long periods of low and high catches of garfish is characteristic. It is probable that the periodicity depends on the stock size of the species and on interannual variations in its migration route. In the first half of the 1950s, the yearly catches of garfish in Estonian waters constituted 20–30 tonnes. The landings increased to 100–150 tonnes in the late 1950s. In 1964, the catches exceeded 800 tonnes, and during1974–81, they varied between 396 and 1081 tonnes. In 1983–1991, the annual Estonian garfish catch varied between 0 and 23 tonnes. Since 1993, the catches have again increased and have constituted from one hundred to four hundred tonnes per year (Ojaveer 2014). In the ICES area, the yearly garfish catches in 1973–1980 constituted up to 2566 tonnes (Draganik and Kuczyński 1983).

Garfish is usually fished as bycatch in herring trapnets and other gear in coastal waters. A smaller amount of the fish is taken with fykes, gill nets, trawls (usually together with herring) and hooks.

3.9.3.7 River Lamprey *Lampetra fluviatilis* (L.)

River lamprey does not belong to fish but rather cyclostomates having no ossified skeleton. However, we address the species here together with fish, as is customarily done.

This is an anadromous species reproducing in rivers falling into the European Atlantic from Ireland, Scotland and South Norway to the North Mediterranean. Spawning migration of freshwater populations from large lakes (Onega, Ladoga) or seas to rivers takes place from late summer to the following spring, preferably in dark nights at the high water level in the river. At the beginning of the spawning migration, river lampreys stop feeding. In the river, a number of changes occur in the body of the lampreys (degeneration of intestines, maturation of gonads, decrease in fat content, etc.). Spawning grounds are on sandy or gravelly bottoms at a depth of 0.2–1.5 m. For spawning, a depression with a diameter of about 40 cm in the bottom of a river of relatively high flow rate is prepared by the males. One/several females and several males spawn simultaneously in the same depression/nest. Spawning begins at 9–10 °C, the water temperature during spawning is usually

12–18 °C, and spawning is most intense in darkness. Data from North Latvia and Estonia show that the fecundity varies from 4800 to 54,000 eggs per female. The color of the somewhat elliptic eggs is white or creamy, and their diameter is about 1 mm. Generally, lampreys die within two weeks after spawning. The length of hatched larvae is about 4 mm. Newly hatched larvae lack pigmentation and live on the endogenous resources distributed passively with water flow. On some days, their activity increases, dark pigmentation appears and the larvae (ammocoetes) screw into the soft bottom.

The ammocoetes live in the river bottom sediments for usually 3–6 years, feeding on detritus. The river lamprey larvae live mostly in bottom areas where the depth of the water is no more than 50 cm. In northern Latvia, metamorphosis begins in late summer at the length of 8–13.5 cm and lasts until the following spring. After metamorphosis, lampreys head for the sea, where they spend up to 3 years (Saat et al. 2003). During this period, the river lampreys take up a parasitic way of life. They fix onto the prey species and feed on its soft tissues. Their common prey species are herring, smelt, sprat, cod, vimba bream, salmon, sea trout, whitefish, pikeperch, etc.

Females are somewhat larger than males. In the rivers of the basin of the Gulf of Finland, the average length of lampreys (both sexes) was 32–35 and the weight was 65–75 g in 1990. In 1993, in the Pärnu River, the mean length of female lampreys was 36.6 cm and the weight 91 g, while males were 35.1 cm and 81 g. The largest female river lamprey caught in the Pärnu River was 45.2 cm long and weighed 190 g.

In the Latvian and Estonian rivers, the parasitic load in river lampreys is rather low and the number of specific parasites is limited. Burbot and perch are found to prey on lamprey larvae in rivers. The following parasites have been found: of cestodes, *Eubothrium* spp. (in the intestines) and *Proteocephalus* spp., of trematodes, *Diplostomum spathaceum* (in the eyes) and *D. petromyzi-fluviatilis* (in the brain), of nematodes, *Cysticola farionis* and *Cucullanus truttae* (both in the intestines), of acanthocephalans, *Echinorchynchus gad*i (in the intestines), of Hirudinea, *Piscicola geometra*, and of crustaceans, *Argulus foliaceus*.

River lamprey is a valuable resource. It is fished mainly during its spawning migration. The landings and the size of resource vary over broad limits. Assessment and forecasting of lamprey resources is almost impossible, as the species has no bone structures to determine the age of specimens. It is assumed that the strength of lamprey year class depends on the temperature and water level in the river in the repreduction period and during the first year of life of the ammocoetes. The year class abundance is also positively related to solar activity. However, a negative correlation exists between the lamprey catches and the abundance of salmon and cod. It has been assumed that the stock size of river lamprey has been negatively affected by the construction of dams and other human activities changing natural conditions in rivers.

In Estonia, Latvia, Finland and Sweden, the river lamprey stocks populate a number of rivers, and the tradition of lamprey fishery goes back to the Middle Ages. Therefore, fishing for this species has been allowed to continue in these countries, despite the species being referred to in Annex 2 of the EU Habitat Directive. The landings are highest in Latvia. During the current century, the annual landings have amounted to about 100 tonnes. In the 1970s, the annual catches were higher, amounting to 400 tonnes, but the landings diminished with the increase in cod stock. After the disappearance of cod from the NE Baltic, the river lamprey stock gradually recovered (Bizarks and Abersons 2011). In Estonia, the yearly catches of lamprey have recently varied between 38–45 tonnes. The main landings come from the Narva River. The catches have been lower than the average catches in the period 1928–1938 (67 tonnes). At the beginning of the 1970s, the annual catch of river lampreys in Finland was about 2.7–3.0 million individuals or about 130 tonnes. In 2003–2004, up to 1 million lampreys were caught per year. The intensity of lamprey fishing has increased. Construction of hydroelectric power stations has reduced the spawning areas of river lampreys in most rivers. So spawning migration has been prevented or spawning areas have been destroyed (Lehtonen 2014; Tuunainen et al. 1980). In the catches of Swedish fishermen, river lamprey is a common species. However, the recent annual catch in commercial fishery has been rather moderate, varying between 1 and 15 tonnes. In Poland, the annual catches of lamprey landed in the period 1930–1939 in the Free City of Danzig frequently exceeded 100 tonnes. It was caught in the Vistula River system and the estuary. After World War II, the catches were still good, reaching 60 tonnes and more. They had diminished to less than 2 tonnes in this river system by 1990 (Bartel 1993). In the Darss-Zingst-Bodden Chain, river lamprey is regularly but also rarely found, as the species has no spawning places in this area (Winkler 1996).

3.9.4 Freshwater Fish

Freshwater fish are of greater importance in areas of lower salinity, especially in the coastal areas of the northern and easten parts of the Baltic Sea, mainly anywhere in the estuarine areas and bights where the limiting influence of salinity on the freshwater organisms is lower. This group consists mainly of warm-water species (burbot excluded), and therefore they prosper chiefly in the warmer and shallower parts of the sea. Of freshwater fish populating the coastal zone of the Baltic Sea, the most important economically are perch, pikeperch and pike. Today, the fishery for these species is important both as a separate branch of the economy and as a constituent of tourism and recreational fishery.

The catch of freshwater fish is comparatively high in Finland, and is also important in Sweden, Estonia, Latvia and Lithuania.

The management of freshwater fish resources has fallen under the competency of the member countries of the EU. However, the state of numerous local populations of varying exploitation rate is very different. Basic data for estimation of the state of stocks are of low quality. Statistics on catches and discards, as well as on catches per unit effort, are missing or incomplete. Differentation of local populations has not been performed to a degree that would allow for their analytical assessment and management. Due to a rather large number of species and local populations, this task is complicated and requires resources, time and competent scientists. As the greater part of freshwater fish is rather cheap, attempts have been made to gain indications on the changes in their stocks on the basis of estimates of the relative abundance of year classes in the experimental fishery. In Estonia, Finland and Sweden, the basic data on the abundance of year classes and the total mortality of a number of freshwater species are collected from the experimental net catches at permanent study sites. The results of the work at this level cannot yield data for the analytical assessment and proper management of resources. However, the work has been intended for the creation of a basis for future joint analysis of variations in year class strengths and stock prognoses (Karås et al. 1997).

3.9.4.1 Perch *Perca fluviatilis* L.

Perch is widely distributed in European freshwater bodies and brackish waters (excluding the Pyrenean, Apennine and Balkan Peninsulas), including the northern part of Scotland, a greater part of Norway and the northern Asia up to the Kolyma River. Up to the present, delimitation of perch unit stocks in the Baltic aimed at assessment and management of the resources has not been performed, though some preparations in this direction have been made.

Perch is well-adapted for life and reproduction in a brackish-water environment. The species is able to reproduce in such areas where spawning of other freshwater species has commonly failed, e.g., in the coastal zone of Vaindloo Island in the central part of the Gulf of Finland (Ojaveer 2014). However, strong year classes of perch develop regularly in such areas where favourable environmental conditions exist for spawning and growth of young stages.

Perch attains sexual maturity at the age of 3 years. Its length at maturation may exceed 20 cm. The fish spawns in May. The peculiar ribbon of eggs is shed on water plants and other objects in bottom water layers. The fecundity of perchs over 30 cm is more than 100,000 eggs. The best temperature interval for embryonic development is from 8 to 18 °C. The newly hatched larvae are 5–7 mm long. Perch can feed on variable pelagic and demersal organisms in the coastal zone. Younger age groups live on copepods (especially *Acartia* spp. and Harpacticoida), Chironomidae, isopods, amphipods, mysids, fish eggs, etc. At a length of 12–15 cm, perch transfers to a fish diet. It uses gobies, small herring, small perch, sticklebacks, roach, ruffe, eelpout, bullhead, etc. The year class abundance and growth rate vary considerably by area.

Feeding intensity is the highest in summer. In winter, its feeding is rather weak.

Perch as a widespread, numerous and rather small fish is an important food item for predatory fish, especially for burbot, pike, pikeperch, cod and large perch. Sea birds and seals also prey on perch.

Perch is the host of a number of parasites. In the NE Baltic, the largest number of parasites belongs to the protozoans (mainly infusorians) and helminths (mainly trematodes and cestodes). Crustaceans, acanthocephalans and fungi (*Ichthyophonus hoferi*) are also identified as perch parasites.

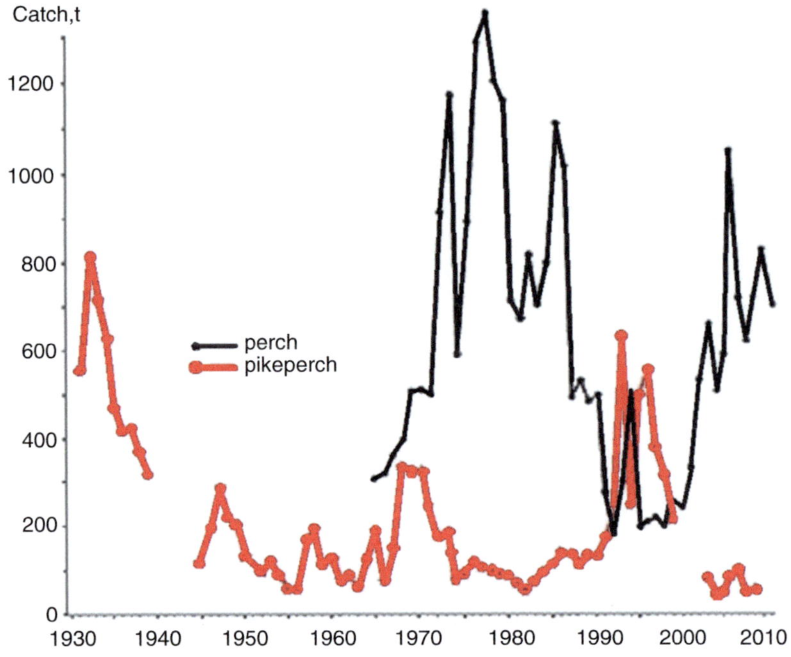

Fig. 3.51 Catches of pikeperch and perch in the Gulf of Riga, 1930–2012 (Ojaveer 2014)

Variation of perch year class strength in different populations in the Baltic Sea has been considerable. A common pattern of the population abundance was found in the perch of the Archipelago Sea's Northern Quark and the east coast of Sweden in the Baltic Proper. Similarities between areas of this nature, although following a different pattern, also appeared for populations within the Estonian and Latvian coastal waters (Karås et al. 1997).

Perch is an important fishing object for professional and freetime fishermen in Finland and Estonia, with the annual average catch amounting to about 1500–1600 tonnes. In the period 2001–2012, the yearly perch catches in the Estonian coastal waters varied from 386 to 1117.2 tonnes, and in Finnish coastal fishery, from 633 to 1020 tonnes. In Fig. 3.51, Estonian and Latvian catches of perch and pikeperch in the Gulf of Riga are shown (the Estonian landings predominated in both perch and pikeperch catches). In perch fishery, two periods of very good catches can be seen, divided by moderate landings from 1990 to the early 2000s. In Sweden, perch is fished mainly in the northernmost part of Bothnian Bay, in the Northern Quark, in the coastal areas of the Bothnian Sea and the Baltic Proper, in the Stockholm Archipelago, and southwards from there, up to Kalmar (Ask and Westerberg 2005). Perch landings by professional Swedish fishermen reached about 160 tonnes in 1995, but decreased to about 100 tonnes in 2003. However, it has been estimated (Thoresson 1996) that the landings by the professional fishermen tend to be less than half of the total perch catches in Sweden. Perch also belongs to the commercial fish species along the coasts of the Russian Federation in the Gulf of Finland and of Latvia and Lithuania. In Latvia, perch landings constituted about 30–50 tonnes (chiefly in the Gulf of Riga) in the period 2000–2004. In Lithuania, perch is fished mainly in the Curonian Lagoon, where annual catches varied from 2002–2012 between 30 and 67 tonnes. In the coastal zone of Poland, perch belongs to the common fish species. In the mid-1990s, the species was abundant in Pomeranian Bay. Its landings varied both by period and fishing area. The largest catches were obtained from the Sczecin Lagoon, where they exceeded 500 tonnes per year in the 1980s (Skora 1996). According to Winkler (1996), perch is rather abundant and is fished in the Darss – Zingst area, especially in the estuarine environments with oligo- and mesohaline districts available. Perch is mainly fished with fish traps and nets during its spawning time in spring.

3.9.4.2 Pikeperch *Sander lucioperca* (L)

Pikeperch supposedly immigrated into the Baltic Sea from the Caspian–Black Sea basin, probably during the Ancylus period. Nowadays, the species' area reaches from the Ural Mountains and the Aral Sea in the east to the Elbe River and the SW coast of Norway in the west, from the polar circle in the north to the Caspian and Black Seas in the south. The area of the species has recently widened to England, France, Denmark, Turkey, Central Asia and West Siberia. Pikeperch is a freshwater species, but also occurs in brackish water. In the Baltic Sea, pikeperch populates coastal areas of lower salinity in the Bothnian Sea, especially in the southern part, in the Stockholm Archipelago, in the northern and eastern parts of the Gulf of Finland, in the Gulf of Riga (mainly Pärnu Bay), and in the Gulf of Gdansk and the Szczecin Bay area (Ojaveer 2014). Pikeperch has a number of ecological varieties: brackish-water, lake and semi-migratory forms; the type of development of oocytes in these forms can be different. The species has a weak sexual dimorphism – during the breeding season, most of the male fish have bluish-grey marble-patterned coloration around their genital opening and belly area (Erm 1981). As a rule, the belly of females is white. Male pikeperch attains sexual maturity at the age of 3–4, females at 5–6 years. The fecundity of larger (over 60 cm) females may be more than 1 mill. eggs. Spawning starts at 12–15 °C and finishes at 20–21 °C. Pikeperch prefers hallow (2–4 m) places with sand or gravelly bottom for spawning. Males guard spawned eggs. In Pärnu Bay, spawning takes place at the end of May and in June, in the Gulf of Finland, in June and early July.

Important preconditions for the development of abundant year class are a uniformly warm reproduction period and the subsequent summer months, as well as a rich occurrence of gobies, which are the main food item of young pikeperch. In the case in which the natural spawning substrate is absent, pikeperch may shed eggs on artificial spawning substrate (plants, coarse netting, etc.). The length of newly hatched larvae is 3.9–4.5 mm; they lack pigmentation. They start exogenous feeding 3–4 days after hatching. The first food animals are copepod nauplii and small cladocerans. A rather large variety of food animals is used by the larvae: *Eurytemora* spp., *Acartia* spp., *Podon* spp., in their young stages and other invertebrates of suitable size. At the age of 1 month, young pikeperch start catching fish larvae, including pikeperch larvae. 2–3 cm long pikeperch live mainly on goby larvae also consuming amphipods and mysids. Larger pikeperch prey on herring, smelt, gobies, bleak, sticklebacks, eelpout, perch, ruffe, roach and other cyprinids. Under good environmental conditions, pikeperch grow rapidly and reach 40-cm length in the fifth year of life. The $L\infty$ for the Pärnu Bay pikeperch is 83.8 cm. In this bay, 15-year-old pikeperches have been caught (Erm 1981).

The only predators dangerous to pikeperch older than 3 years are seals, cod and pike. The composition of parasites varies with area. A very dangerous parasite capable of annihilating pikeperch eggs (especially in thick egg layers) on the spawning ground are the hyphae of *Saprolegnia* spp. The main parasites for young and adult pikeperch are helminthes, but also crustaceans and infusorians.

Pikeperch is a very highly appreciated fishing object, both for professional and amateur fishermen. The highest catches are in Finland (the average annual landings during 2001–2012, 450 tonnes). In Finland, pikeperch landings compose, by size, about 11% of coastal catches (excluded herring), but by value, the percentage is 23%. Estonian coastal fishery landed, on average, about 100 tonnes of pikeperch a year during the period 2001–2012, caught chiefly in Pärnu Bay. Landings of pikeperch from Pärnu Bay were largest in the 1930s, at the beginning of the heavy exploitation of this stock (Fig. 3.51). Commercial pikeperch landings in the Swedish coastal zone constituted 65 tonnes in 1994, but decreased to about 30 tonnes in 2004 (Ask and Westerberg 2005). On the Lithuanian and Polish coasts, pikeperch is a common but not abundant species in catches. In the Szcecin Lagoon, its landings amounted to 200–300 tonnes in the 1980s, while in other coastal areas, they were lower and considerably fluctuated (Skora 1996). In the Darss-Zingst-Bodden area of estuaries, pikeperch is estimated to be the most important species for the fishery. The species has benefitted from the increase in biological productivity and improved conditions for hunting their prey. The abundance and landings of pikeperch in this area increased in the mid-1960s, and since then, the average annual landings varied by around 100 tonnes up to the1990s (Winkler 1996).

3.9.4.3 Pike *Esox lucius* L.

Pike is widely distributed in the fresh and brackish waters of Europe (except its southern part and the greater part of Norway), North Asia and North America. The species is common and also sporadically abundant in the coastal waters of the Baltic Sea. It is an important commercial fish, both at the eastern and western coasts of the Gulf of Bothnia, the Aland Sea and along the coasts of the Baltic Proper from the Stockholm Archipelago up to Hanö Bay, in the Gulfs of Finland and Riga, and in the Curonian Lagoon. In the early part of the twentieth century, pike was abundant in the Darss-Zingst shallow-water area in the Southern Baltic. Due to worsening of spawning and hiding possibilities in this part of the sea, the conditions for pike deteriorated, but after Winkler (1996), this species still constituted some percentage of commercial catches in this district during the late twentieth century. In the Polish lagoons, the importance of pike in landings drastically decreased in the 1970s, mainly because of considerable deterioration of environmental conditions (Skora 1996).

Males attain sexual maturity mainly at the age of 3–4, females at 3–5 years. Spawning of pike commonly starts at the time of ice breaking, at 4–6 °C. Pike spawns in flooded shallow places (up to 0.5 m), mainly on dead vegetation. Spawning usually lasts 3–4 weeks. Smaller pikes spawn somewhat earlier than bigger ones. The results of spawning depend on the water level on the spawning grounds (on the availability of acceptable spawning places). Therefore, in low-water springs, some part of the females cannot spawn. Pike cannot reproduce in salty water (sea). The eggs are rather big. The fecundity of the specimens over 70 cm in length can amount to 200,000. The embryonic period lasts 12–25 days, and the length of the larva at hatching is 6–9 mm. At the length of 13–17 mm, the larvae transfer to exogenous feeding, consuming insect larvae, invertebrates, and larvae of roach and other fish. Metamorphosis takes place at the length of 25–40 mm. After metamorphosis, pike live on fish of appropriate size. Pike is a useful predator, since its diet consists mostly of small and, to a large extent, economically undesirable fish.

Pike is a predatory fish of rapid growth rate. It has already reached a length of 50 cm at the age of 4–5 years. The main parasites for pike belong to the helminths, with infusorians and sporozoans also being important. The most prevalent parasite species are *Myxidium lieberkuehni* (in the kidneys and urinary bladder), *Henneguya psorospermica* (in the gills), *Diphyllobothrium latum* plerocercoids (freely in the viscera, rarely in the musculature), *Rhapidascarus acus* (in the intestines), *Azygia lucii* (in the stomach), a number of acantocephalans (in the intestine), the ectoparasite crustacean *Argulus foliaceus*, etc. The composition of parasites changes with the age of the pike. In the coastal sea, pikes frequently suffer from skin tumors (Ojaveer 2014).

Pike is an extremely popular object in amateur fishery. As with other fish species of limited abundance and sporadic occurrence, regular stock assessment has commonly not been performed. Pike stocks have usually been protected by national regulations (minimum legal length in catches, closed seasons and areas, limitation of catches).

3.9.5 Relict Fish Species

The relict species immigrated into the Baltic during the earlier stages of development of the sea. Cold-preferring relict species have populated the Baltic Sea since the first stages of its development. Some of them disappeared during the subsequent stages of development, while others were able to adapt to and withstand the changes in salinity, temperature and other conditions, as well as the pressure of later immigrants, abiding in the composition of the Baltic ecosystems up to the present. Quite few of them have increased their abundance to a level influencing ecosystems as basic members and also causing interest from an economic point of view. This has, however, been a rather specific case, as present conditions usually do not facilitate high abundance level of species which had their flourishing periods in an earlier time. The relict species of comparatively larger abundance and economic importance today are smelt and eelpout. For management of the stocks of these species, their condition has been assessed, and in some countries, national management regimes introduced. The economic interest towards lumpsucker, fourhorned sculpin and other relict species is less pronounced, but nevertheless, they cannot be ignored as important long-term members in the food chain of their area that also attract consumers today. They are caught and processed, including for human consumption.

3.9.5.1 Smelt *Osmerus eperlanus eperlanus* (L.)

Smelt populations of the Baltic Sea have been intensely studied since the middle of the recent century.

Smelt is a morphologically-variable polymorphic species. The European smelt (*Osmerus eperlanus eperlanus* L.) occurs in the coastal waters of the Atlantic Ocean from the south of Norway to NW Spain. In the Baltic Sea, smelt has survived as an Ice Age relict and has formed local populations differing from one another in morphometric features, growth rate, age of sexual maturation, etc. (Shpilev et al. 2005; Fig. 3.52).

Larger smelt populations occur in the Gulf of Bothnia, the eastern Gulf of Finland, the Gulf of Riga and the Curonian Lagoon, where well-aerated water of low temperature persists year round and no permanent halocline exists. Smelt also exists in the Gulf of Gdansk, Pomorze Bay and in the shallow Darss-Zingst area of low-salinity estuaries. Dwarfed freshwater smelt *Osmerus eperlanus eperlanus morpha spirinchus* Pallas occurs in lakes and other water bodies in the Baltic area and the Volga and Petchora River watershed areas. In the Gulf of Finland, two forms of smelt have been differentiated: the shallow-water smelt that lives permanently in the shallow coastal zone and the deepwater smelt that migrates into deeper areas after spawning. These two forms differ in their time of sexual maturation, length of life and growth rate. In the Curonian Lagoon, one smelt group migrates into the open sea after spawning and feeds there in the coastal zone. Another group lives permanently in the lagoon. It has been argued that the smelt living in the lagoon, in fact, represents the freshwater smelt and that the groups considered are morphologically different.

Reproduction The sexual maturation of smelt takes place at the age of 2–3 years, but some part of the population has already matured by the time they are 1-year-olds (Shpilev et al. 2005). Spawning shoals consist mainly of 3–6-year-old specimens. The fecundity varies widely and depends chiefly on body weight (in the Gulf of Riga, $F = 677\ W + 1250$, in which F denotes fecundity and W body weight). Smelt spawn in rivers, river estuaries and the freshened inner parts of gulfs. The beginning

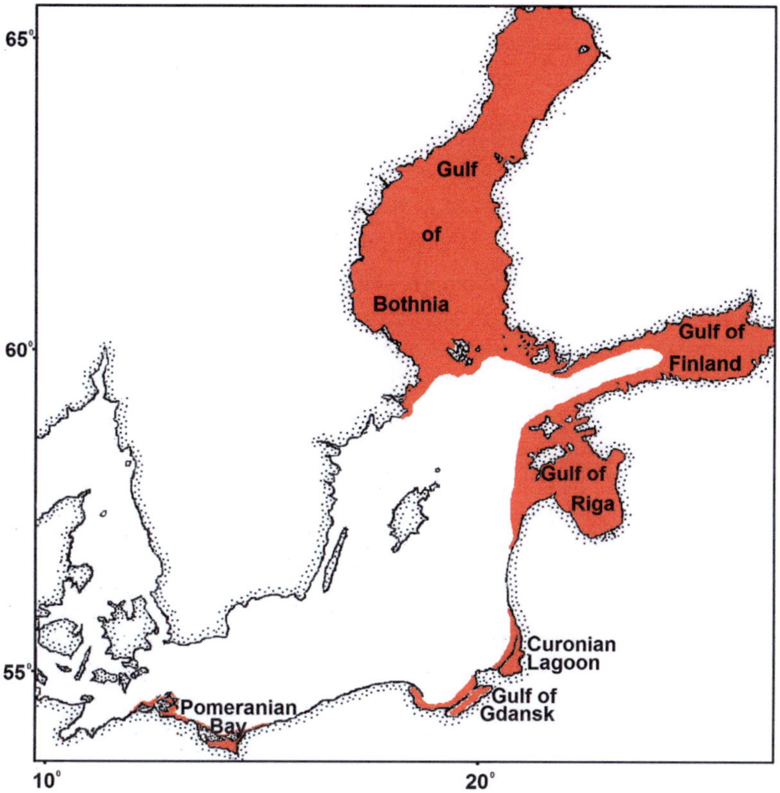

Fig. 3.52 Smelt populations in the Baltic Sea (Ojaveer 2014)

of spawning depends on water temperature. Maturing smelt usually form prespawning concentrations in November. At the end of November and the beginning of December, they start migration to the inner parts of bays, close to the spawning grounds where they winter. In spring, after the temperature rises to 1–2 °C, the smelt move to the spawning places. The duration of the spawning time of populations may last 1–6 weeks. Because of the dependence of spawning on temperature, in the Baltic Sea, the reproduction of smelt populations begins and finishes earlier in the southern regions.

In the Curonian Lagoon, smelt spawning starts in the second week of March and lasts usually until the first week of April. Temperature optimum for spawning is 2.5–5.6 °C. In Pärnu Bay, spawning starts at 1–2 °C, usually in late March or early April, in some years, below ice. Eggs are spawned at a salinity of 1–4 psu on the vegetation-free sand. Spawning intensity is highest at 3–6 °C. Spawning lasts up to the beginning of May and ends when the temperature in the spawning grounds has increased to 9–10 °C. In the Pärnu River, smelt spawn earlier than in Pärnu Bay. Spawning in the river is facilitated by strong currents and a high water level. In rivers, smelt spawn in rapids on sand or stony bottoms at a depth of 1–2 m.

In the Gulf of Finland, spawning starts at a temperature of 5–6 °C and lasts up to mid-June. For shallow-water-smelt, the optimum spawning temperature varies between 4 and 7 °C, for deepwater smelt, from 8 to 11 °C.

In Pärnu Bay, embryonic development of smelt takes place at temperatures from 1.2 to 11.4 °C and lasts 30–40 days. Rapid changes in the temperature reflect a significant drop in the survival rate. At hatching, the length of the embryo varies from 5 to 6 mm. The yolk sac resorbes in 7–9 days. The embryos keep to shallow water with sandy bottoms. In July, their length reaches 2–2.5 cm (Shpilev et al. 2005).

Feeding Smelt larvae live on copepods of early development stages. The common food of smelt smaller than 6 cm in length is *Eurytemora affinis*, *Acartia bifilosa* and *Bosmina coregoni maritima*. At the body length of 6–7 cm, smelt transfers to benthos-feeders. At this length, smelt mainly consumes mysids and amphipods. A very important food item in the Gulf of Finland is *Limnocalanus grimaldii*. Adult smelt also consumes fish eggs and fry, herrings, gobies, eelpouts, smelts and other small fish, and, in the Gulf of Riga, the alien cladoceran *Cercopagis pengoi* as well. The daily ration constitutes 0.82% on average of smelt's body weight in spring, 0.49% in summer, and 1.71% in autumn.

Growth and Age Smelt grows best in autumn, from September to November. In December, when temperature decreases, growth rate drops or growth ceases altogether. In the Baltic Sea, the growth from south to north decreases. The length at the end of the first year is 5–9 cm in the Gulf of Riga and 5.5–6.5 cm in the Gulf of Bothnia. The biggest smelt measured in the Gulf of Bothnia was 247 mm long and weighed 68 g, while in the Gulf of Finland, the longest smelt reached 40.0 cm; in the Gulf of Riga the corresponding figures were 31.3 cm and 195.9 g, and in the Vistula Firth, 27.0 cm and 153 g.

Investigations on long-term growth variations have shown that from 1969–2002 the growth of younger smelt accelerated. However, since the beginning of the 1990s, the average length and weight in older age groups decreased both in the Gulf of Riga and the Gulf of Finland. It is probable that these long-term changes are related to periodic fluctuations in the corresponding food chains induced by climate. In this period, the abundance of the eurybionts (*Mysis mixta*, *Neomysis integer*, *Eurytemora hirundoides*), which are the main food animals of young smelt, increased. Simultaneously, compared to the 1960s, the importance of the coldwater relict species *Limnocalanus grimaldii*, *Mysis relicta* and *Monoporeia affinis* in smelt's food significantly decreased. Abundance of *Pontoporeia femorata*, a valuable food item for older smelt, seriously decreased both in the environment and as a component in smelt's food. In the 1960s, it constituted 20–25% of smelt's food, but in the 1980s, the animal only occurred at depths below 30 m and its share as a component in smelt's food had diminished to 0.3%.

Enemies and Parasites Smelt eggs on spawning grounds are preyed upon by benthos-feeding fish – vimba bream, sticklebacks, etc., while their larvae are consumed by coastal fish. Adult and young smelt are a common food item for pike-perch, cod, perch, eelpout and a number of other fish. Smelt occurs mainly in gulfs, sounds and other near-coast areas, and therefore, it is included in the diet of the predators in these environments.

Smelt's parasites have been studied chiefly in the Gulfs of Finland and Riga. In these gulfs, a total of 38 parasite species for smelt have been found. 71% of the parasites belong to the helminthes. The most important of these are *Cystidicola farionis* (mainly in the air bladder), *Rhapidascaris acus* (in the inner organs), and *Hysteriothylacium aduncum* (larvae in the liver, imagoes in the intestines). These parasites occur in more than 15% of smelts, while other parasites are found less frequently. The occurrence of certain parasite species may only be considerably higher in certain smaller isolated areas. Ectoparasites belonging to crustaceans (e.g., *Caligus rapax* and *Lepeophtheirus salmonis*) can be found in one smelt in twenty or more, and this may cause symptoms related to illness. Ocurrence of the dangerous fungus disease caused by *Ichthyophonus hoferi* (in the liver, etc.) began in these gulfs in the early 1990s.

Abundance Dynamic There are large variations in smelt year class abundance, which mainly depends on the mortality rate in the embryonic and larval stages. It has been found that significant negative correlation exists between the water temperature during the embryonic development and the abundance of the year class. Smelt's local populations differ in their year class abundance. However, the periods of appearance of abundant year classes from the 1960s to the beginning of the 1970s and from the end of the 1970s to the middle of the 1980s, as well as the periods of moderate/weak year classes in the 1970s and from the late 1980s to the 1990s, are common for the smelt populations both in the Gulf of Riga and the Gulf of Finland. It is probable that the periods of abundant and weak year classes in the Gulfs of Riga and Finland have a general background of large-scale changes in the marine ecosystems under the impact of long-term fluctuations in climate.

Assessment and Management In the Baltic Sea, smelt is important in the food web, especially in the ecological subsystems of the Gulf of Riga, the Gulf of Finland, the Bothnian Sea and Bothnian Bay, as well as in the Curonian Lagoon. In these areas, rather abundant smelt stocks transfer the production of cold-water zooplankton and nektobenthos to salmon, sea trout and other top predators. Available smelt catch figures do not allow for correct long-term comparison of exploitation of the stocks throughout the Baltic Sea. However, it can be stated that the catches largely vary both in time and by sea area, probably due to differences in the abundance dynamics and exploitation rate of populations.

In the Gulfs of Bothnia, Finland and Riga, smelt is mainly fished as bycatch in herring and sprat trawl fishery. Commonly, in bycatches in trawl fishery and catches in other gear types, smelt occurs together with other fish (sticklebacks, roach, eelpout, lumpsucker, etc.) and the bycatches are not always landed. Therefore, the catch statistics probably do not reflect smelt catches correctly. The available catch statistics (Shpilev et al. 2005) show that the largest annual smelt landings occurred in the Gulf of Bothnia in the early 1990s (nearly 1500 tonnes), in the Gulf of Finland in the late 1960s (about 4000 tonnes), in the Gulf of Riga in the late 1960s (about 7000 tonnes), and in the Curonian Lagoon in 1975 (about 2000 tonnes). Catches depend on the abundance of year classes and landings on the economic situation. Smelt is not a generally accepted and sought-after food fish. Smelt landings are mostly used for animal fodder, production of fish meal, etc. But it has a stable circle of consumers (in Estonia, Russia, etc.) who appreciate smelt as having certain merits as human food. Therefore, smelt's resources are being investigated with the goal of their correct assessment and management. Hudd (1985) has calculated the total mortality of the smelt of the Gulf of Bothnia (Z = 0.70) and divided it into the natural mortality (M = 0.15) and fishing mortality (F = 0.55). In some countries, smelt is protected by national fishing regulations, including minimum allowed legal length in catches, etc.

3.9.5.2 Eelpout *Zoarces viviparus* (L.)

In the coastal waters of Northern Europe, eelpout is a rather common species. Its area reaches from the White Sea to the English Channel, the Irish Sea and the Shetland Islands. The species also occurs in the coastal zone of the Baltic Sea, mainly in stony environments in the vegetation zone, with older and larger specimens also appearing in deeper sandy and muddy grounds. The species is abundant in archipelagoes and the Gulfs of Riga, Finland and Bothnia. Distribution is influenced by temperature – during feeding periods, eelpout mainly populates deeper and colder (1–3 °C) water layers, while its numbers are still moderate below the 60 m isobath. During the restoration period, eelpout is found chiefly at salinities of 5–6 psu. Comparatively sedentary eelpout forms regional intraspecific groups in its area (Ojaveer and Lankov 1997). In various areas/age groups, the numerical importance of males and females in the population may differ. Based on differences in growth parameters, the appearance of sagittal otoliths with the size of annual increments (Fig. 3.53), number of vertebrae, reproduction pattern and feeding habits, eelpout of the Gulf of Riga has been divided into two phenotypically distinct and

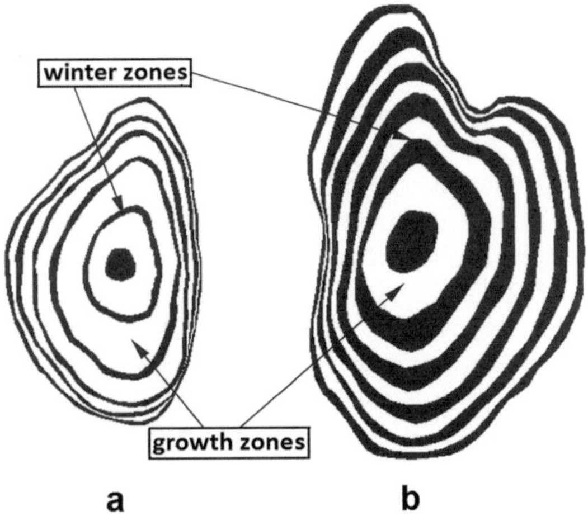

Fig. 3.53 Otoliths of eelpout from the Gulf of Riga: (**a**) deepwater type; (**b**) coastal type

spatially separated ecological groups: one is distributed in cold deep water below the seasonal thermocline (below the 20 m isobath) and the other in variable environments in the shallow coastal zone in the Pärnu River estuary.

Reproduction Sexual maturity is attained in the second year of life when the fish is commonly 13–15 cm long. Spawning shoals are mainly composed of 3–4-year-old fishes. Eelpout gives birth to live descendants. Fertilization takes place from July to September, and the embryos occur in the ovaries beginning in September

Young eelpouts of the length of 3.5–5.5 cm hatch from December to March. In connection with this event, the mother may undertake some migrations in the population area.

Feeding During its whole life, eelpout lives in the bottom water layers. Its food organisms have a nektobenthic or demersal way of life. In the Gulf of Riga, eelpouts of a length below 11 cm feed on benthic organisms in shallow water – *Monoporeia affinis, Saduria entomon* and amphipods. In older eelpouts, the number of possible food animals is larger. In shallower water, especially in the vicinity of estuaries, their food consists of *Saduria entomon, Macoma baltica, Mytilus edulis, Nereis diversicolor, Corophium volutator*, etc. In deeper water, the list of prey species is shorter, consisting mainly of Monoporeia, Pontoporeia, Saduria, etc. In the Gulf of Riga, eelpouts' stomachs constitute herring eggs, herrings, gobies, *Mytilus edulis*, mysids, polychaetes, etc.

The feeding activity of eelpout is highest in spring and early summer. In this time, its stomach mainly contains amphipods, *Nereis diversicolor, Saduria entomon, Macoma baltica*, etc. The choice of food animals decreases during summer and in autumn, when mainly only *Saduria entomon* remains. After September, feeding activity increases again only to drop to the minimum in December–January.

Growth and Age It has been found that at different depths, the length and weight growth of eelpout significantly differs. On this basis, corresponding biological groups of this species have been differentiated.

Parasites 30 parasite species have been found in eelpout in the Gulf of Riga. The core part of them is constituted by ten species of eumarine or brackish-water origin. From unicellular species, *Pleistophora typicalis* has infected over 10% of eelpouts. Helminths are comparatively frequent parasites of eelpout, too. For a number of parasites – *Diplostoma baeri, Contracaecum osculatum, Corynosoma strumosum, C. semerme, Hysteriothylacium aduncum, Rhapidascaris acus, Echinorchynchus gadi*, and *Pomphorchynchus laevis* – eelpout is the final or transitional host. Spores and granulomas of the parasitic fungus *Ichthyophonus hoferi* have been found in the liver, gonads and other eelpout organs. The most widespread eelpout parasite is *Hysteriothylacium* spp. which had infested 48–100% of the eelpouts investigated in the years 1978–1995. The main parasite for the eelpouts investigated in the southern Gulf of Finland in1983–1988 was *R. acus*, and in the early 1990s, *Corynosoma* spp. and *Contracaecum* spp.

In the eelpouts populating Kiel Bay, some parasitic taxa have been found (*Echinorchynchus gadi, Podocotyle atomon, Contracaecum aduncum, Contracaecum* spp., *Cryptocotyle* spp., *Ascarophis* spp. and larvae of *Rhapidascaris* spp.) with *varying* infestation rates. The main parasitizing period was early summer, when the parasites did not cause substantial harm to the host organism.

Indicator Species of Environmental Pollution Because of their comparatively sedentary way of life, rather medium-ranked abundance, acceptable size, wide distribution and length of life, the species is a suitable indicator for the assessment of results of environmental pollution. The characteristic method of reproduction of this species has a positive sense in the treatment of the species as an indicator for environmental pollution. Based on the reaction of eelpout, the impact of the aggregation of polycyclic aromatic hydrocarbons, dangerous compounds (PCBs), various concentrations of zinc, copper, cadmium, tin, etc., in various organs of eelpout (including embryos) and their influence upon its life processes have been estimated.

Assessment and Management Eelpout is a standard member of Baltic ecosystems from Bothnian Bay to the southern parts of the sea. Exploitable concentrations of this fish occur mainly in the gulfs. In the coastal areas throughout the Baltic Sea, the species is landed as bycatch. A comparatively long tradition in eelpout exploitation exists in the Gulf of Riga. The fish is highly appreciated by the inhabitants of Latvia. In the 1960s, the eelpout stock condition in the Gulf of Riga notably improved.

Intense eelpout trawl fishery during 1964–1978 resulted in high landings; during this period, eelpout constituted 54% of the catches of demersal fish in this gulf. The catches were composed of the 8–9 age groups, while in the NW and eastern parts of the gulf, 2–5-year-old eelpouts dominated, and at the slopes of banks in the central area of the gulf, older fish were in the majority. During the 1960–1970s, the total eelpout landings (Estonian and Latvian catches) continuously reached over 5000 tonnes, and in 1974, over 20,000 tonnes (including bycatch). The exploitation rate was too intense and resulted in very high mortality of the stock. The situation was complicated by a rapid worsening of the quality of marine environments at this time and the immigration into the Gulf of Riga of huge cod shoals, which fed intensely on all fish species in the gulf. Consequently, in addition to very high fishing mortality, eelpout's natural mortality also increased rapidly in this period. After stock collapse, fishery for eelpout in the Gulf of Riga was closed from 1980 to 1989. The stock therefore got the possibility of recovering. The fishery was reopened in 1990, and the possible catch was limited to a comparatively low quota.

3.9.5.3 Lumpsucker *Cyclopterus lumpus* L.

Lumpsucker is widely distributed in the Baltic Sea, but it also occurs in the North Sea and the eastern part of the Atlantic Ocean from the Kara Sea to the Gulf of Biscay, in the coastal zone of western Svalbard and Iceland, and also in the western part of the Atlantic Ocean. Lumpsucker is a slow demersal fish that spends the greater part of its time fastened to the substrate. It is probable that in the Baltic waters, lumpsucker represents an Ice Age relict population adapted to the Baltic Sea environment, having smaller body dimensions and markedly slower growth (Ojaveer 2014). In the NE Baltic, adult specimens can reach a length of 18 cm and a weight of 80–230 g, in the Southern Baltic, 24.4 cm and 345 g (Heese 1998). In the Transition Area and at the west coast of Sweden, the weight of lumpsuckers can be as high as 1–5 kg. The fish finds consumers in Denmark, Germany and Iceland, and its eggs are used for the production of artificial caviar (Muus and Dahlström 1966).

3.9.5.4 Fourhorned Sculpin *Triglopsis quadricornis* (L.)

The species has predominantly arctic distribution. It populates the coastal arctic waters of North America, Greenland and Eurasia up to 83°N. It was probably one of the first fish of marine origin to immigrate into the Baltic Sea basin after the establishment of a connection between the Baltic Ice Lake and the ocean and the origination of the Preboreal Yoldia Sea (Munthe 1956). Since then, it has populated the Baltic Sea and evidently survived its freshwater stage.

Presently, fourhorned sculpin occurs mainly in the Gulfs of Bothnia, Finland and Riga, as well as in the northern part of the open sea, preferably in the neighbourhood of gulfs. In the Gulf of Riga, it can be found in its largest numbers between the Irbe Strait and Ruhnu Island and on the slopes of the Ruhnu Deep. In the Gulf of Finland, it occurs mainly in the central and eastern parts, from Naissaar Island to Narva Bay, chiefly in the area of islands.

The area of the species reaches down to a depth of 60 m. The fish can be caught in bottom water layers where it seems to live rather sparsely. However, both in the gulfs and in the open Baltic, some specimens have been caught by pelagic trawl in the 50–65 m layer above greater depths. Different subspecies of fourhorned sculpin also dwell in some lakes in Scandinavia (Mälaren, Vänern, Vättern etc.), Finland (Saimaa, Päijänne) and Russia (Ladoga, Onega).

In the Baltic, fourhorned sculpin is confined to the area with a salinity below 6 psu and temperatures from 2 to 8 °C. Being rather stenothermic, it seldom occurs at temperatures exceeding 9–10 °C (Westin 1968a).

Fourhorned sculpin undertakes regular annual migrations. In April–June, when the temperature near the coast increases, it migrates into deep water below the thermocline, where it feeds on relict crustaceans (mainly *Monoporeia affinis*) from July to September. In October–December, after the cooling of surface layers and the establishment of the homothermium, it returns to shallower areas. The males dig nests from November onwards and guard developing eggs from January to March. The activity of the males is highest in late March, in the period of the hatching of the embryos. Fourhorned sculpin changes its activity pattern twice a year: from nocturnal to daytime in November and vice versa in spring (April), in accordance with changes in the light intensity.

Reproduction In general, fourhorned sculpin matures at the age of 5 years. In Swedish waters, the fecundity of fourhorned sculpin varies from 792 to 5900 eggs (Westin 1968b). The oocytes are 2.4–2.9 mm in diameter. Roe colour varies greatly, usually from yellow to bluish green. The males guard the roe after spawning. The length of the parental cycle can be up to 3 months. The larvae hatch in March and spend their first months in pelagic water layers.

Feeding The 4.5–10-cm long young fourhorned sculpins distribute in rather deep water, where they feed on *Monoporeia affinis*. The food of adult fourhorned sculpin in the Northern Baltic Proper, the Gulf of Finland, the Gulf of Riga and the Estonian Archipelago consists of *M. affinis*, *Saduria entomon*, *Pontoporeia femorata*, *Crangon*, *Harmothoe*, *Cardium*, *Lymnaea*, mysids, gammarids, herring, smelt, sticklebacks, sprat, eelpout, sand goby, fish eggs, etc. (Westin 1970; Ojaveer et al. 2003; etc.).

Age and Growth The growth of fourhorned sculpin varies considerably in rather small areas, which can be associated with a mosaic ecosystem structure and the fish's limited movements, which do not evidently involve long migrations. The oldest fourhorned sculpin found had 13 winter rings in its otolith. One-year-old fish caught in the Gulf of Riga had an average length of 10.6 cm and weighed 12 g, the 13-year-old fourhorned sculpin, 30.6 cm and 436 g, respectively. The largest specimen caught in the Gulf of Riga was 34 cm long and weighed 690 g (Ojaveer et al. 2003).

Enemies and Parasites Because of its large, spinned and tuberculated head, fourhorned sculpin can obviously only be preyed upon in the Baltic by large cod and seals. The fish is heavily infested with parasitic nematodes found in its stomach. According to Westin (1970), in Swedish waters, the infestation rate varied from 10% in November to 40–50% in summer. In addition to nematodes, fourhorned sculpin suffers from Protozoa, Cestodea, Trematoda and Acanthocephala. The main parasites are *Bothriocephalus scorpii*, *Contracaecum osculatum*, *Corynosoma* sp., *Rhapidascaris acus* and *Eubothrium crassum*.

Assessment and Management Abundance of fourhorned sculpin has considerably fluctuated. During the 1960s, its numbers clearly increased, probably thanks to rich food resources (fish injured in trawl fishery). At that time, up to 100 kg of fourhorned sculpin was caught per hour with experimental trawls in the central part of the Gulf of Riga. In 1977, the total catch of sculpins in the Gulf of Riga was 2040 tonnes, but in 1984, only one tonne was caught. However, in recent years, the abundance of the species has somewhat improved in the northern Baltic. The flesh and liver of fourhorned sculpin are acceptable as human food, but are not widely sought after. As unwanted bycatch in bottom trawls, the species is usually discarded.

3.10 Marine Birds

Birds, together with marine mammals and large fish, compose the ultimate link in the marine food chain. Generally, marine birds are more numerous in seas of high latitudes, where the water is colder and richer in food. These seas contain more potential food for birds: planktonic organisms and also the animals that feed on plankton. The Baltic Sea, with its diverse coasts and islands, is rich in birds. The majority of them perform their seasonal activity in their adapted Baltic Sea environments, and some may come to the Baltic area by chance, as strays. We cannot imagine our sea without picturing its many variable bird populations.

Bird species are related to the Baltic Sea through various ties and multiple corresponding backgrounds. The air space over the Baltic Sea has been particularly important for a much larger number of birds than actually breed or winter in the Baltic Sea area. Two tracks constituting part of the main Western Palearctic bird migration route between Eurasia and Africa cross the Baltic Sea. In autumn, birds migrate thousands of kilometres from the coasts of the Arctic Ocean/Northern Russia and Scandinavia, but also from the northernmost Baltic areas, over the islands of Saaremaa and Hiiumaa, the Gulf of Riga, Gotland, Öland, the large Polish estuaries, Schleswig-Holstein and Rügen southwards, heading to warmer wintering places. These may be situated considerably southwards of the Baltic Sea. In spring, they undertake a migration of the same duration, now in the opposite direction, to their breeding and feeding areas northwards. The migration of water birds mostly proceeds along water bodies or above a free surface of water to render possibilities for resting and feeding. The culmination of the spring migration takes place towards the end of May and the beginning of June. The migration of a number of species concentrates into a comparatively narrow passage in the area of the Gulf of Finland, where, on some days, hundreds of thousands of migrating birds can be seen. Every year, this journey is performed by over one and a half million waterfowl: dunlins, scoters, geese, ducks, divers, terns, mergansers, eiders, guillemots, etc. A number of species take resting breaks at the Baltic Sea during migration. The barnacle goose has resting places chiefly in northern Germany, on Gotland Island and in western Estonia. This species has now started nesting at the Baltic Sea.

Some 30 species of water birds breed in the area of the Baltic Sea. The birds in the Baltic Sea area can be divided into a number of ecological groups, for example, marine and freshwater birds. In nesters, i.e., the species that generally nest in sea areas, the marine component constitutes a smaller part. The majority of marine species prefer to be active in the areas of sea coasts. In the Baltic Sea area, notably in its northern and eastern regions, freshwater birds constitute an important part. These species chiefly nest at inland water bodies, but also populate coastal areas and islands of the Baltic Sea, especially shallower coves, thickets of reeds, rocky islets, etc. The species diversity of birds in the Baltic area is by no means low, as the shortage in marine species has been compensated for by abundant freshwater species.

A number of comparatively large areas of the Baltic Sea are widely known as important wintering places for waterfowl, including in the southwestern, southern and even central areas of the sea.

As a wintering area, the Baltic Sea can be compared with the North and Wadden Seas. Here, about nine million birds belonging to seagulls, ducks, eiders, scoters, divers, terns, guillemots, razorbills, etc., spend the winter. The importance of this area can be stressed by the fact that mild winters are spent here by a substantial part of a number of species of western Palearctic populations. These wintering places are chiefly situated on ice-free coastal areas with a depth up to 40 m. This is characteristic of the bottom-feeding birds, which constitute at least three-fourths of the number of species wintering at the Baltic Sea. The most important wintering areas are the Gulfs of Szczecin, Pomorze and Riga, the Danish Sounds, the west coasts of Saaremaa and Hiiumaa, Hoburgs Bank, Midsjö Bank, Odra Bank and other shallower areas in the Baltic Sea. The number of birds wintering in these areas depends on the severity of the winter and ice conditions. Ice formation in northern wintering areas induces birds to migrate from the northern to the southern wintering areas at the Baltic Sea. In milder winters, birds are distributed in the wintering areas of the sea more equally. Skov et al. (1997) estimated that in 1992 and 1993, in the Gulf of Riga, the Gulf of Pomorze, the Odra Bank, the Hoburgs Bank and the Midsjö Bank, the consumption rates of bivalves by the black scoter, velvet scoter and long-tailed duck in kilograms per square kilometre per day had a considerable spatial variation: in the centre of the distribution area, it generally exceeded 30 kg, and the highest consumption rates were 100 kg. The total flux to seaducks on the Hoburgs Bank and the Midsjö Bank was estimated at 80 tonnes d^{-1}, in the Gulf of Pomorze, at 185 tonnes d^{-1}, and in the Gulf of Riga, at 234 tonnes d^{-1}. On the Hoburgs and Midsjö Banks, bivalves were mainly used within the depth range of 20–40 m.

The Baltic Sea area is an important reproduction area for over 30 bird species. In the composition of Baltic nesting birds, the number of freshwater species is rather high, and among the species of marine background, a number of relict endemic species can be found. In addition, within the limits of the sea, one can see significant differences in the distribution of nesting bird populations. In the southern part of the sea, nesting birds are concentrated into large populations in comparatively narrow areas, whereas northwards, they are distributed much more evenly. This phenomenon is probably due to the higher diversity of conditions in the northern coastal areas, including the isles.

At the end of the twentieth and the beginning of the twenty-first century, both increases and decreases have occurred in the abundance dynamics of bird populations in the Baltic area. The growth in the productivity of coastal waters following eutrophication, starting in the 1950s, improved nutrition possibilities for waterfowl, especially for those feeding on fish. Also, discards of unwanted fish catches have improved feeding conditions for piscivorous birds. However, the overloading of the marine ecosystem with human-produced compounds has also exerted pressure upon marine birds. Beginning in the 1950s–1960s, under the impact of high concentrations of DDE and other pollutants, eggshells of some bird species started to thin. Other problems related to reproduction have also cropped up. It has been found that the problems are mainly related to deviations in the hormonal system of birds, which manifest themselves in various fashions. The limitation of dangerous pollution during the subsequent decades generally weakened the anthropogenic pressure upon the environment (Fig. 3.54). This resulted in a decrease in mortality in birds and facilitated an increase in their abundance. However, a number of causes of mortality of aquatic birds are still topical, among them water pollution with petroleum (including large oil spills), mortality in fishing gears, etc. (HELCOM 1996). The

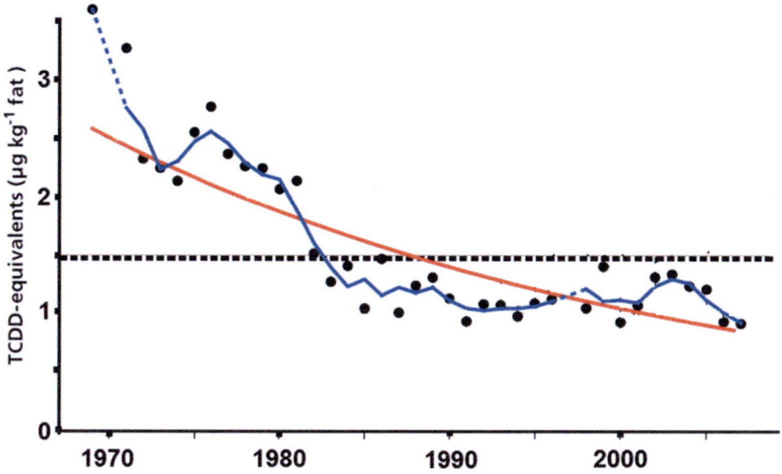

Fig. 3.54 Temporal trends of 2,3,7,8,-Tetrachlorodibenzo-p-dioxin (µg kg^{-1} fat) in common guillemot (*Uria aalge*) eggs from Stora Karlsö in the Western Gotland Basin. The horizontal line represents the geometrical mean, the *red line* is the trend line and the *blue line* the running mean smoother of the time series (HELCOM 2010)

effect of fishing for cod, flatfish and salmon with nets was estimated as being an important cause (about 10–20%) of the mortality of local wintering populations of velvet scoters, long-tailed ducks, common eiders, guillemots, etc., in the Gulf of Gdansk, the Kiel Bight and the Kattegat (HELCOM 2002, 2010).

Ecosystem-based management of sea areas and assessment of the impact of eutrophication, fisheries, climate change, etc., on the natural systems include birds. Problems in the development of reasonable management of some bird species need fast solutions. They are related to the problems concerning the following species.

Cormorant (*Phalacrocorax carbo sinensis*). The cormorant has two sub-species in the Baltic Sea area. *Phalacrocorax carbo carbo*, which generally nests in the area of the Atlantic Ocean, is often found in the Baltic area in winter. The other subspecies, *Phalacrocorax carbo sinensis*, breeds in the Baltic Sea area. At the end of the nineteenth century to the beginning of the twentieth century, the Baltic population of the species was hunted down to near extinction. During 1930–1950, the species returned to the Baltic coasts, beginning with its SW part. Starting in the 1980s, the area for cormorant has rapidly widened, and in the 1990s, its abundance doubled (HELCOM 1996). The notable increase in the abundance of cormorants coincided with the growth in the numbers of some fish species (e.g., cyprinids, which constitute the main food for cormorants) after the decrease in toxic pollution in the water of the Baltic Sea. Cormorants do not compete with fishermen, as their food is composed mainly of fish species classified as second-rate food fish of minor economic value. Nevertheless, piscivorous cormorants are considered to be a pest for fishermen and a nuisance for coastal inhabitants. In some countries specific cormorant management measures have been applied (HELCOM 2010).

White-tailed eagle (*Haliaëëtus albicilla*) feeds on fish. In the 1950s, the populations of the species (particularly their reproduction) throughout the Baltic Sea were strongly influenced by DDTs, PCBs and disturbances by man, including illegal hunting. After the DDTs and PCBs were banned, the reproductive success of the species improved. Today, the population has increased and has regained its old areas.

The barnacle goose (*Branta leucopsis*) has usually bred on coastal lowlands and valleys in eastern Greenland, Spitzbergen and Arctic Russia. To migrate from the southern areas there, it is necessary to fly over the Baltic Sea. Colonies of barnacle geese have been found at the Baltic Sea beginning in 1975, and their number has increased since then. The expansion of the breeding area of this species has been based on the increase in its numbers, probably due to the cessation of hunting in Denmark in 1954 and in all other EC countries in 1977 (required by the EU Birds Directive). In 1993, it was estimated that the total number of breeding pairs in Denmark, Sweden, Finland and Estonia together was about 2000.

Dunlin (*Calidris alpine*) is of circumpolar distribution and has a stable breeding population in northern temperate and Arctic latitudes. Like most wader species, typical and important coastal birds, dunlin has its main breeding sites in coastal habitats – beaches, islets, and coastal wetlands. These habitats have dramatically reduced, or their suitability has deteriorated due to various drainage projects in wetlands, disturbances related to tourism, the pollution of estuaries, increase in predators, etc. Therefore, the situation of the population around the Baltic Sea is either endangered or vulnerable.

Caspian tern (*Sterna caspia*) has breeding areas on small isolated sandy islands or cliffs in the outer parts of Swedish, Finnish and Estonian archipelagos and in the Ladoga area in Russia. Caspian terns face threats both in breeding areas (parasites, etc.) and wintering areas in West Africa (chiefly hunting) (HELCOM 1996).

Common eider (*Somateria mollissima*) has its main breeding sites at the coast. It has been estimated that between 1949 and 1985, the abundance of the population increased by about ten times. Probable reasons for this were an improvement of the feeding conditions resulting from eutrophication, the ban on egg collecting in all Baltic Sea states, and severer hunting restrictions for this species. The current trend in population abundance varies by area. In the Gulf of Bothnia, the abundance of common eider is increasing, but a mass mortality of young eiders took place in the Kattegat in the 1990s and in the Northern Baltic in 1996. According to HELCOM (2007a, b) estimates, in the Baltic Sea, the number of animal species endangered by man is 61; specialists are of the opinion that this number includes 13 species of birds. To protect Baltic fauna, including its rich bird populations, a number of protection areas with various regimes have been formed (Fig. 3.55). Endangered, vulnerable and threatened species are the main protected organisms there. Chief attention is directed at the small endangered local populations that are noticably decreasing. HELCOM (2010) has claimed that its aim is to decrease the mortality of bird species due to human activity to zero by 2015.

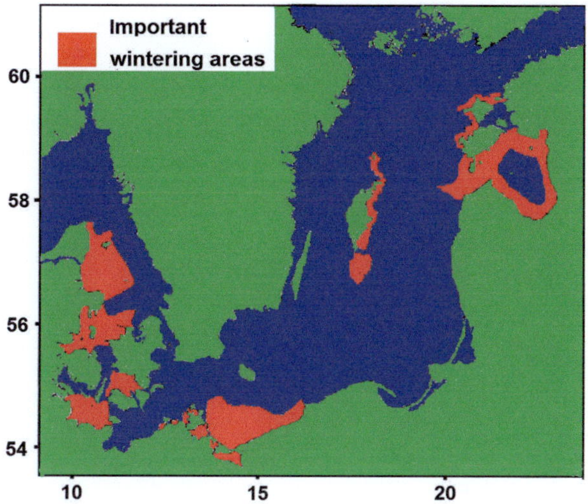

Fig. 3.55 Important wintering areas of marine birds in the Baltic Sea (HELCOM 2007a, b)

3.11 Marine Mammals

Marine mammals populating the Baltic Sea belong to two orders: the true seals (Phocidae) belong to the order Pinnipedia, whereas the family Delphinidae or dolphins belong to the order Cetacea.

It has been supposed that seals immigrated into the Baltic Sea at the latest during the beginning of the Litorina period. The first to come was possibly the ringed seal *Phoca hispida*. However, the grey seal (*Halichoerus grypus*) and the common seal (*Phoca vitulina*) also already lived in the Baltic Sea in the early Litorina period. Presently, they populate different areas of the sea (Fig. 3.56; HELCOM 1996; Harding and Härkönen 1999; Härkönen et al. 2013; etc.). The fourth seal species, the harp seal (*Phoca groenlandica*), immigrated into the Baltic in the Litorina period and was present for several millennia. In the present day, the species has disappeared from the Baltic, the causes being disputable.

The Ringed Seal *Phoca hispida botnica* (Gmelin) has been treated as a sub-species of the species *Phoca hispida hispida* of arctic distribution. No interspecific groups have been found in this sub-species in the Baltic. The body length of the adult ringed seal is up to 1.3–1.5 m and the weight generally 50–60 kg, in rare cases reaching up to or even more than 90 kg. The back and sides of the animal are dark brown, with elongated bright annular spots. Throughout the history of the area, the limit of the reproduction area of ringed seals in the Baltic Sea has shifted northwards. In the mid-Neolithic, the species reproduced in the Gotland area. Today, ringed seals are connected with the area of fast ice, where they have openings for breathing and passages in the snow. Therefore, it is important for them that the temporal duration of the ice period enable reproduction. Ringed seals live in the north-

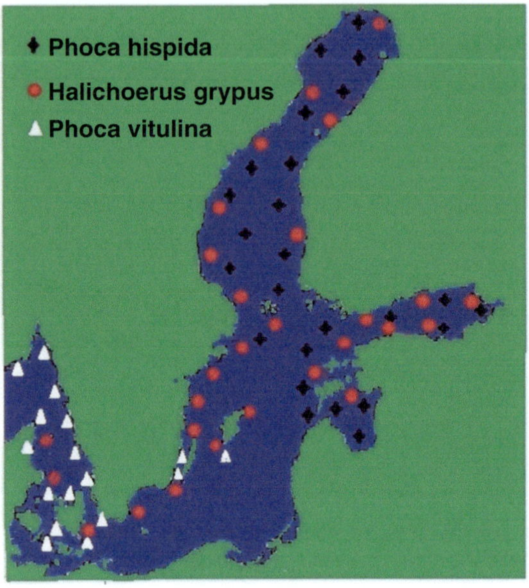

Fig. 3.56 Areas of seal populations in the Baltic Sea (Ojaveer 2014)

ern Baltic and have formed three reproductive populations: in the Gulfs of Bothnia, Finland and Riga. No reproductive population has been found southwards, where the occurrence of ringed seals is presently rather scarce. Generally, the species likes coastal areas. In autumn, ringed seals migrate into gulfs to find ice. In winter and spring, they live mostly in the Gulfs of Finland and Riga, mating, reproducing and shedding their hair. Ringed seals have 1–2 white long-haired pups, already capable of swimming some days after their birth. After ice break-up, ringed seals leave the coastal waters, and in summer, they live at a certain distance from the coast. The greater part of a ringed seal population is composed of 12–15-year-old animals; the age of the oldest specimens is over 25 and can even reach 35 years (Harding and Härkönen 1999). The bulk of the food of ringed seals in the Gulf of Finland is composed of herring, three-spined sticklebacks and smelt. The remaining part includes eelpout, sprat, fourhorned sculpin, river lamprey and the benthic isopod *Saduria entomon*. The average daily food requirement is estimated at 1.7 kg per animal. Härkönen et al. (2013) estimated the ringed seal subpopulation in Bothnian Bay at 6038 animals, and the population is increasing. However, the subpopulations in the Gulf of Finland (<50 animals) and the Archipelago Sea (150–300 animals) are declining, and the abundance in the Gulf of Riga (1400 animals) shows no indications of increasing either.

The Grey Seal *Halichoerus grypus* is presently the most abundant seal species in the Baltic Sea. This species is the largest seal in the Baltic. The length of an adult male reaches 2.5–2.7 m, in some cases, up to 3 m, and its weight is up to 250–300 kg. Females are smaller and weigh 180–250 kg. The back of the males is a dull light brown with dark spots, while females are brighter. Genetic studies have shown that in the Baltic Sea, we have to make do with an isolated population of the grey seal. The population is distributed in the central and northern parts of the sea; the bulk of the species populates the area north of the 58°N. The usual living area of grey seals is high seas around islands and islets. The species arrives in the Gulf of Riga in late autumn, in the time when the gulf freezes. It spends the winter in the vicinity of fast ice on the broken ice. In February–March, the female has one or two pups, which start swimming at the age of about 1 month.

In historic and also prehistoric times, reproduction areas of the grey seal were probably situated in the southern Baltic, the Danish Sounds and the Kattegat. The relevant sub-fossil findings are relatively scarce in the northern Baltic. Therefore, it is supposed that the area of the grey seal has shifted northwards in the Baltic. During this process, the population had to adapt to bringing forth young ones on ice, which may have taken time. Nowadays, the grey seal is regarded as a typical animal reproducing on ice (Jüssi et al. 2008). However, in the 1990s, grey seals were observed bringing forth their pups on the north coast of the Gulf of Riga, on skerries and islets in the Stockholm Archipelago, and in SW Finland. Reproduction of grey seals at coasts has even been mentioned in ethnographical sources, thus this is not a new feature in the biology of this species. Commonly, reproduction of the species takes place on drifting ice in the central and northern Baltic. Reproduction on ice is accompanied by a clearly smaller mortality of the offspring of grey seals than is the bringing forth of their young on the coast. After cold winters (in 1952, 1969, 1971, etc.), a number of reproduction communities were found in the Irbe Sound and the Saaremaa area. When the ice conditions are poor, the lounges are preferably situated at stony coasts. About a third of the food (by weight) of the grey seals is composed of herring, about the same amount is constituted of cod, river lamprey and eelpout, and the remaining part consists of perch, flounder, sprat, roach, white bream, etc. The average daily food requirement has been estimated at 3.2 kg fish.

Härkönen et al. (2013) estimated that the abundance of grey seals in the Baltic Sea is close to 28,000 animals (in the Mid-Swedish Archipelago, 10,224, in the Archipelago and Åland Sea, 8285, in the West Estonian Archipelago, 3385 animals, etc.) and both its distribution and abundance is increasing.

The Harbour Seal *Phoca vitulina* is a stocky, 1.6–1.8 m-long yellowish grey marine animal weighing 60–80 (rarely up to 115) kg. It lives near the coast and comes right up to the coast from time to time to rest. It breeds at the beginning of summer; the female usually has only one pup, which is able to swim. The offspring of the harbour seal depend on water temperature (the lowest limiting value is 3 °C). This may be one of the factors limiting the area of the species in the Baltic Sea.

A relatively large harbour seal population can be found in the Kattegat and the Belt Sea – according to estimates by Härkönen et al. (2013), it constitutes about 8500 specimens. Simultaneously, in the Danish and Swedish marine areas in the neighbourhood, only about 800 animals have been registered. Archeo-osteological data indicate that the harbour seal has never belonged to the fauna of the northern part of the Baltic Sea.

Recently, the harbour seal populations suffered from a viral disease, which caused their heavy mortality in the Baltic. Afterwards, the abundance of harbour seals has increased both in the Kattegat and the SW part of the Baltic Sea. The northernmost harbour seal population presently lives in the Kalmar Sound situated between the Swedish mainland and Öland Island. The small Kalmar Sound population is genetically different from the far more abundant Kattegat population. Like other seals, the harbour seals feed on fish; their average daily food requirement is estimated at 2.1 kg.

In the Baltic Sea, only one cetacean species – the **harbour porpoise** (*Phocoena phocoena*) – lives continuously and reproduces. The habitat of the harbour porpoise is near the coast, where it enters into gulfs and bays and also into the estuaries of large rivers. It follows ships and plays around them. Harbour porpoise feeds on herring, salmon and other fish, but its relations with fishermen are not tense. It has its pup (of a length of 60–80 cm and weight of 6–8 kg) in spring. The harbour porpoise is distributed mainly in the southern part of the sea, where the water is ice-free year-round. In severe, cold and icy environments, harbour porpoises can perish. Solitary harbour porpoises have been found as strays in the Gulf of Finland up to Kunda and Narva Bay, in the Gulf of Riga, up to Ruhnu, and in the Gulf of Bothnia, up to Vaasa.

Härkönen et al. (2013) estimate that the abundance of the harbour porpoise populations in the Baltic Proper is decreasing (23,000 animals in 2005) and that the subpopulation in the Baltic Proper is critically endangered. As with seals, the health of harbour porpoises is endangered by water pollution, they can be entangled in fishing gear, and they are also disturbed by human activity. Protection of harbour porpoises is executed based on the 'Agreement on the Conservation of Small Cetaceans in the Baltic, North East Atlantic, Irish and North Seas' (ASCOBANS). An Action Plan for the recovery of the harbour porpoise has been developed. Bycatch in the commercial fisheries, although very low, is identified as a prime obstacle to recovery, and hence the main management initiative is to reduce the bycatch in the fisheries inter alia through the introduction of acoustic transponders. Driftnetting was banned in the Baltic Sea from 2008, but other passive gears are still in use, and pingers (acoustic deterrents) have been introduced in some fisheries when fishing with passive gears. Especially in low-density areas, Marine Protected Areas do not have the potential for significant conservation benefits.

At the beginning of the nineteenth century, the ecosystems of the Baltic Sea were rather close to the natural situation. The intensity of exploitation of natural resources was generally low and the changes in the environment induced by man were observable mainly in the coastal zone and in the estuaries of larger rivers. Marine mammals exploited fish resources together with man, who caught fish and hunted seals, but did not substantially influence the marine environment. According to rough indirect estimates (Harding and Härkönen 1999; HELCOM 2002; ICES 2007), around about 1900 in the Baltic Sea (together with the Danish Sounds), there were approximately 100,000 grey seals and nearly 200,000 ringed seals, while in the Baltic Proper, the number of harbour seals was approximately five thousand. Naturally, a considerable fraction of the fish production in the Baltic Sea was consumed to feed such large mammal populations. In addition to competition in the exploitation of fish resources, seals damage fishing gear. Therefore, seals were considered to be the enemies of fishermen. Seal bounties were paid during the periods 1889–1927 and 1941–1977 in Denmark, 1903–1967 in Sweden, and 1909–1918 and 1924–1975 in Finland. Sealing remained one of the cornerstones of the local economies of coastal settlements in the north, as during the period from 1300 AD to 1800 AD, the Baltic was the most important producer of seal oil in Europe (Harding and Härkönen, 1999).

References

Adjers K, Appelberg M, Eschbaum R, Lappalainen A, Minde A, Repecka R, Thoresson G (2006) Trends in coastal fish stocks of the Baltic Sea. Boreal Environ Res 11:13–25

Altukhov KA (1959) Materialy po biologii salaki Litovskogo poberežja Baltijskogo morya (Materials on biology of herring of the Lithuanian Baltic coast). Trudy AN Litovskoi SSR, Ser B 2(18):197–224

Aneer G (1989) Herring (*Clupea harengus* L.) spawning and spawning ground characteristics in the Baltic Sea. Fish Res 8:169–195

Apine SO (1984) Tšhislennost i produktsiya bakterioplanktona v Rižskom zalive (The abundance and productivity of bacterioplankton in the Gulf of Riga), Hydrobiology of the Gulf of Riga. Zinatne, Riga, pp 28–39

Arntz WE (1974) A contribution to the feeding oecology of juvenile cod (*Gadus morhua* L.) in the Western Baltic. Rapp P-v Réun Cons Int Explor Mer 166:13–19

Aro E (1989) A review of fish migration patterns in the Baltic. Rapp P-v Réun Cons Int Explor Mer 190:72–96

Aro E (2000) The spatial and temporal distribution patterns of cod (*Gadus morhua callarias* L.) in the Baltic Sea and their dependence on environmental variability – implications for fishery management. Dissertation, University of Helsinki

Ask L, Westerberg H (eds) (2005) Fiskbestånd och miljö i hav och sötvatten. AB Danagards Grafiska, Odeshog

Bagge O, Steffensen E (1989) Stock identification of demersal fish in the Baltic. Rapp P-v Réun Cons Int Explor Mer 190:3–16

Bagge O, Thurow F, Steffensen E, Bay J (1994) The Baltic cod. Dana 10:1–28

Bartel R (1993) Anadromous fishes in Poland. Bull Sea Fish Inst 1(128):3–15

Bekkevold D, Gross R, Arula T, Helyar S, Ojaveer H (2015) Outlier loci detect infraspecific biodiversity amongst spring and autumn spawning herring across local scales. PLoS One 11(4):e0148499

Berner M (1968) Einige orientierende Untersuchungen an der Otolithen der Dorsches (*Gadus morhua* L.) aus verschiedenen Regionen der Ostsee. Fisch Forsch 6(1):73–76

Berner M, Vaske BB (1985) Morphometric and meristic characters of cod stock in the Baltic Sea. ICES statutory meeting, paper CM 1985/J:11

Biester E (1979) Der Frühjahrshering Rügens – seine Rolle in der Fischerei der Ostsee und in den Übergangsgebieten zur Nordsee, Rostock

Bizarks J, Abersons K (2011) Anthropogenic influence on the dynamics of the river lamprey *Lampetra fluviatilis* landings in the River Daugava basin. Sci J Riga Techn Univ Environ Clim Technol 7:32–38

Blaxter JHS (1958) The racial problem in herring from the viewpoint of recent physiological, evolutionary and genetic theory. Rapp P-v Réun Cons Int Explor Mer 143(2):10–19

Bylund G, Lönnström L-G (eds) (1994) Diseases and parasitess of flounder (*Pleuronectes flesus*) in the Baltic Sea, Baltic marine biologists publication, vol 15. Abo Academy, Turku

Chrzan F (1962) Pokarm i odżywianie się dorsza w Zatoce Gdańskiej. Prace Morsk Inst Ryb Gdyni t 11/A, Gdynia
Cieglewicz W, Posadzki K (1947) Comparative studies of the spring and autumn spawning herring of the Gulf of Danzig. Cons Perm Int Explor Mer Ann Biol 2(1942–1945):159–165
Demel K, Mulicki Z (1959) Studia ilosciowe nad fauna przydenna południowego Bałtyku. Prace Morsk Inst Ryb Gdyni t 10/A:19–33
Draganik B, Kuczyński J (1983) Garfish from the Puck Bay. ICES statutory meeting, paper CM 1983/J:3
Draganik BA, Yu M, Psuty-Lipska I (2005) The status of the turbot *Psetta maxima* (L.) stock supporting the Baltic fishery. Bull Sea Fish Inst 1(164):23–53
Ehnholm G (1951) Studier över strömmingen i Östra Kvarken. Merentutkimuslait Julk/ Havsforskninginst Skr 149:1–94
Elmgren R (1989) Man's impact on the ecosystem of the Baltic Sea: energy flows today and at the turn of the century. Ambio 18:326–332
Elwertowski J (1957) Szprot. Biologia, polowy, przetworstwo, Gdynia
Elwertowski J (1960) Biologische Grundlagen der Sprottenfischerei in der östlichen und mittleren Ostsee. Fisch Forsch 4:1–20
Erm V (1981) Koha. Pääsuke, Tallinn
Erm V, Rannak L, Sõrmus I, Štšukina I (1970) Väinamere kalastik. In: Kumari E (ed) Lääne- Eesti rannikualade loodus. Valgus, Tallinn, pp 31–62
Ezhova E, Źmudziński L, Maciejewska K (2005) Long-term trends in the macrozoobenthos of the Vistula Lagoon, Southeastern Baltic Sea. Species composition and biomass distribution. Bull Sea Fish Inst 1(164):55–73
Feistel R, Nausch G, Wasmund N (eds) (2008) State and evolution of the Baltic Sea, 1952–2005. Wiley, Hoboken
Florin A-B, Franzen F (2010) Spawning site fidelity in Baltic Sea turbot (*Psetta maxima*). Fish Res 102:207–213
Grauman GB (1969) Nekotorye dannye po razmnoženiju kiljki v južnoi tšasti Baltijskogo morya v 1954–1964 godakh (Some data on sprat reproduction in the Southern Baltic in the years 1954–1964). Trudy AtlantNIRO 21:140–150
Grauman GB (1980) Mnogoletniye izmeneniya tšislennosti ikry i litšinok šprota v Baltijskom more (Long-term variations in the abundance of sprat eggs and embryos in the Baltic Sea). Rybokhoz issled v bass Baltijskogo morya 15:138–154
Grauman G, Yula E (1989) The importance of abiotic and biotic factors in the early ontogenesis of cod and sprat. Rapp P-v Réun Cons Int Explor Mer 190:207–210
Grauman GB, Lisivnenko LN, Sidrevitš LL (1989) Nekotoryje aspekty pitaniya litšinok baltijskikh ryb (On some aspects in feeding of fish larvae in the Baltic Sea). Fisch Forsch 27:7–13
Hällfors G, Niemi Å (1974) A *Chrysochromulina* (Haptophycea) bloom under the ice in the Tvärminne archipelago, southern coast of Finland. Memoranda societatis pro Fauna et Flora Fennica 50:89–104
Halling F (1978) Autumn-spawning herring larvae and the abundance of copepods off the Aland islands in September–November 1977. ICES statutory meeting, paper CM 1978/J:20
Harding KC, Härkönen TJ (1999) Development in the Baltic grey seal (*Halichoerus grypus)* and ringed seal (*Phoca hispida*) populations during the 20th century. Ambio 28(7):619–627
Härkönen T, Galatius A, Bräeger S, Karlsson O, Ahola M (2013) In cooperation with the HELCOM SEAL expert group and the JASTARNIA group. HELCOM Core Indicator of Biodiversity. Population growth rate, abundance and distribution of marine mammals. HELCOM 2013 www.helcom.fi
Heese T (1998) Populations of non-commercial fishes in the coastal areas of the Southern Baltic Sea. Bull Sea Fish Inst (Gdynia) 3(145):21–39
Heincke F (1898) Naturgeschichte des Herings I. Die Localformen und die Wanderungen des Herings in den europäischen Meeren. Abhandl Dtsch Seefisch-Ver, Bd. 2, H. 1, 1–CXXXVI und 1–128

HELCOM (1996) Third periodic assessment of the state of the marine environment of the Baltic Sea 1989–1993. Background document. Baltic Sea environment proceedings, no 64B. Helsinki Commission, Helsinki

HELCOM (2002) Environment of the Baltic Sea area, 1994–1998. Baltic Sea environment proceedings, no 82B. Helsinki Commission, Helsinki

HELCOM (2007a) HELCOM thematic assessment in 2007. Climate change in the Baltic Sea area. Baltic Sea environment proceedings, no 111. Helsinki Commission, Helsinki

HELCOM (2007b) HELCOM Baltic Sea action plan. Baltic Sea environment proceedings. Helsinki Commission, Helsinki

HELCOM (2009) Eutrophication in the Baltic Sea. Baltic Sea Environ Proc 115 B:114

HELCOM (2010) Hazardous substances in the Baltic Sea – an integrated thematic assessment of hazardous substances in the Baltic Sea. Baltic Sea environment proceedings, 120B. Helsinki Commission, Helsinki

HELCOM (2012a) Checklist of Baltic Sea macro-species. Baltic Sea environment proceedings, no 130. Helsinki Commission, Helsinki

HELCOM (2012b) Indicator-based assessment of coastal fish community status in the Baltic Sea 2005–2009. Baltic Sea environment proceedings, no 131. Helsinki Commission, Helsinki

Hernroth L, Ackefors H (1979) The zooplankton of the Baltic proper. Fish Bd of Sweden. Institute of Marine Research report, no 2

Hessle C (1925) The herrings along the Baltic coast of Sweden. Publ Circonst 89:1–57

Hoziosky SA, Shvetsov FG, Gradalev E (1989) Seasonal distribution, migration and mortality component dynamics in sprat of the eastern and southeastern Baltic. Rapp P-v Réun Cons Int Explor Mer 190:22–24

Hudd R (1985) Assessment of the smelt (*Osmerus eperlanus eperlanus* L.) stock in the Northern Quark, Gulf of Bothnia. Finn Fish Res 5:55–68

Hüssy K (2011) Review of western Baltic cod (*Gadus morhua*) recruitment dynamics. ICES J Mar Sci 68:1459–1471

ICES (1987) Report of the Working Group on Assessment of Pelagic stocks in the Baltic, ICES statutory meeting, paper CM 1987/Assess:20

ICES (1989) Report of the Working Group on Assessment of Pelagic Stocks in the Baltic, ICES Statutory Meeting, paper CM 1989/Assess: 14

ICES (1995) Report of the Study Group on assessment-related research activities relevant to Baltic fish resources, Riga, Latvia, 23 February – 1 March 1995. ICES statutory meeting, paper CM 1995/J:1

ICES (2000) The status of fisheries and related environment of northern seas. Nord 2000:10

ICES (2007) Report of the WG on Marine Mammals, 27–30 March, Vilm, Germany. ICES Statutory Meeting, paper CM 2007/ACE 03

ICES (2010) ICES report of the ICES Advisory Committee, 2010. Book 8, The Baltic Sea, International Council for the Exploration of the Sea

ICES (2011) ICES Advisory Committee, 2011. ICES Advice 2011. Book 8, The Baltic Sea, International Council for the Exploration of the Sea

ICES (2013) Report of the ICES Advisory Committee, Book 8, The Baltic Sea

ICES (2015a) ICES advice 2015. ICES, Copenhagen

ICES (2015b) Report of the Salmon and Trout Assessment Working Group (WGBAST), 23–31 March 2015, Rostock

International Baltic Sea Fishery Commission (IBSFC) (1975) Proceedings of the first session, October 28–31, 1974, Warsaw, Poland

Jamieson A, Otterlind G (1971) The use of cod blood protein polymorphism in the Belt Sea, the Sound and the Baltic Sea. Rapp P-v Réun Cons Int Explor Mer 165:55–59

Järvekülg A (1979) Donnaja fauna vostotšnoi tšasti Baltijskogo morya (Bottom fauna of the eastern part of the Baltic Sea). Valgus, Tallinn

Jensen AJC (1950) Amount and growth of herring fry in the Danish waters. Rep Dan Biol Stat Copenhagen 51:19–48

Jüssi M, Härkönen T, Helle E, Jüssi I (2008) Decreasing ice coverage will reduce the breeding success of Baltic grey seal (*Halichoerus grypus*) females. Ambio 37(2):80–85

Kääriä J, Eklund J, Hallikainen S, Kääria R, Rajasilta M, Ranta-aho K, Soikkeli M (1988) Effects of coastal eutrophication on the spawning grounds of the Baltic herring in the SW Archipelago of Finland. Kieler Meeresforsch Sonderh 6:348–356

Kaiser W, Schulz S (1976) Zur Ursache der zeitlichen und räumlichen Differenzen des Beginns der Phytoplanktonblüte in der Ostsee. Fisch-Forsch Rostock-Marienehe 1:77–81

Kändler R (1942) Über die Erneuerung der Heringbestände und des Wachstum des Frühjahrs- und Herbstheringe in der westlichen Ostsee. Monatsh Fisch, 1942 B. 10, H. 2:17–22

Kändler R (1951) Über das Laichen des Frühjahrsherings bei Rügen und die Häufigkeit der Brut des Herbstherings in der Beltsee und südlichen Ostsee. Kieler Meeresforsch B. 8, H. 2:145–163

Kändler R, Pirwitz W (1957) Über die Fruchtbarkeit der Plattfische im Nordsee-Ostsee-Raum. Kiel Meeresforsch 13:11–34

Karås P, Böhling P, Järv L, Lehtonen H, Makarchouk A, Ziliukas V (1997) Year-class strength patterns in Baltic perch and pikeperch populations – a basis for development of systems of prognosis. Bull Sea Fish Inst 2(141):3–17

Kautzky H (1988) Factors structuring phytobenthic communities in the Baltic Sea. Doctoral thesis, University of Stockholm

Khlebovich VV (1974) Krititšeskaja solyonostj biologitšeskikh protsessov (Critical salinity of biological processes). Nauka, Leningrad

Klinkhardt M (1984) Untersuchungen zur Embryonalphase des Laiches Rügenscher Frühjahrsheringe unter besonderen Berücksichtigung natürlicher Mortalitätsraten auf einen aufgewählten Laichplatz des Greifswalder Boddens. Dissertation, Wilhelm – Pieck – University, Rostock

Kosior M, Strzyzewska K (1979) Plodność dorszy i śledzi bałtyckich. Studia i Materiały, Seria B. Wydawnictwo Morsk Inst Ryb, Gdynia, pp 5–32

Köster FW, Möllmann C (2000) Trophodynamic control by clupeid predators on recruitment success in Baltic cod? ICES J Mar Sci 57(2000):310–323

Kostritškina EM, Laganovska PJ (1983) Zooplankton (Zooplankton). In: Davidan IN, Savtšuk OP, Sustavov OV (eds) Ekosystem and its components. Project Baltica, vol. 1, pp 170–184

Kuikka S, Vanhatalo J, Pulkkinen H, Mäntyniemi S, Corander J (2014) Experiences in Bayesian Inference in Baltic Salmon Management. Stat Sci 1:42–49

Kupffer C (1877) Über Laichen und Entwicklung des Herings in der westlichen Ostsee. Jahresber Commn Wiss Unters dtsch Meere in Kiel 1874, 1875, 1876, pp 24–35

Lang T (1988) *Hysteriothylacium aduncum* (*Nematoda*) in Baltic cod (*Gadus morhua* L.). ICES statutory meeting, paper CM 1989/J:20

Lankov A (1988) On the role of the sprat (*Sprattus sprattus balticus* Schn.) in the ecosystem of the Northern Baltic. ICES statutory meeting, paper CM 1988/J:12

Lehtonen H (1983) Scientific basis for fisheries management of vendace, *Coregonus albula* (L.) in the Bothnian Bay. Aquilo Ser Zool 22:77–82

Lehtonen H (2014) Actions taken to protect and restore the river lamprey in Finland. http://www.fdd-fsd.eu/eufish/docs/balticLehtonen-Fin.pdf. Visited 08.04.2014

Lehtonen H, Böhling P (1988) Management of the whitefish (*Coregonus lavaretus* L. *s.l.*) fishery in the Gulf of Bothnia. Finn Fish Res 9:373–387

Leppäkoski E (1975) Macrobenthic fauna as indicator of oceanization in the Southern Baltic. Merentutkimuslait Julk 239:280–288

Leppäkoski E (2001) From *INSECTA, PISCATORIBUS…NOXIA* to ecological models – history of brackish-water biology in Finland. Proc Estonian Acad Sci Biol Ecol 50/2: 66–84

Lindquist A (1971) Contribution to the knowledge of Baltic sprat (*Sprattus sprattus*). ICES statutory meeting, paper CM 1971/H:19

Line R, Sidrevitš L (1995) Zooplankton of the Gulf of Riga. In: Ojaveer E (ed) Ecosystem of the Gulf of Riga between 1920 and 1990. Estonian Academy Publishers, Tallinn, pp 175–186

Lumberg A, Ojaveer E (1991) On the environment and zooplankton dynamics in the Gulf of Finland in 1961–1990. Proc Estonian Acad Sci Ecol 3:131–140
Martin G (1999) Distribution of phytobenthos biomass in the Gulf of Riga (1984–1991). Hydrobiology 393:181–190
Martin G (2000) Phytobenthic communities of the Gulf of Riga and the Inner Sea of the West-Estonian Archipelago. Dissertation, University of Tartu, Tartu University Press
Mayr E (1942) Systematics and the origin of species. Columbia University Press, New York
Mellergaard S, Lang T (1999) Diseases and parasites of Baltic cod (*Gadus morhua*) from the Mecklenburg Bight to the Estonian coast. ICES J Mar Sci 56:164–168
Mikelsaar N (1957) Lesta sigimisbioloogiast Läänemere idaosas. ENSV TA Toim Biol 6:256–264
Mikelsaar N (1984) Eesti NSV kalad. Valgus, Tallinn
Miller A (1969) Balti mere kilu ööpäevasest toitumisrütmist. Eesti NSV TA Toim Biol 18:16–28
Mulicki Z (1947) Odżuwianie się storni (*Pleuronectes flesus*) w Zatoce Gdańskiej. Arch Hydrobiol Ryb 13:221–259
Munthe H (1956) On the development of the Baltic herring in the light of the late quaternary history of the Baltic. Ark Zool 9(10):333–341
Muus BJ, Dahlström P (1966) Merikalat ja kalastus. Helsinki
Neb K.-E (1970) Über die Heringe des Windebyer Noors. Ber dtsch Wiss Kommn Meeresforsch Bd. 21 H 1–4, pp 265–270
Nielsen R, Kristiansen A, Mathiesen L, Mathiesen H (eds) (1995). Distributional index of the benthic macroalgae of the Baltic Sea area. The Baltic marine biologists publication, no 18. Finnish Zoological and Botanical Publishing Board, Helsinki
Nielsen EE, Hensen MM, Ruzzante DE, Meldrup D, Grønkjaer P (2003) Evidence of a hybrid zone in Atlantic cod (*Gadus morhua*) in the Baltic and the Danish Belt Sea revealed by individual admixture analysis. Mol Ecol 6:1497–1508
Nissling A, Westin L (1997) Salinity requirements for successful spawning of Baltic and Belt Sea cod and the potential for cod stock interactions in the Baltic Sea. Mar Ecol Prog Ser 152:261–271
Nissling A, Westin L, Hjerne O (2002) Reproductive success in relation to salinity for three flatfish species, dab (*Limanda limanda*), plaice (*Pleuronectes platessa*), and flounder (*Pleuronectes flesus*), in the brackish water Baltic Sea. ICES J Mar Sci 59:93–108
Nissling A, Müller A, Hinrichsen HH (2003) Specific gravity and vertical distribution of sprat eggs in the Baltic Sea. J Fish Biol 63:280–299
Ojaveer E (1981) Influence of temperature, salinity and reproductive mixing of Baltic herring groups on its embryonal development. Rapp P-v Réun Cons Int Explor Mer 178:407–415
Ojaveer EA (1988) Baltijskie seljdi (Baltic herrings). Agropromizdat, Moskva
Ojaveer E (1989) Population structure of pelagic fishes in the Baltic. Rapp P-v Réun Cons Int Explor Mer 190:17–21
Ojaveer E (2002) The role of ICES in the assessment and management of resources in the Baltic Sea. ICES Mar Sci Symp 215:582–589
Ojaveer E (2014) Läänemeri. TA Kirjastus, Tallinn
Ojaveer E, Kalejs M (2010) Ecology and long-term forecasting of sprat (Sprattus sprattus balticus) stock in the Baltic Sea: a review. Rev Fish Biol Fisheries 20(2):203–217
Ojaveer H, Lankov A (1997) Adaptation of eelpout Zoarces viviparus (L.) to spatially changing environment on the coastal slope of the Gulf of Riga (Baltic Sea). ICES CM 1997: EE/3
Ojaveer E, Pihu E, Saat T (eds) (2003) Fishes of Estonia. Estonian Academy Publishers, Tallinn
Ojaveer H, Jaanus A, MacKenzie BR, Martin G, Olenin S, Radziejewska T, Telesh I, Zettler ML, Zaiko A (2010) Status of biodiversity in the Baltic Sea. PLoS One 5(9):e12467
Ojaveer E, Arula T, Lankov A, Shpilev H (2011) Impact of environmental deviations on the larval and year class abundances in the spring spawning herring (*Clupea harengus membras* L.) of the Gulf of Riga (Baltic Sea) in 1947–2004. Fish Res 107:159–168
Otterlind G (1962) Sillens/strömmingens vandringsvanor vid svenska syd- och ostkusten. Osteologie 34(1):15–21

Otterlind G (1967) Migration of plaice and flounder in the southern Baltic. ICES statutory meeting, paper CM 1967/F:34
Otterlind G (1976) Fish stocks and fish migration in the Baltic Sea environment. Ambio Spec Rep 4:89–101
Otterlind G (1985a) Cod migration and transplantation experiments in the Baltic. Zeitschr angewandte Ichthyol 1:3–16
Otterlind G (1985b) Notes on garfish in the Baltic. Meddelande från Havsfiskelaboratoriet, no 311. Havsfiskelaboratoriet, Lysekil
Parmanne R (1988) Silakan kasvu, morfologinen muuntelu ja vaellukset Itämeren pohjoisosassa. Dissertation, University of Helsinki
Parmanne R, Popov A, Raid T (1997) Fishery and biology of herring (*Clupea harengus* L.) in the Gulf of Finland: a review. Boreal Environ Res 2:217–227
Platpira V (1995) Heterotrophic bacteria. In: Ojaveer E (ed) Ecosystem of the Gulf of Riga between 1920 and 1990. Estonian Academy Publishers, Tallinn, pp 139–159
Plikshs M, Kalejs M, Grauman G (1993) The influence of environmental conditions and spawning stock size on the year class strength of the eastern Baltic cod stock. ICES statutory meeting, paper CM 1993/J:22
Pocwierz-Kotusa A, Kijewskaa C, Petereit C, Bernašc R, Więcaszek B, Arnyasie M, Liene S, Kente MP, Wennea R (2014) Genetic differentiation of brackish water populations of cod *Gadus morhua* in the Southern Baltic, inferred from genotyping using SNP arrays. Marine genomics, vol. 19, pp 17–22. doi:10.1016/j. margen.2014.05.010
Rajasilta M, Eklund J, Kääria J, Ranta-aho K (1989) The deposition and mortality of the eggs of the Baltic herring, *Clupea harengus membras* L., on different substrates in the south-west archipelago of Finland. J Fish Biol 34:417–427
Rannak L, Arman J and Kangur M (1983) Lõhe ja meriforell. Valgus, Tallinn
Rechlin O (1964) Untersuchungen zum Verlauf der Laichperiode des Herbstherings in der Mecklenburger Bucht im Jahre 1963. Fisch Forsch, Rostock – Marienehe 2(1): 25–29
Rechlin O (1967) Heringsfischerei und Heringsbestand in der Ostsee während der Jahre 1963 bis 1965. Fisch Forsch, Rostock – Marienehe 5(2):23–31
Rechlin O (1969) Beziehungen zwischen Larvenaufkommen beim Herbsthering und den Wasserumwälzungen in der Ostsee. Limnology (Berlin) 7(1):167–170
Remane A, Schlieper K (1958) Die Biologie des Brackwassers. Die Binnengewässer, Bd. 22. E. Schweizerbart, Stuttgart
Saat T, Tambets J, Kangur M (2003) Lampern, river lamprey, Lampetra fluviatilis. In: Ojaveer E, Pihu E, Saat T (eds) Fishes of Estonia. Tallinn, Estonian Academy Publishers, pp 48–52
Sandström O (1982) The plankton fauna of the Gulf of Bothnia. In: Müller K (ed) Coastal research in the Gulf of Bothnia. Dr. W. Junk Publications, The Hague, pp 173–196
Seletskaja AV (1970) Morfologitšeskaja kharakteristika kiljki Baltijskogo morya (Morphological characteristics of the Baltic sprat) Rybokhoz issled bass Baltiiskogo morya 5:76–92
Shpilev H, Ojaveer E, Lankov A (2005) Smelt (*Osmerus eperlanus* L.) in the Baltic Sea. Proc Estonian Acad Sci Biol Ecol 54(3):230–241
Sick K (1965) Haemoglobin polymorphism of cod in the Baltic and Danish Belt Sea. Hereditas 54:19–48
Sjöblom V, Parmanne R (1979) Abundance of sprat eggs and larvae in the Northern Baltic in 1978. ICES statutory meeting, paper CM 1979/J:20
Skora KE (1996) A comparison of changes in the composition of fish catches in the Polish lagoons in 1960–1989. Proceeding of Polish-Swedish symposium on Baltic coastal fisheries, 2–3 April 1996, Sea Fisheries Institute, Gdynia, Poland, pp 225–241
Skov H, Durick J, Andell P (1997) The ecological role of wintering seabirds in the Balti Sea. In: Andrushaitis A (ed) Proceedings of the 13th Baltic marine biology symposium, pp 221–225
Stankus S (2003) The peculiarities of turbot (*Psetta maxima* L.) biology and their role in the ecosystem of the Baltic Sea coastal zone of Lithuania. Acta Zool Lithuanica 13(2):217–237

Steemann Nielsen E (1975) Marine photosynthesis. Elsevier, Amsterdam/Oxford/New York

Strodtman S (1918) Weitere Untersuchungen über Ostseefische, III Bericht, Wiss Meeresunters, N. F. 14:33–95

Stroganov AN, Bleil M, Oeberst R, Winkler H, Semenova AV (2013) On differentiation of cod (*Gadus morhua*) groups in Baltic Sea. Russ J Genet 49(9):937–944

Strzyżewska K (1969) Studium porōwnawcze populacji śledzi trących sie u polskich wybrzeży Bałtyku. Prace Morsk Inst Ryback 15, Ser A, Gdynia, pp 211–277

Štšukina I (1970) Pitanie i migratsii retšnoi kambalõ (*Pleuronectes flesus trachurus* Duncker) v raione ostrova Hiiumaa (Feeding and migrations of flounder (*Pleuronectes flesus trachurus* Duncker) in Hiiumaa area. Trudy BaltNIIRH 4:361–376

Telesh IV, Alimov AF, Golubkov SM, Nikulina VN, Panov VE (1999) Response of aquatic communities to anthropogenic stress: a comparative study of Neva Bay and the eastern Gulf of Finland. Hydrobiology 393:95–105

Thoresson G (1996) The Swedish coastal fishery in the Baltic Sea. Bull Sea Fish Inst Gdynia 2(141):67–78

Thoresson G, Kangur M, Repećka R, Saat T, Vitinsh M (1996) Development of a resource assessment system for Baltic coastal fish stocks. Polish–Swedish symposium on Baltic coastal fisheries, October 1996, Sea Fisheries Institute, Gdynia, pp 283–292

Torn K, Krause-Jensen D, Martin G (2006) Present and past depth distribution of bladderwrac (*Fucus vesiculosus*) in the Baltic Sea. Aquat Bot 84(1):53–62

Trei T (1991) Taimed Länemere põhjal. Valgus, Tallinn

Tuunainen P, Ikonen E, Auvinen H (1980) Lampreys and lamprey fisheries in Finland. J Fish Res Board Can 37(11):1953–1959

Uzars D (1975) Peculiarities of feeding and quantitative food consumption of eastern Baltic cod. ICES statutory meeting, paper CM 1975/F:4

Uzars D-AV (1989) Feeding of Baltic cod (*Gadus morhua callarias* L.) and its role in the sea ecosystem. Rapp P-v Réun Cons Int Explor Mer 190:63–66

Uzars D, Plikshs M (2000) Cod (*Gadus morhua*) cannibalism in the Central Baltic: interannual variability and influence of recruit abundance and distribution. ICES J Mar Sci 57:324–329

Valle KJ (1934) Suomen kalat. Otava, Helsinki

Vallin L, Nissling A, Westin L (1999) Potential factors influencing reproductive succes of Baltic cod *Gadus morhua*: a review. Ambio 28(1):92–99

Voipio A (ed) (1981) The Baltic Sea. Elsevier, Amsterdam/Oxford/New York

Wasmund N, Siegel H (2008) Phytoplankton. In: Feistel R, Nausch G, Wasmund N (eds) State and evolution of the Baltic Sea, 1952–2005. Wiley, Hoboken, pp 441–481

Weber W (1971) Laichplätze des Herings (*Clupea harengus* L.) der westlichen Ostsee. Kieler Meeresforsch 27, H. 2:194–208

Westin L (1968a) Lethal limits of temperature for fourhorn sculpin *Myoxocephalus quadricornis* (L.). Inst Freshw Res Drottningholm Rep 48:71–76

Westin L (1968b) The fertility of fourhorn sculpin *Myoxocephalus quadricornis* (L.). Inst Freshw Res Drottningholm Rep 48:67–70

Westin L (1970) The food ecology and the annual food cycle in the Baltic population of fourhorn sculpin *Myoxocephalus quadricornis* (L.). Inst Freshw Res Drottningholm Rep 50:168–210

Winkler HM (1996) Peculiarities of fish communities in estuaries of the Baltic Sea; the Darss-Zingst-Bodden as an example. Limnology 26(2):199–206

Yurkovska VA (1990) Obštšaya tšislennost, biomassa i fiziologitšeskije gruppõ bakterii Baltiiskogo morya v osennii period (the total abundance, biomass and physiological groups of bacteria in the Baltic Sea in autumn). In: Andrushaitis GP, Laganovska PYu, Apine SO (eds) Plankton Baltiiskogo morya. Zinatne, Riga

Yurkovska VA, Apine SO, Platpira VP (1983) Mikroorganizmõ (Microorganisms). In: Davidan II, Savtšuk OP, Sustavov VP (eds) Ecosystem and its components, Project "Baltika", vol 1. Gidrometeoizdat, Leningrad, pp 157–170

Zalachowski W (1985) An attempt to apply Bajkov's formula to estimation of food uptake of cod in the Southern Baltic. Acta Ichthyol Piscat 15(1):3–19

Zettler ML, Schiedek D, Glockzin M (2008) Zoobenthos. In: Feistel R, Nausch G, Wasmund R (eds) State and evolution of the Baltic Sea, 1952–2005. Wiley, Hoboken, pp 517–541

Zuzarte F, Köster FW, Möllmann C, Voss R and Grönkjaer P (1996) Diet composition of cod larvae in the Bornholm Basin. ICES statutory meeting, paper CM 1996/J:19

Chapter 4
Recent Dynamics of the Environment and Biota

Abstract Of natural factors the most important for biota is water salinity. Dynamics of salinity corresponds to the dynamics of the freshwater discharge into the gulfs of Riga and of Finland with the time lag of 2 years. In the Gotland Deep in the water layers which can be inhabited by pelagic fish (O_2 lower level 2 cm^3 dm^{-3}) it was averagely 9.85 psu during 1902–1936, 10.65 in 1936–1980 and 9.36 in the period 1981–1998. Comparing different regions, in the period 1981–1998 the salinity in the main deeps was lower than in 1951–1980. In the Gotland and Northern deeps the salinity was about 6 psu lower than in the Bornholm Deep. Winter temperature conditions in 1981–1998, compared to the period 1951–1980, were the most constant in the Bornholm Deep area but the temperature was relatively notably higher in the Gotland area and especially in the Northern Deep. A clear regime shift in the ecosystems of the Baltic Sea from the domination of marine characters in the environment to the conditions favouring brackish-water organisms was probably induced by a shift from a low to rich river discharge into the gulfs of Riga and of Finland and a decrease in intrusions of saline water into the Baltic. The changes included a rather notable reorganization of the Baltic food chains and ecosystems which started in the 1980s. Beginning with the mid-1980s the abundance of the cod year classes decreased. The diminuation of the cod stock was supported with its high exploitation rate and fishing mortality. The amount of sprat consumed by cod declined. Increase in winter temperature improved reproduction and wintering conditions of sprat and the abundance of this species increased. The freshwater and brackish-water biota extended their area. Biomass of marine species (incl. valuable food organisms of fish) dropped. The resulting deterioration of the food composition, including in pelagic fishes, caused a decrease in their growth rate and weight which reflected also in catches.

Important changes occurred also in the anthropogenic influences. The Baltic Sea retained its clean-water character up to the beginning of the twentieth century. Environmental problems related to the economic development of human societies notably increased since the 1950s. Oxygen deficiency related to eutrophication, toxic pollution with metals and their organic compounds as well as with persistent organic compounds, oil and radioactive substances ever magnified. Impact of chemical warfare agents, intense marine fishery, transportation and tourism complicated the human impact. The tolerance of xenobiotic influences by the rather simple and

E. Ojaveer, *Ecosystems and Living Resources of the Baltic Sea*,
DOI 10.1007/978-3-319-53010-9_4

unstable ecosystems of the Baltic Sea, suffering steady salinity and thermic stress, is different compared to the relatively stable ecosystems in the seas of normal salinity.

4.1 Changes in the Environment and Biota Induced by Natural Conditions

In the natural state, the abundance, structure and other characters of biota are regulated by the physical, chemical and other abiotic parameters, but also by a number of biotic environmental factors. It has been stated that in the Baltic, both under the conditions of the relatively stable regime of the environment and under the extraordinary conditions of regime shifts (which have commonly been understood as environmental changes causing large, persistent, often abrupt alterations in ecosystem structure and functions), the basic natural regime-forming factors are salinity and temperature (cf. Kalejs and Ojaveer 1989). Oxygen conditions of water layers, pH, content of nutrient salts, etc., are also very important for the existence of organisms.

Below, recent changes in salinity, temperature and other important living conditions, which have been reflected in the abundance, distribution, growth and other population parameters of organisms in the Baltic Sea regions, are discussed.

4.1.1 Salinity

Commonly variations in salinity do not influence the condition of ecosystems instantly, but with a time lag. To study the actual impact of variations of salinity on the biota during the last century, it is insufficient only to account for the variations of salinity at the bottom layers of deeps. Fish and other higher organisms may not have the possibility of settling in the deepest layers in all sea areas because of oxygen deficiency (Fig. 2.23). To study the impact of salinity changes on the biota, we defined the changes in salinity in the water layer at the depth of the oxygen concentration 2 $cm^3\ dm^{-3}$, which is the threshold salinity rendering the distribution of Baltic pelagic fish species possible. During the period 1953–1996, the average depth of this water layer in the Gotland Deep was 91 m. We assume that, on average, the salinity at this depth has a substantial direct importance for the distribution, reproduction, feeding, abundance and other aspects of the existence of higher organisms whose distribution is limited by oxygen concentration. In this water layer, salinity displays periodic variations. In the temporal course, the average salinity in this water layer and the condition of fish stocks in the Baltic Sea allows for discrimination of the following periods (Fig. 4.1):

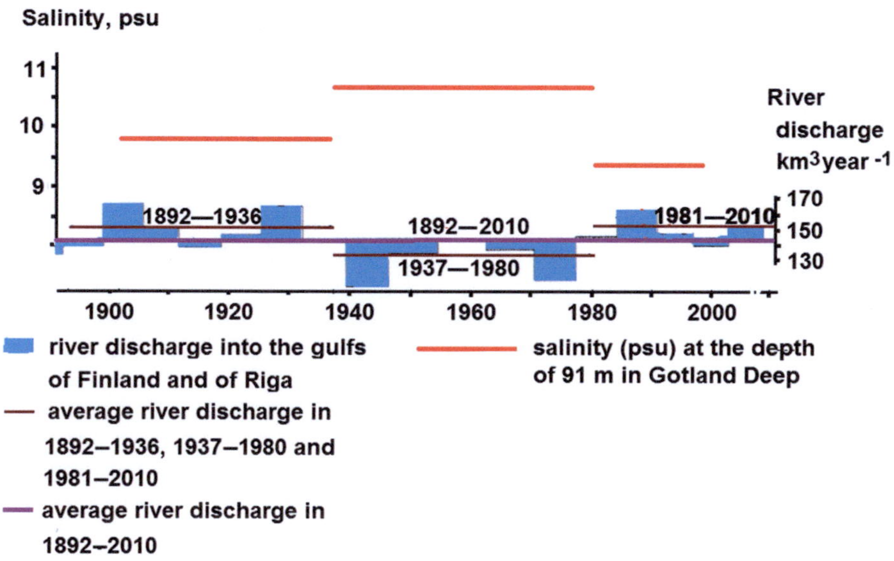

Fig. 4.1 Average river discharge into the Gulfs of Finland and Riga (km^3 a^{-1}) during the periods 1892–2010, 1892–1936, 1937–1980 and 1981–2010 and changes in the average salinity in the Gotland Deep at a depth of 91 m in 1902–1936, 1937–1980 and 1981–1998 (Ojaveer 2014)

1. In the period 1902–1936, the salinity at the depth of 91 m in the Gotland Deep was comparatively low – on average, 9.85 psu. The probable background of this rather low salinity is that the runoff of rivers of the Gulf of Finland and the Gulf of Riga catchment areas during the period 1890–1930 was substantially larger than the average long-term discharge. Such salinity of the ecosystem was reflected in the biota, including in the state of fish stocks. According to Eero et al. (2007), during that period, cod fishery was moderate in the Baltic, while Elwertowski (1957) estimated that sprat catches were good.
2. In the middle of the 1930s, the freshwater discharge into the Gulfs of Finland and Riga decreased. Water exchange between the North and Baltic Seas intensified and the salinity of the Baltic Sea increased. The salinity was rather high from the mid-1930s to the late 1970s (Matthäus 1993). In the period 1937–1980, the average salinity at the depth of 91 m in the Gotland Deep was 10.65 psu. Against the background of rising salinity, the abundance of cod increased beginning in the second half of the 1930s and remained at a good level up to the second half of the 1980s. At the beginning of that period, the resources of pelagic fish (herring and sprat) sharply decreased (Elwertowski 1957), but later improved again due to eutrophication of the marine environment and the resulting amelioration of feeding conditions of pelagic fish. Also, the abundance of some herring populations (notably gulf herring populations) declined (ICES 2010). Both sprat and herring populations were thinned, chiefly due to the mortality caused by the ever-growing cod population (ICES 2010).

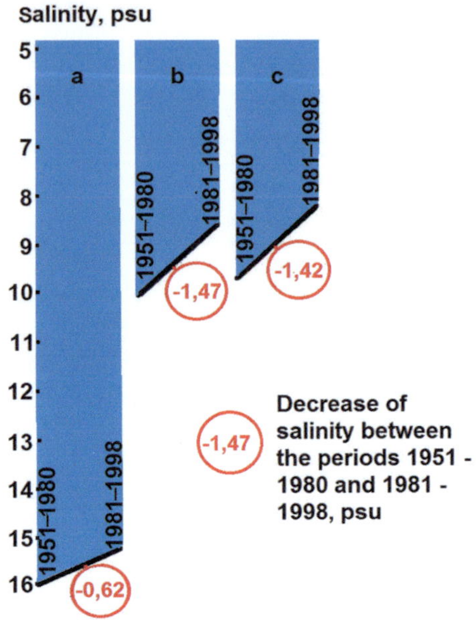

Fig. 4.2 Change in the average salinity at a depth of 85 m (the depth of the O_2 concentration of 2 cm^3 dm^{-3}) in the Bornholm (**a**), Gotland (**b**) and Northern (**c**) Deeps during 1951–1980 and 1981–1998 (Ojaveer 2014)

3. At the end of the 1970s, the period of scanty precipitation and low river discharge finished. After the strong inflow of the North Sea water in 1977, notable saltwater intrusions into the Baltic Sea discontinued for one and a half decades. In the period 1981–1998, the average salinity at the depth of 91 m in the Gotland Deep fell to 9.36 psu. Such a rapid salinity fall was accompanied by a drastic decrease in the abundance of year classes and landings of both the western and eastern populations of cod in the Baltic Sea. Moreover, the year classes and resources of marine herring in the central and southern Baltic also diminished. However, the resources of sprat and gulf herring increased (ICES 2011a).

It is noteworthy that variations in salinity in the central deep of the sea (Gotland Deep) did not influence environmental conditions in the Baltic uniformly: the natural regions were affected differently. To evaluate and compare the changes in the average salinity in different areas, in the Bornholm, Gotland and Northern deeps, we can use the data from the periods 1951–1980 and 1981–1998. The water layer with an oxygen concentration of 2 cm^3 dm^{-3} (Fig. 4.2) situated at about an equal depth (85 m) in the three deeps (cf. Fig. 2.23), was taken as the basis. The changes in salinity between the periods 1951–1980 and 1981–1998 at 85 m in the three deeps allow us to draw the following conclusions:

1. compared to the period 1951–1980, in 1981–1998, the average salinity was lower in all studied deeps;
2. the decrease in salinity was the most moderate in the Bornholm Basin. In this deep, the importance of the decrease had a much smaller ecological effect than in the other deeps. In this deep in both periods, the salinity was higher than

11 psu, below which reproduction of the Baltic cod is not possible (Aro 2000). Therefore, reproduction of cod populations was continuously possible in the area of the Bornholm Deep. In the Gotland and Northern deeps, the drop in the average salinity between the compared periods was almost equal. However, salinity remained under the 11 psu level during both periods. Differently from the SW region, salinity in the areas of the Gotland and Northern deeps commonly stayed below the level required for the reproduction of cod. In these regions, major salinity increases can only occur in the periods of low river discharge and intense saltwater inflows.

Naturally, salinity variation also affects the composition of biota outside the deeps, including in gulfs and the coastal zone. Therefore, with increasing or decreasing salinity, the environment and the character of the biota vary in all Baltic Sea regions.

4.1.2 Temperature

Temperature largely varying both temporally and spatially throughout the Baltic Sea plays a very important role in the life cycle of all organisms. The impact of temperature is most obvious in the organisms of southern background and is expressed most clearly in winter. To compare wintering conditions of marine organisms, one should estimate the thickness of the water layers with acceptable temperature and oxygen conditions for the species in the wintering period. From abundant fish species, sprat is the most sensitive towards low temperature. Usually, wintering layers of sprat are limited from above with the 3 °C temperature isoline. From below, the distribution of pelagic fish shoals is limited with the layer of oxygen concentration of 2 $cm^3\ dm^{-3}$. Comparison of the periods 1951–1980 and 1981–1998 (Fig. 4.3) shows that in the latter period, the thickness of the water layers suitable for sprat wintering increased in all deeps considered.

The comparison indicates that the conditions remained the most constant in the Bornholm Deep area. In this area, the average thickness of the water layer with favourable conditions for sprat wintering increased in the period 1981–1998 compared to the period 1951–1980, but only from 47 to 58 m or by 11 m. In the Gotland Deep, the increase was from 38 to 64 m or 26 m, and in the Northern Deep, from 26 m to 67 m or 41 m. This shows the change in the living conditions of biota in the compared regions between the periods 1951–1980 and 1981–1998, as well as their overall variations in separate basins. The change in the living conditions between these periods in the Northern Deep area (Fig. 4.3) indicates that in the period 1981–1998, the average wintering conditions were substantially more favourable than in 1951–1980.

Figures 4.2 and 4.3 characterize the Bornholm Deep area as the main region under the control of a marine climate in the Baltic Sea area. The variations in water temperature and oxygen content indicate that the environment in the areas of the

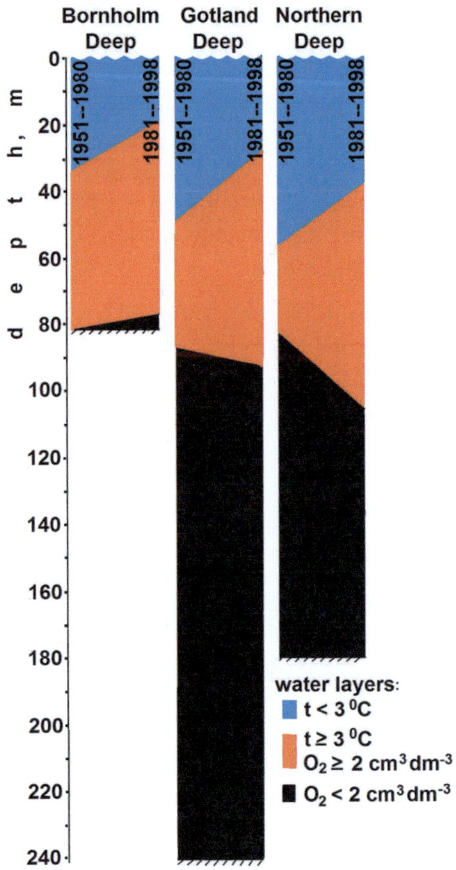

Fig. 4.3 Change in the average wintering volume of pelagic fish (the water layer limited from above with a temperature of 3 °C and from below the oxygen content 2 cm dm^{-3}) in the Bornholm, Gotland and Northern Deeps between the periods from 1951–1980 and 1981–1998 (Ojaveer 2014)

Gotland and Northern Deeps (the eastern and northwestern regions) is much more strongly influenced by continental climate and varies notably more than in the Bornholm area (southwestern region).

Climate Characteristics Salinity and temperature conditions affect biota, especially the distribution, reproduction and feeding of organisms, year-round. The impact of both these regime-forming environmental conditions fluctuates depending on climatic factors. In the Baltic Sea area, the limiting influence of temperature is particularly obvious in winter. The severity of winters in the Baltic area varies on a wide scale. The formation of the winter regime in the Baltic area should be considered in the system of the worldwide processes of the Arctic Oscillation.

It has been found that the climate in the Northern Atlantic – the European sector of the Arctic Oscillation – depends on the oscillation of the difference in the sea-level pressure between the Azores and Iceland, with the core pressures sensitively reacting to changes in the net heat fluxes between the ocean and the atmosphere. Corresponding deviations in air pressure are reflected in the winter conditions in the Northern Atlantic and also in the neighbouring areas of the Baltic Sea (Feistel et al. 2008).

From January to March, two clearly distinct climate situations can mainly be differentiated in this area: the continental type characterized by the flow of dry and chilly air masses from the Eurasian continent to the Baltic Sea space and the marine type characterized by the invasion of humid air masses from the Northern Atlantic under mild western winds and their domination. It has been shown that the character of weather in the Northern Atlantic and also in neighbouring areas of the Baltic Sea depends basically on the differences in air pressure on the sea level between the Icelandic low and Azores high pressure systems and the related direction of winds.

A positive North Atlantic Oscillation (NAO) index is associated with an anomalously low sea-level air pressure in Iceland, strong meridional pressure gradients over the North Atlantic, and intensified westerlies. Strengthened westerlies transport air masses of moderate temperature toward Western Europe. The negative NAO index represents a domination of the continental climate in the Baltic area and the reigning of severe wintering conditions for the biota in the Baltic Sea. This is associated with a reduced westerly air flow and the zone of westerlies displaced southwards. Associated quasi-rhythmic changes between the climate modes show periods of 6–8 years.

The climate in the Baltic Sea area is also influenced by the global fluctuations and by the impact of the continental climate type, the importance of which is increasing eastward. Therefore, to better reflect the situation in the Baltic Sea, Hagen and Feistel (2005) created a new climate index, which they called the Winter Baltic Climate Index (WIBIX). The index is based on winter (January–March) anomalies of air pressure difference between Gibraltar and Reykjavik to preserve the basic relations of the factors in the NAO, adding the sea level anomalies at Landsort to characterize the filling level of the Baltic Proper, as well as the data on the maximum Baltic ice cover to include continentally dominated alignments of atmospheric centres of action. The power spectrum of the WIBIX exhibits quasi-cycles of 2.2, 3, 6, 8 and 14 years, and suggests the existence of periods of about 40 and 100 years.

Regularities in the variations in environmental factors and their impact on the biota cannot be fully transferred from one macro-region of the Baltic Sea to another. In the Baltic Sea, the environment and biota vary from region to region and depend on the climate in the concrete region. The thermic situation in the large gulfs (Gulfs of Riga, Finland and Bothnia) differs significantly from the circumstances in the central and southern parts of the sea, as well as from the situation reigning in the area of the Azores and Iceland in the Atlantic Ocean. The type of atmospheric circulation (the marine type – basically westerly winds and a mild temperature regime from January to March; the continental type – mainly easterly winds and a severe thermic situation in winter) very substantially determines the life conditions of biota in winter. It has been proven that the abundance of year classes in the Gulf of Riga spring spawning herring in 1951–2004 significantly correlated with the monthly average air temperature during January–March in their home gulf, whilst their correlation with the average monthly NAO indices during this period was non-significant (Ojaveer et al. 2011).

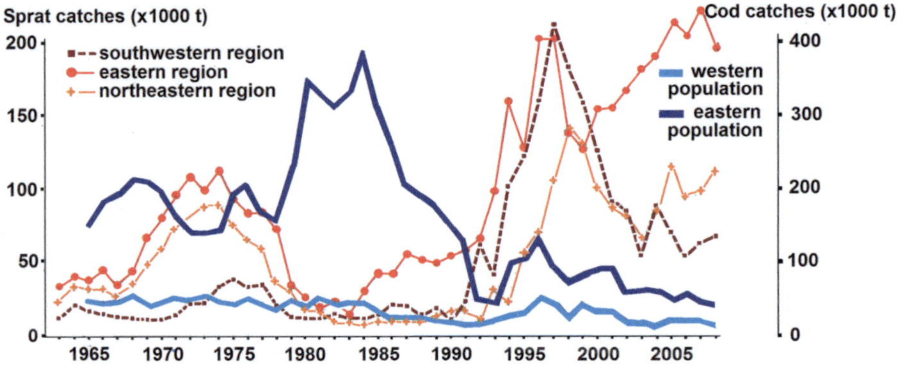

Fig. 4.4 Dynamics of cod and sprat catches under the cumulative effect of climate and interspecific relations (ICES data) (Ojaveer 2014)

The relations in the large gulfs with a substantial impact from the enormous Eurasian continent can basically differ from those in the Baltic Proper. Therefore, in studies on the dependence of biota on climatic processes in large gulfs, the relationships developed for other areas may not yield the expected results.

4.1.3 Interactions of Fish Species

In addition to the direct influence of the most important environmental factors, the Baltic Sea ecosystems are influenced by other, indirect and less noticeable but still substantial natural and complex (natural and anthropogenic) factors. Important problems occur because of interrelations between organisms in the ecosystems, especially concerning feeding relations. Also, contests for habitat or for the reproduction environment and the relations between parasites and hosts are continuously essential. Reproduction of fish is seriously affected by some microorganisms, in regard to both their excretions and the compounds forming at their decomposition after perishing. It has been known long ago that pelagic fish avoid assemblages of cyanobacteria and some other microalgae.

Simultaneous variations in salinity, temperature, oxygen and other conditions reflect on the relations between organisms and organism groups in ecosystems. This has caused great changes in the structure of biota, including exploitable resources. From the economic point of view, the yearly variations in the amount, quality and distribution of cod, sprat and herring resources are of direct and substantial importance. To understand the scales of the impacts, one should consider the catch dynamics of the most productive fish species of the open Baltic – cod, sprat and herring – shown in Figs. 3.15 and 4.4. These species constitute about 95% of the commercial catch of fish in the Baltic Sea. During the recent four decades, the biomass of cod (ICES Subdivisions 25–32), herring (ICES Subdivisions 25–29), and sprat (ICES Subdivisions 24–32) varied between 2.5 and 4 million tonnes. The biomass of other species is not known, but it is not likely to constitute a higher fraction of the biomass than of the commercial catch. Cod preys heavily on herring and sprat. Based on more than 43,000 analyses of cod stomachs from the central Baltic, it has been concluded that about a half of the diet of cod is composed of herring and sprat, which is more than that landed by the fishery. The cod cannibalism concerns almost exclusively 0- and 1-year- old cod. The magnitude of cannibalism suggests a rather strong self-regulatory mechanism of the cod stock. Salmon, trout, pike, pikeperch and perch are other piscivorous fishes in the Baltic. Herring, sprat, juvenile cod and other fish species are parts of their diet. Cod has the highest biomass of the predators in the Baltic Sea.

Rearrangements in the biotic part of ecosystems are initiated with changes in the most important abiotic factors. The decrease in the freshwater runoff, increase in the frequency of saltwater inflows from the North Sea, and the rise in salinity in the Baltic Sea that began in the second half of the 1930s changed the properties of the marine environment. The increase in the salinity and oxygen content in the bottom water layers has been of special importance. High salinity and good oxygen conditions in deeps are the most important preconditions for the formation of strong cod year classes. Moreover, an increase in salinity is also important for the improvement of living conditions of marine invertebrates (including valuable food items of young stages of cod and other fish, such as the copepod *Pseudocalanus minutus elongatus*, etc.) and the increase in their abundance and biomass. A notable eutrophication of the Baltic Sea started in the 1950s. The braking of this process (limitation of anthropogenic eutrophication by the HELCOM and other organizations) took several decades. During that time, eutrophication meant ever-growing biological productivity. This resulted in, among other things, a substantial augmentation of the abundance and biomass of pelagic fish resources. Also, the average individual weight of these fish increased. The situation especially favoured cod, mostly the eastern population of this species, as the feeding conditions particularly improved in the area of this population. Due to favourable reproduction conditions, large cod year classes hatched. Big shoals of young cod migrated from their reproduction areas in various directions, including into the Gulfs of Riga, Finland and Bothnia. Cod consumed an extra rich food resource: herring, sprat, common sand eel, gobies, perch, pikeperch, roach, bleak, smelt, eelpout, etc. The high biomass of pelagic and coastal fish that had formed under the conditions of rich biological productivity during eutrophication was able to sustain a very large cod stock. Particularly abundant cod year classes hatched from the middle of the 1970s to the early 1980s. This caused a deep depression of sprat stocks since the 1980s. Cod resources and landings increased to a maximum height (Fig. 4.4).

These developments took place over a little more than 40 years after the substantial decrease in the freshwater runoff of the rivers discharging into the Gulfs of Finland and Riga (the mid-1930s). Consequently, based on the presumable regular alternation of the abiotic regime-forming environmental conditions in the Baltic area after 42–43-year periods estimated by Hela (1966) and Hagen and Feistel (2005), the subsequent regime shift was expected in the Baltic. After a very large saltwater intrusion into the Baltic Sea in 1977, the abiotic regime-forming conditions altered and such inflows stopped for one and a half decades. Naturally, the Baltic biota reacted to the shift in the abiotic environmental conditions with a certain time lag. The shift from the domination of the marine biota to the conditions favouring brackish-water organisms took place some years later, mainly during the second half of the 1980s (HELCOM 2002; Alheit et al. 2005; ICES 2009, 2011b; Ojaveer and Kalejs 2012; etc.).

This well-documented, very clear regime shift in the Baltic Sea environment, ecosystems and fish stocks was probably induced by the shift from a low to a rich river discharge to the Gulfs of Finland and Riga (which corresponds to the shift from high to much lower water salinity in the area of cod spawning places), with the shift in the winter severity and other climate conditions. The shift created better conditions for the organisms of brackish-water background (Baltic gulf herring populations, invertebrates adapted to low salinity, etc.), but meant a clear deterioration of the living conditions for marine organisms requiring higher salinity (invertebrates of marine origin, cod, flatfish, etc.). The changes included a rather notable reorganization of the Baltic food chains and ecosystems, which started in the 1980s.

Abiotic processes (an increase in river discharge and decrease of intrusions of saline water into the Baltic) and the resulting deterioration of salinity and oxygen conditions in deeper layers brought about a notable fall in the abundance of cod year classes beginning in the mid-1980s. The decrease in the cod stock was supported by a very high exploitation rate and fishing mortality of the population (ICES 2011a; Figs. 3.44 and 4.4). Very good cod catches, which had been landed over a long period of time, decreased. Cod and other organisms requiring comparatively high salinity retreated to the southern and southwestern parts of the sea. The amount of sprat consumed by cod declined. An increase in temperature and improvement of other reproduction and wintering conditions of sprat also favoured a rise in its stock. The abundance and catches of all the three sprat populations increased after the disappearance of large cod shoals (Fig. 4.4). Due to the favourable hydrological regime, sprat year classes that had formed since the early 1990s were generally good or rich. Sprat populations and catches increased and the abundance of pelagic fish species magnified. The freshwater and brackish-water biota extended their area. Marine species (including valuable food organisms of fish) retreated towards the south-southwest and their abundance and biomass dropped. The resulting deterioration in the food composition of pelagic fish caused a decrease in the growth rate and also in the average individual length and weight of pelagic fish in catches. After a gradual restoration of cod populations that began in the mid-1990s, the sprat stock of the SW Baltic once again had higher natural mortality, which resulted in a decrease in its catches (Fig. 4.4).

4.2 Changes Caused by Anthropogenic Impacts

The formation of the Baltic Sea, consisting mainly of melting waters of glaciers, started in the Arctic climate more than ten thousand years ago. It is probable that up to the end of the nineteenth century, the anthropogenic pollution only influenced the sea near some larger towns of that time. The sea retained its clean-water character up to the beginning of the twentieth century. At that time, the Baltic Sea was a relatively oligotrophic water body. Its natural state was controlled by its self-purification capacity. This means that in the regulation of the situation in the system, the leading role belonged to the aquatic organisms, which accumulated organic compounds from water, mineralized them in their body, and facilitated sedimentation of the mist. Around 1940, the amount of nutrient salts discharged into the Baltic Sea was only somewhat larger than that at the beginning of the twentieth century. Environmental problems related to the economic development of human societies notably increased beginning in the 1950s (HELCOM 1996, 2002). From then onwards, exploitation of raw materials from nature, their processing, utilization and the produced wastes created large-scale changes in the Baltic ecosystems. Anthropogenic eutrophication, oxygen depletion, toxic pollution, overexploitation of living resources, etc., induced variations in the state of ecosystems and living resources. The periods and extent of variations in the ecosystems and resources for natural and anthropogenic causes are different. The effect of the anthropogenic impacts in the Baltic Sea differs from that in the seas of normal salinity. In many respects, this is due to the basic characteristics of the Baltic Sea (cf. paragraph 2.1.), first of all, to its shape and bottom profile, the narrow connection and peculiarities of the water exchange with oceanic areas, the related brackish-water character and accumulation of pollutants, but also to climatic conditions and the activity and economic development of people in the catchment area of the sea.

The tolerance of xenobiotic influences by the rather simple and unstable ecosystems of the Baltic Sea, suffering steady salinity and thermic stress, is different compared to the relatively complicated and stable ecosystems in seas of normal salinity. Also, this tolerance differs between the ecosystems of the Baltic Sea. This situation is related to the heterogeneous descent of organisms and their generally insufficient adaptation to the areas of their present habitation.

4.2.1 Eutrophication

Eutrophication – enrichment of a water body primarily with nitrogen and phosphorus, and in some cases also carbon salts – results in an increase in productivity and causes changes in the structure and functioning of the ecosystem. Eutrophication was historically first noticed in lakes. In general, oligotrophic lakes are transformed into nutrient-rich eutrophic lakes over many years, under the influence of nutrient salts carried into the lakes from the land. Most lakes are eutrophic nowadays. The studies on the influence of nutrients on aquatic ecosystems started in the first half of the twentieth century. The studies on the relationship between the transportation of phosphorus and nitrogen into water bodies and the formation of the water quality had already been carried out for management purposes in the 1970–1980s.

Input of nutrients results in a higher nutrient concentration in the receiving water body and an increase in the biomass of phytoplankton and filamentous algae and primary production. As the process proceeds, the changes in physical, chemical and biological qualities of the aquatic environment are followed by obstructions in light penetration and increasing oxygen concentration in euphotic water layers. Moreover, they are also accompanied by a deterioration in oxygen conditions and the related limitation of the distribution of organisms or their elimination in deep layers.

When eutrophication began to cause problems for the Baltic Sea ecosystems, its main source was agriculture. The phosphorus and nitrogen discharged by ditches, rivers, etc., originated, for the most part, in manure stores, fertilizers, pastures, dairies, sewage, etc.

With economic growth, the amounts of nutrients and other pollutants from agriculture, industry, transportation, mining and municipal wastes increased and diversified. The results also became more complicated and their role increased. The natural self-purification capacity of the sea could no longer counterbalance its condition. In the Baltic Sea, the level of nutrients constantly increased, including during vegetation periods when, at the balanced state of the sea, the nutrient concentration decreases to zero. Primary production magnified step by step, and its continuously high level became the ordinary situation in large areas of the sea. In the process of primary production, oxygen is released. Also, organic matter is formed, requiring large amounts of oxygen both alive and when perished, decaying in surface or deep water layers.

As the main part of the nutrients is discharged from the mainland, the results of eutrophication are expressed earliest and most clearly in the coastal sea. The general responses of pelagic ecosystems to nutrient enrichment can, in principle, be a gradual change towards: (1) increase in plankton primary production; (2) dominance of microbial food webs over the "classic" planktonic food chain from small to large organisms; (3) a dominance of non-siliceous phytoplankton species over diatom species; (4) a dominance of gelatineous zooplankton (jellyfish) over crustacean zooplankton increasing sedimentation of organic matter to the bottom of the sea; (5) increased sedimentation of organic matter to the seafloor; (6) near-seafloor oxygen depletion caused by oxygen consumption by the degrading organic matter; (7) loss of higher life forms, including fish and bottom invertebrates owing to poor oxygen conditions (HELCOM 2009).

In the nutrient-rich Danish fjords, the primary production is estimated to be several times higher than in the open part of the Kattegat. The domination of filiform algae has raised a number of problems. Epiphytic algae suffocate hosts (*Zostera marina*, *Fucus vesiculosus*) and cause their stunting. Therefore, the species composition of macroalgae changes and their growth rate decreases. Mass development of *Nodularia spumigena* and other cyanobacteria causes a decrease in water transparency and a cover of foul-smelling algae on the sea surface. Such cases have been

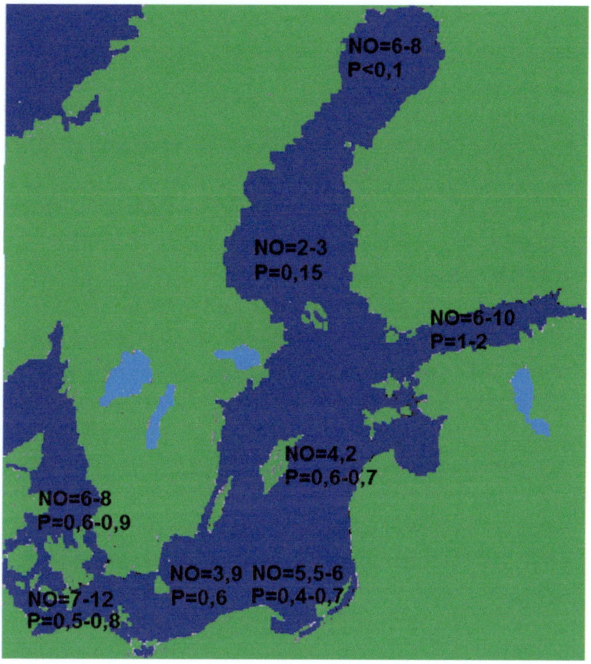

Fig. 4.5 Concentration (μmol dm^{-3}) of nitrates and nitrites (NO) and phosphates (P) in the surface layer of the Baltic Sea in winter 1989–1993 (HELCOM 1996)

interpreted as the result of the discharging of nutrient salts into the sea (Forsberg 1991). However, extensive blooms of cyanobacteria (including *Nodularia spumigena* and *Aphanizomenon flos-aquae*) in the open Baltic were reported as early as before the 1900s (Ojaveer 2014). In the condition of excessive nutrients, such circumstances are becoming customary. Nowadays, large algal assemblages are common, and have been registered by satellites. As both *A. flos-aquae* and *N. spumigena* are able to assimilate nitrogen from the atmosphere, they have the advantage over other algae in developing successfully despite nitrogen scantiness if the quantity of phosphorus is not limited. *Nodularia* spp. produce hepatoxin, which can degenerate liver cells, cause tumours, etc. In Sweden, Denmark and other Baltic Sea countries, blooms of *Nodularia*, *Chrysochromulina* and other toxic algae have affected/killed horses, cows, birds, fish and other vertebrates (including humans) and invertebrates (Forsberg 1991; etc.). Obviously, the nutrient conditions are excellent in the Baltic for many bloom-forming species. Therefore, it is reasonable to assume a continuation of the blooms.

Although the continuous addition of nutrients has increased their amount in the Baltic Sea, in different regions, the concentrations are still different (Fig. 4.5). Since the 1960s, nitrogen concentrations have magnified in all regions. The same can be assumed for phosphorus concentration, except in the Gulfs of Bothnia and Riga (HELCOM 1996). According to Forsberg (1991), during the twentieth century, the addition of nitrogen into the Baltic Sea increased by four times and the addition of phosphorus by eight times. This has considerably increased the concentration of these nutrients, except in Bothnian Bay. In that bay, phosphorus plays the most important role as the limiting nutrient. The large surplus of nitrogen compared to phosphorus has been explained by the fact that in the rivers flowing through forests into Bothnian Bay, phosphates are coupled to iron. It is assumed that low solubility and precipitation may cause sedimentation of phosphorus. Primary production is low, which means that only a small proportion of inorganic nitrogen is bound in organic matter settling to the bottom.

In the twentieth century, the changes in the Baltic ecosystems due to human impact were estimated based on the supposed changes in the major energy flow expressed as organic carbon (Forsberg 1991):

(1) pelagic primary production increased by 30–70%; (2) zooplankton production increased by 25%; (3) sedimentation of organic carbon rose by 70–190%; (4) the macrobenthic production above the halocline approximately doubled; (5) oxygen deficiency in bottom waters wiped out the macrobenthos over nearly 100,000 km^2 of the deeper bottoms of the Baltic Proper and the Gulf of Finland; (6) fish catches grew substantially (this was only partly related to fish productivity, a very important factor also being the increase in the exploitation rate).

One of the most crucial problems in the management of ecosystems of the Baltic Sea has been the tense oxygen conditions in deep water. In the surface layers, contrastingly, the oxygen concentration has been close to saturation due to the oxygen dissolution from the atmosphere and the availability of the oxygen released in photosynthesis. However, at a depth of 20 m, the amount of oxygen formed in the primary production may be less than is required for the respiration of organisms and bacterial degradation of sunken organic material. If oxygen-rich surface waters do not mix with deep waters, the deep layers can run into serious oxygen deficiency. Such mixing can potentially take place in gulfs without any or with a weak halocline, but is hardly possible in the open part of the sea, where the halocline is an unpassable barrier to substantial mixing. As deeper layers receive large amounts of organic material (dead organisms sinking from higher water layers), the oxygen consumption in bottom layers is higher than the existing oxygen amount. When the oxygen reserves begin to expire, bacteria start using other compounds for respiration. Nitrates are transformed into atmospheric nitrogen, and thereafter sulphates are converted into hydrogen sulphide (H_2S), poisonous for higher life forms. The dead bottoms originating in deeper areas of the Baltic Sea have seized a large part of the deeps in the Baltic Sea. This means that oxygen deficiency has become continuous and remains so until the next inflow of saline and oxygen-rich water from the North Sea.

It was estimated that in the middle of the twentieth century, before the notable increase of eutrophication in the Baltic Sea, fish productivity of this sea would have allowed for catches of about 400,000 tonnes (Demel 1970). In recent decades, the real catches have been much higher (Fig. 3.15). The productivity of higher water layers of the sea and fish catches there have increased together with eutrophication. However, compared to the earlier times, the catches have lost their stability.

A eutrophication-based increase in fish catches was gained on account of a substantial deterioration in living conditions of a number of valuable fish species (especially salmonids and coregonids requiring water of high oxygen concentration). The

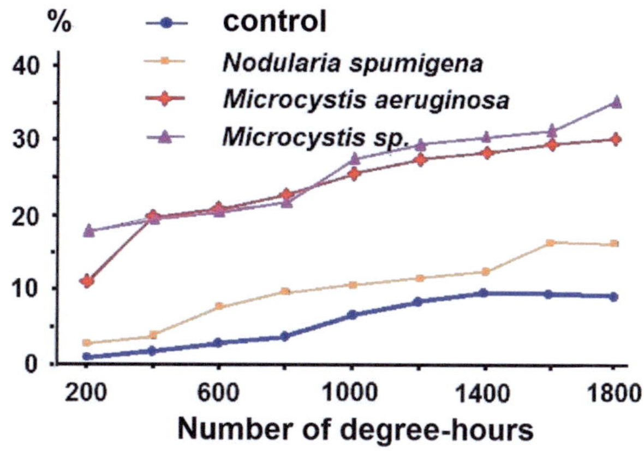

Fig. 4.6 Cumulative deviation of spring spawning sea herring embryos from normal development during the embryonic life under the influence of excretions by *Nodularia spumigena, Microcystis aeruginosa* and *Microcystis* sp. (Ojaveer 2014)

result was the disappearance of these species from many areas of the sea and the pauperization of ecosystems of the Baltic Sea.

The majority of important exploited resources suffer under the negative influence of eutrophication. Sprat resources are vulnerable during the wintering time, as sprat is sensitive to low temperatures and insufficient oxygen concentrations. Therefore, it cannot use deeper (and warmer) water layers for wintering if the oxygen conditions are unacceptable there. A number of fish populations are suffering under the shrinking of food resources, because a part of the potential food-producing volume of the sea cannot fulfil its purpose for food production due to oxygen deficiency. Eutrophication is limiting the reproduction possibilities of all important commercial fish. In cod and flatfish, some spawning areas (where the acceptable salinity occurs in the area of oxygen deficiency) cannot be used for reproduction. This has a very important negative impact on these fish stocks and fisheries. On the spawning grounds of the autumn spawning herring, short-term oxygen deficiencies can most probably annihilate all the demersal eggs/embryos. This can result in the failure of a whole year class.

Cyanobacterial blooms related to eutrophication/pollution negatively affect embryonic development of fish. Specific experiments (Ojaveer et al. 2003b) have showed the negative impact of nodularin (the excretion of *Nodularia spumigena*) and the excretions of *Microcystis aeruginosa* taken from the Baltic Sea and the cultivated *Microcystis* spp. upon the embryonic development of the spring spawning herring of the NE Baltic (Fig. 4.6). Also, under the influence of the excretions of *Nodularia spumigena*, the zooplankter *Eurytemora hirundoides* (an important food for herring, especially in its larval stages) will die in a few days, males first (Fig. 4.7). Cyanobacterial poisons negatively affect embryos of herring and most likely those of other fish species.

Nowadays, the times of natural development in the Baltic Sea are in the remote past. For some time now already, the self-purification capacity of the Baltic Sea has been unable to fulfil its task. The Baltic Sea is one of the most polluted areas of the World Ocean. Pollution of the bottom layers and the oxygen deficiency under conditions of stagnation have grown into important ecological factors exerting crucial

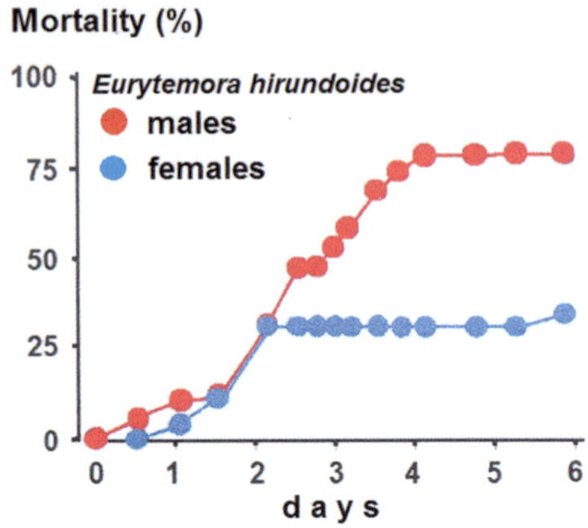

Fig. 4.7 Mortality of the male and female *Eurytemora hirundoides* under the excretions of *Nodularia spumigena* (Ojaveer 2014)

influence from time to time upon the composition and biomass of the biota in the bottom layers of the sea. Therefore, living conditions in the near-bottom layers have changed: the gas regime has deteriorated, and the amount of sediments (including decaying sediments) has increased. This limits the abundance of a number of species up to elimination. Today, H_2S at the bottom and planktic microalgae in pelagic layers can seriously limit the composition and distribution of biota in the Baltic. Eutrophication is perhaps the single greatest threat to the Baltic Sea environment (HELCOM 2010).

4.2.2 *Toxic Pollution*

The 'industrial revolution' started in the countries around the Baltic Sea in the 1850s–1860s. Production has considerably diversified, particularly since the 1950s (Backlund et al. 1992). Many organic and inorganic substances needed by consumers are developed each year. The raw materials for their production are extracted from nature, converted into products by means of chemical or mechanical processes, and used, and after use, a large part of these compounds is emitted and discharged into the sea as waste. Not all of these compounds are necessarily dangerous to life. However, some disposed products constitute a huge burden to the marine ecosystems. Inorganic acids, cyanids, compounds of mercury, lead, cadmium, copper and other heavy metals, toxic organic substances, e.g., halogenated organic compounds, creosol, phenols and their derivatives, aromatic hydrocarbons and a number of synthetic washing materials are hazardous for nature.

A substance is dangerous to the natural environment if it is (1) persistent in the environment, meaning the compound influences the environment over a long period, only breaking down very slowly; (2) if the compound is lipophilic (fat soluble), meaning it easily penetrates cell membranes and accumulates in fatty tissues; or (3) if it is toxic to the environment, i.e., causes elimination of the organisms (acute toxicity) or manifests itself through physiological or behavioural deviations of organisms. Particularly harmful are geno-toxic substances inducing alteration of genes and carrying their influence over to the next generation.

The influence of chemicals depends on their concentration and the duration of exposure, as well as on the characteristics of the ecosystem (salinity, temperature, pH, oxygen supply, content of other pollutants, etc.). The Baltic Sea includes regions with considerably differing environmental conditions, therefore the effect of a compound can notably vary by parts of the sea. In addition, the present-day knowledge on the impact of various pollutants and their complexes is far from exhaustive.

The majority of pollutants belong to one of the following three groups: (1) metals, especially heavy metals and their organic compounds; (2) organic compounds with one or more of their hydrogen atoms substituted with chlorine or some other halogen atom (halogenated hydrocarbons); (3) polyaromatic hydrocarbons (PAHs) with molecules composed of a number of aromatic rings. A certain amount of these compounds can originate from natural processes. However, the vast majority of these substances are industrially produced.

Metals Metals occur in marine water in dissolved (ionic) form, being bound to particulate matter or inorganic–organic complexes. A number of metals are important for the fulfilment of biological functions in living organisms (molybdenum, manganese, cobalt, copper, zinc), but lead, cadmium and mercury do not seem to have any biologically important function. At high concentrations, all metals have negative impacts.

The initial developmental stages of organisms are especially sensitive to pollutants (Kihlström 1992). Generally, the reproduction of organisms is influenced by very low concentrations of pollutants. Specific experiments arranged in the 1970s indicated that very low concentrations of copper, cadmium (0.005 mg dm^{-3}) and zinc (0.005 mg dm^{-3}) decrease the percentage of fertilization, exert a negative impact on embryonic development, and limit the normal hatching of embryos (Ojaveer 1988). It was found that with increasing copper and cadmium concentrations, the percent of the total hatch, that of normal embryos, and the average length of newly hatched larvae considerably decrease (Fig. 4.8).

Persistent Organic Compounds In the Baltic Sea, halogenated organic compounds constitute a very significant group of pollutants. The majority of them belong to the chlorinated organic compounds, which are generally persistent, lipophilic and toxic (Backlund et al. 1992). Part of the chlorinated hydrocarbons can be of natural origin (e.g., some dioxins), but a great majority of these substances are human-made and used in industry (e.g., PCBs, DDT, DDE, toxaphene, lindane, etc.). The hydrocarbons and ethers constituting over four chlorine atoms are especially dangerous to the environment. Special attention has been paid to dioxins, which are possibly the most dangerous and toxic chemical substances known today. Due to their toxic and bio-accumulative properties and persistence, dioxins are a major environmental concern. In particular, the dioxin contamination of fatty fish in the Baltic Sea is a hazard to the health of the human population around the Baltic Sea.

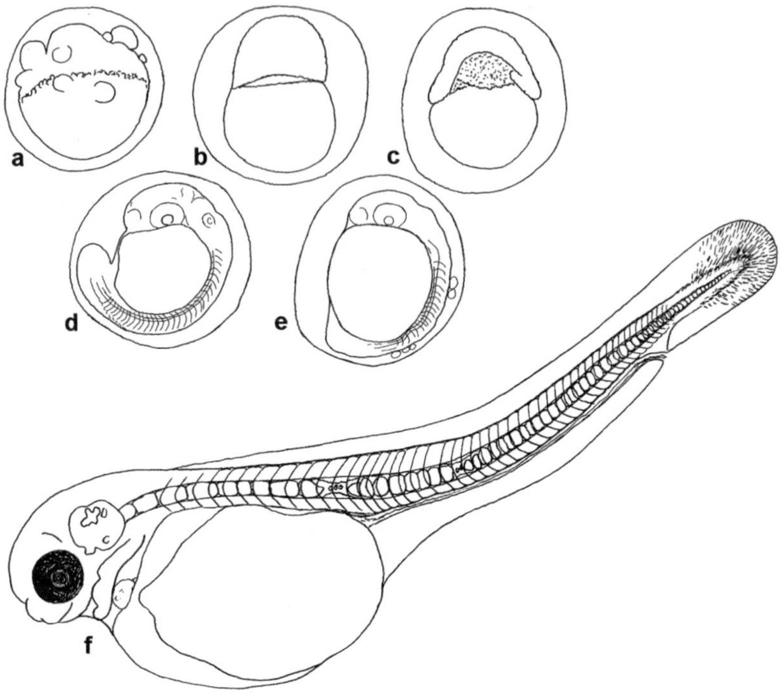

Fig. 4.8 Embryos of spring spawning herring deviated from the normal development or perished in the copper, cadmium and zinc solutions in different developmental stages: (**a**) abnormal cleavage; (**b**) abnormal blastula; (**c**) abnormal gastrula; (**d**, **e**) aberrations at the stage of mesoderm segmentation (myotomes of irregular size and shape, optic vesicles and chorda can be abnormal, the auditory capsule can be absent, the tail bud aberrant, etc.); (**f**) hatched abnormal prelarva (heart, auditory capsule and olfactory capsule aberrant, yolk hardened, body curved, secondary chorda cells abnormal, etc.); author's data

Dioxins are a group of polychlorinated aromatic compounds with similar structure and physical-chemical properties. They are not produced intentionally, but are formed as byproducts of natural events or human-made processes, such as pulp and paper bleaching and the incineration of waste. Dioxins are colourless, odourless organic compounds and highly soluble in fat. This means that they bind to sediment and organic matter and are absorbed in animal and human tissue. They are persistent and not biodegradable, and therefore accumulate in the food chain. Of the 210 different dioxin compounds (congeners), about 17 are of toxicological concern. The most toxic form is 2,3,7,8-tetrachlorodibenzo-*p*-dioxin (2,3,7,8-TCDD). Dioxins are among the 12 chemicals banned by the Stockholm Convention on Persistent Organic Pollutants. The European Commission's policy and strategy to reduce exposure to dioxins has resulted in a range of Community legislation on dioxins in food and animal feed. Dioxins cannot be prohibited, as they are not produced intentionally. Nevertheless, the production and use of PCBs has been discontinued in most industrialized countries. In the EU, their use has been prohibited since 1978. Existing PCB-containing equipment was taken out of service by the end of 2010.

Dioxins and dioxin-like PCBs create problems today, regardless of the decreasing trends and the efforts to eliminate and reduce the polluting sources. Fish caught in the Baltic Sea, in particular, fat-rich fish like salmon, herring and sprat, exhibit high concentrations of these pollutants in their bodies. The health impact of eating food contaminated with dioxins encompasses several possible effects (Backlund et al. 1992). Some dioxins are classified as known carcinogens, but they are also known to be responsible for developmental and neurobehavioral (learning disabilities), reproductive and immune-toxic effects as the result of experiments on laboratory animals. The use of the contaminated fish for other purposes (e.g., fish feed) also poses serious concerns.

Polychlorinated biphenyls (PCBs) are chlorinated aromatic hydrocarbons synthesized by direct chlorination of biphenyls. Some PCBs have toxicological properties similar to dioxins ('dioxin-like PCBs'). They are still present today in some technical installations. These pollutants were mainly produced starting in the 1930s and 1940s and were phased out in the 1970s.

Polyaromatic hydrocarbons (PAHs) consist of several fused aromatic rings with several chlorine atoms. They are often persistent and bioavailable. Most PAHs are produced in incomplete combustion, including in natural fires. PAHs can be found in mineral oil, coal tar, wood tar, etc.

Oil Pollution Petroleum causes a large number of problems, as it includes nearly a thousand different chemical compounds. Petroleum hydrocarbons are released into the Baltic Sea from many sources: runoff from land, ships (including oil spills), atmosphere, off-shore oil platforms, etc. Increased traffic, ignorance of requirements for oil transportation, oil prospecting in the Baltic, etc., augment the risk of petroleum pollution of the sea. Finding a reasonable estimate for the annual input of oil into the Baltic Sea area has caused problems. According to Kihlström (1992), the estimate varies between 21,000 and 66,000 tonnes. It is probable that in the surface layers of the Baltic Sea, petroleum concentrations do not vary very much. However, in the Baltic Sea areas connected to the North Sea and the NE Atlantic, the petroleum concentration is, respectively, about two and three times higher than the averages for the North Sea and the NE Atlantic.

The largest amount of the compounds polluting the sea is emitted from point sources: factories, workshops, power stations, mines, towns, etc. However, a substantial amount of the pollution is also discharged by non-point sources – agriculture, transportation, traffic, etc. A variable portion of anthropogenic pollution arrives in the Baltic Sea region via the atmosphere, transported by winds. Pollutants can be transferred from one individual to another, in particular, from prey to predator in the food web. In these cases, the concentration increases at higher levels in the food web (so-called bio-magnification) (Kihlström 1992).

The transfer from mother to child can occur via the placenta in mammals and via the egg in birds and reptiles. Thereby, at birth, the young have already gotten the same concentration of pollutants as the mother has. In general, in the case in which the previous generation has lived in a polluted environment, the subsequent generation is incapable of beginning its life clean (Kihlstöm 1992).

The reproduction and growth of organisms are extremely sensitive to toxicants. High concentrations of pollutants cause deviations in hormonal and reproduction systems from their normal state and the development of tumours. Also, muscular and nervous systems can fall under pressure. Continuous suppression of reproduction possibilities can actually ruin the whole population concerned. From the 1950s to the 1970s, an immense rise in the pollution level in the Baltic Sea put the Baltic seal populations into a very unsafe situation. Only the rapid reduction of the pollution level in the aquatic environment saved the marine mammal populations and several marine bird species in the Baltic Sea area.

4.2.3 Influences Related to the Storage of Dangerous Substances

Radioactive Substances The Baltic Marine Environment Protection Commission has repeatedly discussed and evaluated the impact of radioactive pollution and the chemical and other warfare agents dumped into the Baltic Sea after WW II (HELCOM 1996, 2002, 2013, etc.). The main radioactive isotopes of anthropogenic origin found in the Baltic Sea are caesium-137 (^{137}Cs) and strontium-90 (^{90}Sr). It has been estimated (HELCOM 1996) that until 1991, the Baltic Sea received 4850–5750 Tbq* ^{137}Cs and 720 Tbq ^{90}Sr (*Tbq = terabecquerel =10^{12} Bq), the main part of which was the ^{90}Sr deposition caused by the Chernobyl accident in April 1986, plus the fallout from atmospheric weapon tests during the 1960s and discharges from the reprocessing plants at Sellafield and La Hague (which somewhat influence the Baltic Sea ecosystems through the inflowing of contaminated saline waters through the Danish Straits). After the Chernobyl catastrophe, the level of radioactive compounds in the Baltic Sea increased up to the beginning of the 1990s. Then, after part of the radioactive material had decomposed and some quantity of it had left the Baltic Sea, the concentration decreased. Because of the comparatively limited water exchange between the Baltic Sea and the oceanic areas, the concentration of artificial radionuclides is higher in the former than in the latter. However, the maximum annual dose to individuals from any critical group in the Baltic Marine Area during the period 1950–2000 was estimated to be 0.2 Sv/year**, which is below the dose limit for the exposure to the general public set out in the Basic Safety Standards (HELCOM 2002) (**Sv = sievert, unit of individual dose). Concerning the level of ^{137}Cs in the sediments of the Baltic Area, in 1998, it was found in the highest concentration in the Bothnian Sea (1370–1560 Tbq) and the Gulf of Finland (230–255 Tbq) against 23 Tbq in the Gulf of Riga. In 2015, there were 7–8 active nuclear power plants around the Baltic Sea. The predominant radionuclide in the liquid discharges from the nuclear power plants and research reactors in the Baltic Sea region is tritium. After tritium, the most abundantly discharged nuclides are ^{137}Cs, ^{90}Sr and ^{60}Co, which contribute more than 98% of the cumulative decay corrected discharges of the other radionuclides (except tritium) in the long term. The importance of tritium in the radiation burden of people is minor because of the low energy emitted by the radionuclide. Nonetheless, most of the radioactivity in the sediments of the Baltic Sea originates from naturally occurring radionuclides. At present, the radioactivity in the sediments is not expected to cause harmful effects to the Baltic Sea wildlife.

Chemical Munition Chemical munition was used in WW I, but in WW II, it was not applied in Europe. However, the stocks of chemical munition were available for application. Liquidation of the stocks was started before the end of the war. Mainly between 1945 and 1948, in the southern part of Skagerrak (outside the Helsinki Convention area), over 30 ships carrying a large amount of chemical munition from the preserved stock of the German army, along with conventional ammunition, were sunk. According to the HELCOM data (HELCOM 1996, 2013, etc.), in the Baltic Sea, at least 40,000 tonnes of chemical weapons, containing some 15,000 tonnes of chemical warfare agents, was dumped altogether. The dumping was carried out in a number of stages during the years 1945–1965 over a comparatively large area, mainly east of Bornholm Island, south-west of Gotland Island and south of the southern entrance of the Little Belt. The main dumping locations are indicated in Fig. 4.9 (HELCOM 2016). The containers with aircraft bombs, artillery shells and grenades holding chemical warfare, and conventional munitions, including mines, were dumped, mainly item-by-item in designated dumping areas, as well as en route from the loading harbours (e.g., Wolgast, Flensburg). Therefore, the dumping sites were scattered along the transport routes from the harbours to the designated dumping areas.

The main dumping areas in the Baltic Sea were the deeps east of Bornholm Island and the 4–6 m thick clayey mud bottoms in the 93–137 m deep area southeast of the Gotland Basin. Chemical warfare agents such as sulphur-mustard, tabun, adamsite and others were dumped. In the mid-1940s, in the area south of the entrance to the Little Belt, at a depth of 25–31 m, chemical weapons containing tabun and phosgene and conventional ammunition, brought chiefly from Wolgast, were dumped. Similar dumping occurred in several areas, chiefly in the Southern Baltic, with the exact locations of chemical warfare materials being uncertain. One of them is situated in the Gdansk Deep at a depth of 80–110 m (HELCOM 2013). There is no exact overview of the dumped chemical weapons and conventional munition. Examinations have indicated that the contemporary status of the dumped materials is variable. Under the influence of currents, water turbulence and other natural processes, but also due to anthropogenic impacts (fishing and other activity), a part of the dumped material has been transferred from its original locations. The bombs, shells and containers are corroded, and the compounds inside have changed their original condition under the impact of marine water, the oxygen contained in the water, hydrogen sulphide, bacteria and other factors. Their influence upon the organisms and the characteristics of the bottom sediments has not been sufficiently studied. Up to now, no decision to remove the dumped chemical warfare and conventional munition from the seabed has been made, as its implementation would be very complicated and connected with risks. Presently, the sulphur-mustard type agents are probably the most dangerous, as these can cause damage to the naked

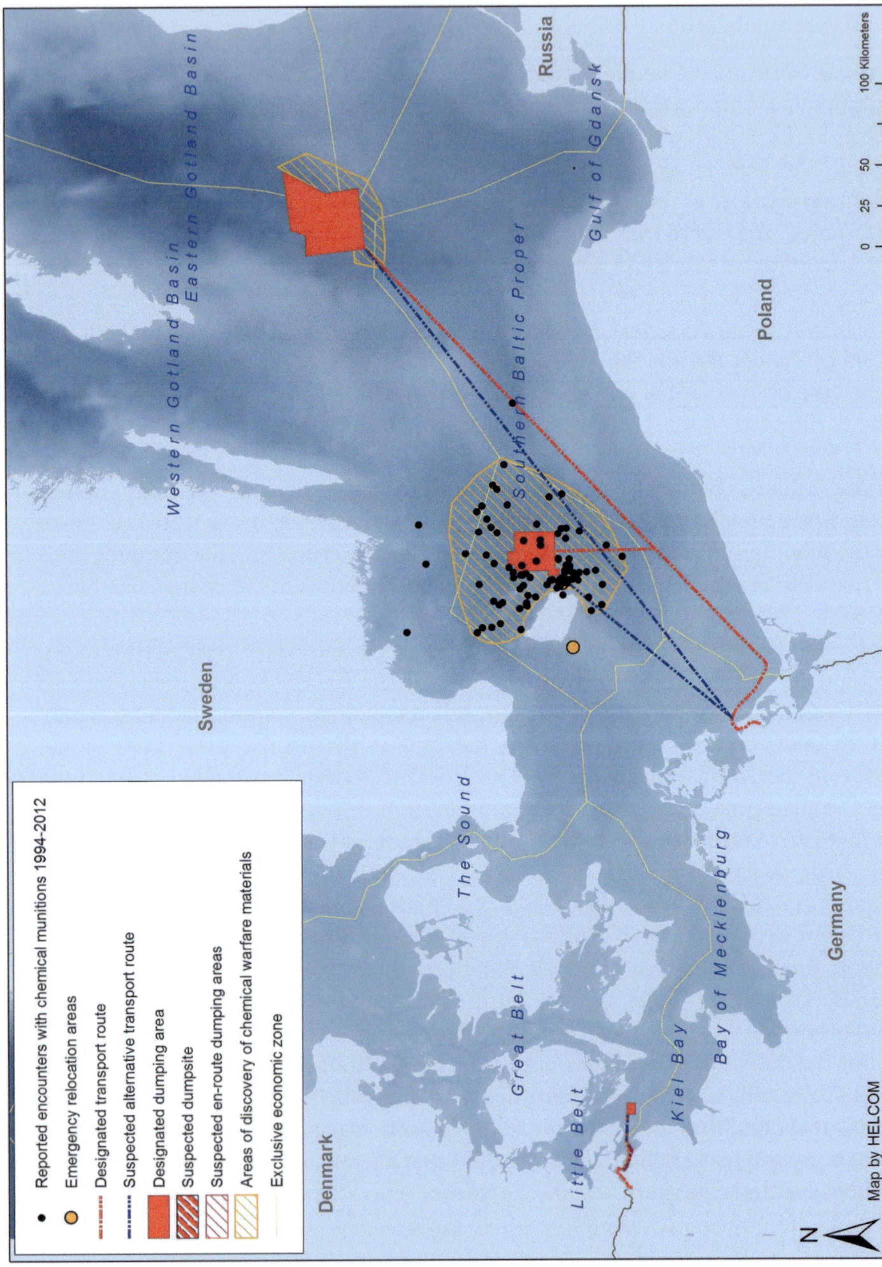

Fig. 4.9 Dumping sites for chemical munitions in the Baltic Sea and the transport routes to the dumping sites – HELCOM 2016 (http://helcom.fi/Lists/Publications/BSEP142.pdf)

skin. Hundreds of fishermen have been injured by chemical weapons, in most cases, in the southern part of the sea. The numbers were highest at the beginning of the 1990s, but incidents have also been reported later. It has been shown that the compounds contained in the dumped chemical weapons exert a harmful influence on fish, some phytoplankton species, microscopic crustaceans, etc. Also, injuries have been caused by lumps of white phosphorus that have drifted to the coasts (including beaches). Generally, the compounds of chemical warfare agents do not dissolve much in water. Information on the amount of damages to the biota and possible needs for investigations is insufficient. However, chemical warfare agents were and still are dangerous to biota. These weapons are a serious limiting factor in using the sea bottom for the installation of cables and pipelines, as well as for recreation (fishing, diving, etc.).

4.2.4 Other Impacts of Human Activity

In addition to the dangerous substances, human activity influences marine ecosystems with organic and inorganic substances that are either nontoxic or are of low toxicity (e.g., faeces, lignin, cellulose, garbage, litter, etc.), but still influence aquatic organisms, mainly by changing general ecological conditions. Inorganic nontoxic substances (sand, clay, larger lumps of soil) can also exert mechanical impacts on organisms by injuring or burying them, changing water transparency, etc. The status of the ecosystems and living resources of the Baltic Sea also depends on other aspects of human activity: extermination of forests, which has lasted for centuries, and construction of dams, bridges and other structures. A life under constant stress factors (unfavourable salinity, severe wintering conditions) increases the vulnerability of Baltic organisms of limited genetic variability to diseases, parasites, and anthropogenic impacts, especially to toxic pollution and oxygen deficiency related to eutrophication. In addition to environmental pollution, the impact of lively, ever-intensifying marine transportation, a wide array of shipping-related wastes, disturbance of organisms by noise and vibration accompanying fishing, extraction of minerals, offshore wind farms, pipelines and cables on the sea bottom, magnetic fields of high-voltage cables, anthropogenic impacts related to tourism and recreation, and other economic activities put extra burden on organisms in the Baltic Sea.

Aquaculture Contrary to global trends, the aquaculture production in the Baltic Sea catchment area stagnated or even slightly declined during 2005–2015, with an annual yield of approximately 100,000 tonnes. The main species reared are rainbow trout (*Oncorhynchus mykiss*) and common carp (*Cyprinus carpio*). Significant amounts of the production are land-based in fresh water; for rainbow trout, this constitutes about 50% of production. Mainly Finland, Sweden and Denmark have hydrological conditions suitable for the sea cage production of rainbow trout. Aquaculture has the potential for future development in the Baltic Sea, and this potential is not restricted to fish; algae and blue mussels are also considered. Around

2015, rainbow trout dominated in the production (about 70% of the total), of which about 50% was reared in marine/brackish waters, and carp in fresh waters (about 20%). The main producers are Denmark and Finland, followed by Sweden and Poland. However, there are environmental effects to consider, such as increased nutrient loads (Saikku and Asmala 2010), but also diseases and parasites, escapements and the possible establishment of alien species.

Primary Ecological Problems The main ecological problems related to human impact in the Baltic Sea are (1) deterioration of the water quality (colour, taste, smell, sediments, changes in the pH, toxicity related to the occurrence of certain algal species, decrease of oxygen content in bottom layers, increase in H_2S concentration); (2) deterioration of the aesthetic quality and recreational characteristics of the sea, and elevated risks to health; (3) limitations for fisheries: diseases, enemies, extra mortality, bad smell/taste of fish, decrease of valuable and increase in inferior fish stocks. Data by HELCOM indicate that the efforts in the direction of limitation of eutrophication and toxic pollution and the attainment of better functions of Baltic systems have yielded some results. However, it should be stressed that (1) the supply of new artificial compounds, untested for the reactions of Baltic organisms, ecosystems and environment, is continuous. The necessary related information for the protection of the Baltic systems is very extensive and its elaboration takes time; (2) information on the compounds already in use is deficient, especially concerning their long-term impacts and their effect on genetic factors. In addition, up to now, the pollution situation in the Baltic Sea has been compared with that of other seas without consideration of their ecophysiological differences (e.g., the differences in water salinity and temperature conditions).

4.3 Immigration into the Contemporary Baltic Sea

On the geological time scale, the Baltic Sea is a very young sea. Nevertheless, the living places in this sea are divided between groups of organisms. However, changes and redistributions of habitats between organisms are possible, and happen rather frequently. With the disappearance of Atlantic sturgeon and a notable decrease in salmonid and coregonid fishes and seal populations, their dwelling areas as well as their places in the food web have been redistributed between other populations. The Baltic Sea is a changing and developing system, with both the disappearance of some species and the appearance of new ones taking place continuously.

Moving northwards and eastwards from the Kattegat, the importance of organisms with a marine background decreases, whereas the share of brackish-water and freshwater fauna and flora increases stepwise. Marine organisms are constantly adapting to lower and freshwater organisms to higher salinity. One of the most important circumstances favouring immigration has been the ample discharge of nitrogen and phosphorus compounds into the Baltic Sea, thereby increasing primary productivity. This has directly and/or indirectly improved the feeding conditions of organisms in the Baltic Sea.

The migration of marine and freshwater organisms to the brackish-water areas of the Baltic Sea and their adaptation there may have been supported both by natural processes (carried by currents into the Baltic Sea, migrations due to climate changes) and human activity.

Such species or intraspecific taxa that have been distributed in the new living places with human help are known as alien (non-indigenous) species. Human aid in immigration can be unintentional (incidental) or intentional. The main reason for the wilful transfer of species by man has usually been the wish to increase the profitable resources of the water body. This type of immigration may have been accompanied by incidental/unintentional invasions by other organisms. Man can facilitate immigration (invasion) of alien species to areas in which they have not previously been found through a number of means: by marine transportation (with the hull, reservoirs, hold, bilge water of the vessel, etc.), through activities related to fishing and aquaculture (with nets and other fishing gear, aquatic organisms, packing material, equipment used in aquatic sports, trading with aquaria and living fish, with the materials related to biodefence), etc.

One of the most important factors/ways related to incidental immigration has been the migration of aquatic organisms along with transportation, which has lasted for centuries and will probably continue. In recent decades, shipping has rapidly developed, and ever bigger vessels with larger ballast water reservoirs are being constructed. When the transport vector is ballast water, the time which species spend in it may limit their survival and the success of the invasion. The possibility of the survival and reproduction of a species in the new area depends on the similarity of the hydro-climatic conditions between its natural habitat and that area.

The bulk of the invasive migrants are microscopic species, with some exceptions, including fish, e.g., round goby (*Neogobius melanostomus*), channel catfish (*Ictalurus punctatus*) and bighead carp (*Aristichthys nobilis/Hypothalmichthys nobilis*). The most important attributes of the species favouring their successful adaptation to the new area are a wide ecological resilience to the variability of environmental conditions, existence of resistant resting stages, high fecundity, high growth rate, short reproduction cycle, ability of vegetative or hermaphroditic reproduction, a wide ecological amplitude, high adaptability, and absence of enemies, parasites and diseases in the new area.

In addition to anthropogenic impacts, supplementation of the Baltic ecosystems also has natural causes. Incorporation of organisms of both marine and freshwater backgrounds into the Baltic brackish-water systems is clearly facilitated by periodic or aperiodic variations in the natural conditions, particularly by the large-scale changes in climatic conditions. During the periods of continuous salinity increases, a number of invertebrates, as well as the fish species that consume them – anchovy, four-bearded rockling, garfish, etc. – expand their presence in the Baltic. Periods of higher salinity are followed by a stronger drive of marine species to the Baltic Sea, but warmer periods are followed by the intense penetration of boreal marine and southern freshwater species into the Baltic ecosystems.

The Baltic ecosystems continuously see new species. The present-day situation of the process should be understood as a stage of complementation of the ecosys-

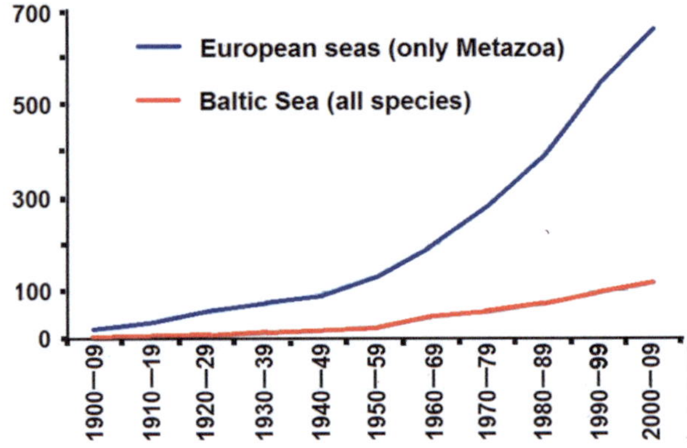

Fig. 4.10 The cumulative number of alien species belonging to Metazoa, in the European seas and the cumulative number of all alien species by decades in the Baltic Sea since the beginning of the twentieth century (Naršćius 2013)

tems under the conditions of strong anthropogenic impact. Since the beginning of the twentieth century, 119 alien species have been registered in the Baltic Sea (the total number for the European seas is 662 species – Fig. 4.9– Narscius 2013). Only some of the alien species that have immigrated to the Baltic Sea have adapted there and been able to create vigorous populations. The number of such species is 37 (AquaNIS 2013-06-23), including, e.g., *Acartia* (*Acanthacartia*) *tonsa*, *Amphibalanus improvises*, *Carassius gibelio*, *Cercopagis (Cercopagis) pengoi*, *Chara connivens*, *Cyprinus carpio*, *Dreissena polymorpha*, *Elodea canadensis*, *Evadne anonyx*, *Gammarus tigrinus*, *Marenzelleria neglecta*, *Mnemiopsis ledyi*, *Mya arenaria*, *Neogobius melanostomus*, *Oncorhynchus mykiss*, *Rhithropanopeus harrisii*, etc. (Fig. 4.10).

Bioinvasions are one of four main dangers to the ecosystems of the World Ocean. They change the ecological quality on several levels (infection with parasites/enemies, changes in the structure of populations/communities and habitats, genetic variations, etc. They can be directly pathogenic to man or the members of existing ecosystems. They can also influence existing biota through competition for food, living space, etc. (Olenin et al. 2007). In the Baltic Sea, the most widely distributed alien species are *Acartia tonsa*, *Cercopagis pengoi*, *Amphibalanus improvises*, *Dreissena polymorpha*, *Gammarus tigrinus*, *Mya arenaria*, *Neogobius melanostomus*, *Oncorchynchus mykiss*, etc. Their impact on the Baltic ecosystems is the most notable. However, it has not been stated that alien species have caused disturbances in the ecosystems in the Baltic Sea. In this sea, nonindigenous species have mostly enriched both the biota and the functions of the food web, including for the exploited resources (Ojaveer et al. 2010). It is important that both the species adapted in the Baltic Sea ecosystems and the immigrants not be simultaneously and equally adapted to all the Baltic Sea ecosystems. Every ecosystem is different from the others and requires corresponding adaptation. Presently, populations belonging to 12 fish species have adapted to the brackish-water conditions to a level that they are probably capable of reproduction in all parts of the Baltic Sea. These include spring spawning and autumn spawning herring, smelt, fourhorned sculpin, bullrout, straight-nosed pipefish, broad-nosed pipefish, eelpout, snake blenny, common sand eel, sand goby and three-spined stickleback. A number of species are developing in this direction. In the formation of the unique brackish-water ecosystems of the Baltic Sea, both mighty natural and anthropogenic processes have their role.

References

Alheit J, Möllmann C, Dutz J, Kornilovs G, Loewe P, Mohrholz V, Wasmund N (2005) Synchronous ecological regime shifts in the Central Baltic and the North Sea in the late 1980s. ICES J Mar Sci 62:1205–1215

AquaNIS (2013) Information system on aquatic Non-indigenous species. World Wide Web electronic publication. www.aquanis.org, version Version: 2.2. Accessed 23 June 2013

Aro E (2000) The spatial and temporal distribution patterns of cod (*Gadus morhua callarias* L.) in the Baltic Sea and their dependence on environmental variability – implications for fishery management. Dissertation, University of Helsinki

Backlund P, Holmbom B, Leppäkoski E (1992) Industrial emissions and toxic pollutants. The Baltic University Secretariat, Uppsala University, Uppsala

Demel K (1970) Naša Baltika (Our Baltic), Petrozavodsk (trans: Brõnskii ES). Polish Edition: Demel, K. (1967) Nasz Baltyk, Warszawa

Eero M, MacKenzie BR, Karlsdottir HM, Gaumiga R (2007) Development of international fisheries for the eastern Baltic cod (*Gadus morhua*) from the late 1880s until 1938. Fish Res 87:155–166

Elwertowski J (1957) Szprot. Biologia, polowy, przetworstwo. Wydawnictwo Morskie, Gdynia

Feistel R, Nausch G, Wasmund N (eds) (2008) State and evolution of the Baltic Sea, 1952–2005. Wiley, Hoboken

Forsberg C (1991) Eutrophication of the Baltic Sea. The Baltic University Secretariat, Uppsala University, Uppsala

Hagen E, Feistel R (2005) Climate turning points and regime shifts in the Baltic Sea region: the Baltic winter index (WIBIX) 1659–2002. Boreal Environ Res 10:211–224

Hela I (1966) Fluctuations in the degree of continentality of Northern Europe in 1866–1965. Geophysica 8(4):331–334

HELCOM (1996) Third periodic assessment of the state of the marine environment of the Baltic Sea 1989–1993; Background document, Baltic Sea Environ Proc no 64B. HELCOM, Helsinki

HELCOM (2002) Environment of the Baltic Sea Area, 1994–1998. Baltic Sea Environ Proc no 82B

HELCOM (2009) Eutrophication in the Baltic Sea. Baltic Sea Environmental Proceedings no 115 B

HELCOM (2010) Ecosystem Health of the Baltic Sea 2003–2007. HELCOM Initial Holistic Assessment, Baltic Sea Environ Proc no. 122. Helsinki Commission, Helsinki

HELCOM (2013) Chemical munition dumped in the Baltic Sea. Report of the ad hoc expert group to update and review the existing information on dumped chemical munitions in the Baltic Sea (HELCOM MUNI), Baltic Sea Environ Proc (BSEP) no 142. Helsinki Commission, Helsinki

HELCOM (2016) http://helcom.fi/Lists/Publications/BSEP142.pdf

ICES (2009) Report of the ICES/HELCOM Working Group on Integrated Assessment of the Baltic Sea (WGIAB), 16–20 March 2009, Rostock, Germany. ICES Statutory Meeting, paper CM 2009/BCC:02

ICES (2010) Report of the ICES Advisory Committee 2010. Book 8, Baltic Sea, ICES Advice 2010. International Council for the Exploration of the Sea, Copenhagen

ICES (2011a) Report of the ICES Advisory Committee 2011. Book.8, Baltic Sea, ICES Advice 2011. International Council for the Exploration of the Sea, Copenhagen

ICES (2011b) Report of the ICES/HELCOM Working Group on HELCOM (2009) Eutrophication in the Baltic Sea. Baltic Sea Environment Proceedings no 115B

Integrated assessment of the Baltic Sea (WGIAB), 4–8 Apr., 2011, Mallorca, Spain. ICES Statutory Meeting, paper CM 2011/SSGRSP.03

Kalejs M, Ojaveer E (1989) Long-term fluctuations in environmental conditions and fish stocks in the Baltic. Rapp P-v Rèun Cons int Explor Mer 190:153–158

Kihlström JE (1992) Toxicology – the environmental impact of pollutants. The Baltic University Secretariat, Uppsala University, Uppsala

Matthäus W (1993) Major inflows of highly saline water into the Baltic Sea – a review. ICES Statutory Meeting, paper CM 1993/C:52

Narščius A (2013) Development and application of the information systems on aquatic non-indigenous species. PhD thesis.,Klaipeda University, Klaipeda

Ojaveer EA (1988) Baltijskie seldi (Baltic herrings). Agropromizdat, Moskva

Ojaveer E (2014) Läänemeri. TA Kirjastus, Tallinn

Ojaveer E, Kalejs M (2010) Ecology and long-term forecasting of sprat (*Sprattus sprattus balticus*) stock in the Baltic Sea: a review. Rev Fish Biol Fish 20:203–217

Ojaveer E, Kalejs M (2012) Long-term prediction on Baltic fish stocks based on periodicity of solar activity. Rev Fish Biol Fish 22:683–693

Ojaveer E, Pihu E, Saat T (eds) (2003a) Fishes of Estonia. Estonian Academy Publishers, Tallinn, 416 pp

Ojaveer E, Simm M, Balode M, Purina I, Suursaar U (2003b) Effect of *Microcystis aeruginosa* and *Nodularia spumigena* on survi val of *Eurytemora affinis* and the embryonic and larval development of the Baltic herring *Clupea harengus membras*. Environ Toxicol 18:236–242

Ojaveer H, Jaanus A, MacKenzie BR, Martin G, Olenin S, Radziejewska T, Telesh I, Zettler ML, Zaiko A (2010) Status of biodiversity in the Baltic Sea. PLoS ONE 5(9):e12467

Ojaveer E, Arula T, Lankov A, Shpilev H (2011) Impact of environmental deviations on the larval and year class abundances in the spring spawning herring (*Clupea harengus membras* L.) of the Gulf of Riga (Baltic Sea) in 1947–2004. Fish Res 107:159–168

Olenin S, Minchin D, Daunys D (2007) Assessment of biopollution in aquatic ecosystems. Mar Pollut Bull 55(7–9):379–394

Saikku L, Asmala E (2010) Eutrophication in the Baltic Sea the role of salmonid aquaculture, consumption and international trade. J Ind Ecol 14(3):482–495

Chapter 5
Assessment and Management of Ecosystems and Living Resources in the Baltic Sea

Abstract The most important commercial living resource in the Baltic Sea is fish. Since the 1970s the total annual fish landings in the Baltic Sea have been reported at 700000–1100000 tonnes. Cod herring and sprat constitute presently about 95% of the total catch. Trawl fishery is the most important way of resource exploitation. Coastal fisheries with gill, pound and trap-nets are conducted along the entire Baltic coastline.

The fishing technology has been developed to a point where a significant proportion of the total fish population could be harvested in a rather short time and it may economically be feasible to do so. Management of the exploitation should: (1) restrict the impact on the ecosystems within the boundaries defined by society and (2) assure that we use the living resources in an optimal way. Basic principles of the assessment and management of marine ecosystems are sustainability, precautionary approach and the single stock perspective. The estimates can be divided: (1) estimates of the conditions of ecosystems and the quantitative forecasts on the size and composition of the existing fish stocks 1–3 years in advance; (2) long-term qualitative prognoses on the state of ecosystems and fish stocks 10–20 years in advance. Fisheries management regulates the fishing effort or catch that is removed from the sea. Reference points for the advisory use in fishery management are based on two main considerations: (1) the maintenance of the reproductive potential of the stock and (2) the maintenance of the growth potential of the stock. The reference points represent long-term averages and do not account for natural variability and the uncertainties in assessment and in practical fishery management. The safe biological limits are a key concept in relation to the state of a fish stock and fisheries: a stock is considered inside safe biological limits if it will be able to sustain its production under the present exploitation. These limits are defined according to reference points which are quantifiable indicators of the state of a stock and the fisheries exploiting it. A reference point is associated with a management objective. The reference points base on single species models and these presently in use represent the spawning stock biomass (SSB) and the fishing mortality (F). The level of the SSB below which a reduced recruitment has been observed, is used as a biological reference point for stock sustainability. This spawning stock reference point was referred as the Minimum Biologically Acceptable Level (M_{BAL}). The fishing mortality was limited with the F_{lim} point to sustain the SSB higher of the M_{BAL}. Also a system of complementary reference points has been developed to achieve the sustainable precautionary management of fish stocks. As the environmental conditions

E. Ojaveer, *Ecosystems and Living Resources of the Baltic Sea*,
DOI 10.1007/978-3-319-53010-9_5

in the Baltic Sea cannot be identified with the conditions in the seas of the salinity of 33–37 psu, the result of the adaptation of ecosystems and living resources to the natural and human-made conditions in the brackish Baltic Sea is different of the same process in the seas of normal salinity. HELCOM estimated in 2010 that the ecosystem capacity to deliver goods and services is hampered. The major concern is eutrophication and contamination, the biodiversity status is unfavourable.

5.1 Composition and Exploitation of Living Resources in the Baltic Sea

The focus in the management of fish stocks is on the supply of human food and the production of feed for aqua- and agricultural production (fish meal and fish oil). There are, however, other uses, although ones not currently of great importance, for example, in pharmaceuticals and cosmetics. We may also consider using fish biomass to replace certain raw materials in the chemical industry. One example is the possibility of harvesting or growing algae for various uses. In addition, there are resources that belong to other phyla (mussels). These, however, are of minor importance and are not discussed here.

In the present Baltic Sea, the main fish catches are produced by the descendants of the marine populations (herring, sprat, cod, flatfish species) of rather wide ecological amplitude that have immigrated into the Baltic Sea earlier. They are readily capable of using the possibilities within the fluctuating environmental conditions of the Baltic for the effective expansion of their reproduction and growth areas. The abundance fluctuations in freshwater, diadromous and relict populations of the sea are relatively more moderate.

A number of fish species are important objects of commercial interest. Fish is the most usable biological resource of the Baltic Sea. Fishery has been an economically important occupation for coastal folk over the centuries. However, in the Baltic Sea, only a fraction of fish species are of substantial direct commercial interest – Baltic spring spawning and autumn spawning herring, Baltic sprat, Baltic cod, flounder, plaice, dab, turbot, salmon, sea trout, eel, etc. – chiefly the populations with a marine background having adapted to the Baltic conditions and migratory species. Presently, these species compose the largest part of the biomass landed by fishing vessels (Fig. 3.15) and also of the food of the coastal population. They all have had long periods of adaptation to the Baltic Sea's environmental conditions during that sea's development. In the periods of historical development of the Baltic Sea, the role of these species has varied. Due to the economic importance, large efforts are presently expended towards investigations of the main fish resources. Therefore, knowledge of these fishes is much wider and better than of the species of lower significance.

Today, fishery (including aquaculture) is important both as a separate branch of the economy and as a constituent of tourism and recreational fishery. In addition to the resources of the commercial fishery, the local fish fauna includes a number of species that are only occasionally exploited by fisheries, or not at all. These species may have importance for other segments of the natural food webs. Some species may be numerous during certain periods, for instance, mackerel *Scomber scombrus*, which does not reproduce in the Baltic, but whose young stages are, in some periods, carried together with water masses from the North Sea into the Baltic, where they develop, grow and are fished. A limited number of fish species are regarded as alien species. Some of them (e.g., round goby *Neogobius melanostomus*) have found a convenient niche for their reproduction and life, and are increasing their abundance in the Baltic.

One of the peculiarities of the Baltic Sea is the relatively high importance of coastal fish of generally rather low biomass. In some ecosystems of the Baltic Sea, freshwater fish, roach, ide, bream, carp, bleak, etc., of the cyprinids, perch, ruffe, etc., of the percids, sticklebacks, gobiids, sand eels and a number of other species constitute a sufficiently ample fraction (Draganik 1996; Thoresson 1996, etc.). The main ecological role of this group is supporting the Baltic ecosystems on the level of fish stocks. These species constitute the main food resource of cod, flatfish, salmon, trout, eel, garfish, pikeperch and other species exploited by commercial fisheries. Their composition varies from place to place. In some areas, the stock size of freshwater and other coastal species justifies their exploitation on a commercial level. They (especially pikeperch, pike and perch) are also interesting as the objects of amateur fishery. Despite the economic revenue for the commercial fishery of coastal fish species representing only a fraction of the offshore pelagic and demersal fishery in the Baltic, the target coastal species are of high socio-economic importance in being highly valued in both recreational and small-scale coastal fishery. They are of great importance to human society, both from socio-economic and cultural points of view.

The most important commercial living resources are Baltic herring, sprat, cod, flounder, plaice, dab, turbot, brill, salmon, sea trout, perch, pikeperch, pike, European/common whitefish, European eel, vendace, garfish, smelt, eelpout, lumpsucker, river lamprey, etc. Cod, sprat, herring, flounder, salmon and trout occur over most of the Baltic Sea, while the other species are localized, often related to certain specific coastal areas.

Fish catches in the Baltic Sea since 1930 are shown in Fig. 3.15. In the period 1974–1984, the total fish catches in the Baltic were reported at 850,000–990,000 tonnes. The highest nominal catches were recorded during 1996–1998, when total landings peaked at 1,100,000 tonnes as a result of a very high landing of sprat. The present annual level of the total catch is approximately 700,000 tonnes (Table 5.1).

The main target species in commercial fishery – cod, herring and sprat – presently constitute about 95% of the total catch. Sprat and herring landings have dominated total catches since the beginning of the 1990s. Trawl fishery is the most important method for the exploitation of these species. In cod fishery, nets and long lines are used, while on herring spawning grounds, trapnet and net fishery are also applied.

Table 5.1 Landings ('000 tonnes) by country and species groups

Source Eurostat/ICES database 2006–2012

Country	Crustaceans and molluscs	Freshwater fish	Diadromous fish	Marine fish	Total	%
Denmark	12	0	1	72	84	11.5
Germany	0	1	0	54	55	7.6
Estonia	0	1	1	72	74	10.1
Latvia	0	0	1	75	76	10.4
Lithuania	0	0	0	20	20	2.8
Poland	0	3	0	108	111	15.2
Finland	0	8	2	114	125	17.1
Sweden	0	0	2	140	142	19–5
Russia	0	4	1	40	44	6.0
Total	12	17	8	695	731	100.0

Coastal fisheries are conducted along the entire Baltic coastline. Coastal fisheries target a variety of species with a mixture of gears, including fixed gears such as gill, pound and trapnets, weirs and Danish seines. Trawling is forbidden in the coastal zone in most countries.

The distribution of fish catches has changed considerably over the past decades. The distribution of the Eastern Baltic cod, in parallel with the decrease in its stock size, has contracted to the southern areas since the mid-1980s. The abundance of sprat, on the other hand, has increased mostly in the northern areas of the Baltic Proper.

5.2 Main Goals of the Management of Ecosystems and Living Resources

As fishing technology has developed to a point at which a significant proportion of the total fish population could be harvested in a rather short time, e.g., within a year, and because fisheries are such that it may be economically feasible to do so, we need to manage the exploitation of the living resources. Fishing also has a number of side effects besides the removal of the target species, such as bycatch of non-commercial species and damage to the bottom fauna by hauled gears with bottom contact. Management therefore is regulating the conflict between the effort that would exist if there was free fishing and what society might accept in terms of impact. Furthermore, fish resources are renewable, and to maintain a healthy stock, a certain degree of fish survival is required for spawning. This latter consideration has led to the Maximum Sustainable Yield (MSY) concept, which, to put it briefly, dictates a limitation on fishing mortality – or the equivalent fishing effort – to maintain full reproduction of the target species.

The basic aim in management is to limit human activity within the boundaries considering the sustainability of the activities. Sustainable development 'meets the needs of the present without compromising the ability of future generations to meet their own needs' (*Our Common Future*, the Brundtland Commission 1987), and we must balance the need to conserve a healthy and productive environment with the need to use nature's resources. The sustainable development concept was formulated as the first principle of the 1992 Rio Declaration: '*Human beings are at the center of concerns for sustainable development. They are entitled to a healthy and productive life in harmony with nature*'. Ecological sustainability is a condition for social and economic sustainability.

Declarations are global in scope, but must be implemented on a regional scale. As has repeatedly been mentioned in this book, the Baltic Sea ecosystems are unique in many aspects, not the least of which is the scale of the brackish-water area. We are therefore considering management actions that are specific to the dynamics of the Baltic Sea ecosystems and effective in relation to the human activities influencing the Baltic Sea. Management should consider concrete conditions and their dynamics.

Management is required, because the fishing fleet capacity is able to affect the stocks beyond what is accepted by society, but also because overexploited stocks are providing suboptimal economic return: we can get more from the ecosystems if we fish less. We need to manage the exploitation of the living resources for two principal reasons: (1) to restrict the impact on the ecosystems within the boundaries defined by society, and (2) to assure that we use the living resources in an optimal way. The meaning of 'optimal' in this context usually has two definitions – Maximum Economic Yield (MEY) and Maximum Sustainable Yield (MSY). In both cases, the emphasis is on the long-term return that can be gained from the ecosystems. However, although not accepted by international society, there are considerations and objectives that could be defined by taking the short term into account. When defining long-term objectives, we need realistic assumptions about the economic future typically formulated in a model on price development, technological development (costs per unit effort and selectivity) and the productivity of the fish stock. We also need to consider whether the current mix of species and size groups that we extract from the ecosystem will be stable over time or whether the present markets will change and demand a different mix. This means that there could be a possibility that non-commercial species will become valuable at some later time. The point to be made is that the concepts that are being put forward are by no means cast in stone and what should be included when calculating MSY is not necessarily constant over time.

Practical management of fisheries is based on the control of an area and the activities taking place in that area. A management unit is defined by the species and the area where it occurs. This is in some cases supplemented by the temporal occurrence of the species in an area. Clearly, this definition is not identical to the biological unit stock. In practice, much effort is devoted to how best to merge the two considerations, and the structure of the fishery is an element in this evaluation, in particular, for fisheries that have marked seasons. As fish stocks may change the area of occurrence and fisheries may change their selectivity and fishing grounds, the definition of management stock is subject to change over time. Practical assessment work is nearly always based on the management definition of a stock, because the assessment results are required in a management context, while the theoretical basis for the assessment is the unit stock. There are exceptions to this in the Baltic Sea, for example, the salmon in the Baltic Proper is a mixture of some 30 stocks, and assessment and management take account of the stock mix. Within a management area, the assessment scientist might map out a further breakdown to account for separate reproduction patterns, growth and mortality.

Ecosystems and living resources should be assessed and managed within their natural boundaries. To reach the aim of management, one should concentrate on the environmental conditions that influence the structure of the ecosystem and limit its functioning, diversity and natural productivity. The scales of ecosystem treatment may be variable. The boundaries of management are determined by users, managers, scientists and the local society.

Based on the standpoints of ICES and HELCOM (ICES 2000; HELCOM 2007a, b), the overall main goals of the management of ecosystems and fisheries may be formulated as follows:

- Ensuring sustainable, sound and healthy ecosystems:
  - Baltic Sea unaffected by eutrophication
  - Baltic Sea life undisturbed by hazardous substances
  - Favourable status of Baltic Sea biodiversity
- Achieving sustainable exploitation of the living marine resources:
  - Economically viable fisheries within sustainable limits
- Maritime activities in the Baltic Sea carried out in an environmentally-friendly way.

The concept of Ecosystem Services is used in many contexts to communicate that mankind uses marine ecosystems. For example, the ecosystems are the basis for fish production, and marine ecosystems both receive and process waste water and provide space and fresh water for aquaculture, to mention but a few. Ecosystem services are typically classified as supporting, regulating, cultural and provisioning services. These services are interrelated and influence human welfare both directly, through human use, and indirectly, via impacts on supporting and regulating services. However, they are increasingly under threat from widespread and growing pressures for the overfishing of marine and coastal resources, water contamination, coastal habitat destruction and a general loss of biodiversity. Impacts on ecosystem services can be examined in qualitative terms, by quantitative measurements or through economic valuation. The latter is introduced with the aim of improving the understanding of the importance of coastal and marine ecosystems to humans, to better inform decision-makers, and thereby to support attempts to influence human behaviour (Garpe 2011).

Management of environment and fisheries is aimed at assuring sustainability of the whole ecosystem; therefore, environmental and fisheries management should be treated together. As presented in Chap. 2, because of geographical differences in salinity, temperature and other environmental conditions, there are local populations of fish and other organisms that form local ecosystems. Each of these ecosystems needs to be considered separately, while management considerations dictate the use of larger units. Furthermore, for fisheries, the sampling of the catches provides key information, but the detail that can be obtained is limited by practical constraints.

Management is based on national sovereignty, which is area-based, but as many of the effects are trans-boundary, management needs to be internationally coordinated. Cooperation takes place primarily through the relevant international organizations. Ecosystem treatment includes the requirement that treatment of every ecosystem and population should be related to concrete conditions and their dynamics.

5.3 Basic Principles in the Assessment and Management of Ecosystems and Living Resources

Basic principles of the assessment and management of marine ecosystems are sustainability, a precautionary approach, and the single stock perspective (ICES 2000).

1. Sustainability in fisheries can be considered from three perspectives:

   (1) Resource perspective, in which the reproductive capacity of the fish stocks is the core issue;
   (2) Ecosystem perspective, in which the continued functioning of the ecosystem as a productive and healthy environment is important;
   (3) Production chain perspective, which includes social and economic sustainability.

   This general political conception is in conformity with the theory of maximum sustainable yield, which was a component of the theory of fishery management in the 1920s and gained the position of the leading principle of management in the 1950s.

2. A precautionary approach in the exploitation of fish resources is also a leading principle. Application of this principle should minimize the risks arising from the limitations of our knowledge. The possibility of reproduction of the stock biomass as the basis for its exploitation should be kept in mind. Treatment of ecosystems and their components cannot always be safely calculated with models; certain limits exist in our understanding of the behaviour of ecosystems and our ability to forecast their responses.

3. Management of fisheries should grant sustainability to the whole ecosystem. Management of fisheries and ecosystem protection should be treated together. A very important principle of ecosystem treatment is considering regional ecosystems separately. This principle is based on the requirement that biological systems be considered separately. Due to different salinity, temperature and other environmental conditions, during historical development, local populations of fish and other organisms have formed in different parts of the Baltic Sea. Each of these populations has its own area. The conception of panmictic mixing cannot be applied to the members of these populations. Therefore, to obtain correct data in assessments, every population should be treated separately.

5.3.1 Assessments and Management Recommendations

On the basis of their goals and the methods of composition, the assessments and prognoses for the management of fisheries can be divided as follows.

1. Short-term changes in the distribution of fish related to environmental variations (mainly deviations in temperature, currents, etc.) or to the biological or seasonal cycle of fish, and the prognoses based on these deviations. The estimates on the short-term changes in the distribution or behaviour of fish concerning variations in their movements, distribution or activity (spawning, feeding or wintering migration, etc.) are not related to enforcing principles of better management of ecosystems or the rationalization of the exploitation of living resources. Such estimates are usually made for the better organization of fisheries.
2. Estimates of the condition of ecosystems and the quantitative forecasts on the size and composition of the existing fish stocks 1–3 years in advance. The objective of this work is the quantitative assessment of resources through the established stock units. Such prognoses can be composed on the basis of the data characterizing the condition and exploitation of resources (the size, species and age composition of catches, average weight of individuals by age groups, maturity stages of fish, materials indicating catch intensity, etc.). The numerical strength of the recruiting age group, which is of very high importance for the future catches, is taken into account in accordance with a previously agreed-upon method. The chief objective of such assessments is the short-term forecasts used in operational management, e.g., setting annual quotas and other regulations that affect the exploitation of the fish stocks to prevent their overexploitation.
3. Long-term qualitative prognoses on the state of ecosystems and fish stocks in the future. The long-term prognoses are composed 10–20 years or more in advance. These are based on periodic fluctuations of ecosystems in the Baltic Sea and can be used for the preparation of strategic decisions related to the impact of changes in ecosystems and fish catches upon society.

5.4 Assessment of Existing Fish Resources of the Baltic Sea

Below, the basic principles of making assessments and prognoses on ecosystems and the existing fish stocks (1–3 years in advance) and the long-term forecasts (10–20 years in advance) are discussed. The sustainable exploitation of ecosystems and fish stocks requires composition of both of them.

The assessment procedures forming the basis for assessment and sustainable management of existing fish resources are performed by the ICES Working Group on the Assessment of Fish Stocks in the Baltic. The necessary material is submitted by the countries surrounding the Baltic Sea. The members of the WG represent the riparian countries, and additional members are possible. The methods in use are checked by the ICES. Management of fisheries regulates human activity and exercises influence upon fishing operations. Consequently, this activity affects the viability of resources indirectly. In the long perspective, the sustainability of resources is of basic significance for achieving economic and social goals. The basis of fishery resources is multiple. Therefore, sustainability is related both to fisheries and ecosystems.

For the assessment of fish stocks today, a method of analysis of virtual populations and corresponding complementary methods are widely applied (ICES 2000, 2011a).

Safe Biological Limits and Reference Points The agreement of the UN Conference on Straddling Fish Stocks and Highly Migratory Fish Stocks stated that management should adopt measures to ensure the long-term sustainability of fish stocks, promote the objective of their optimal utilization, and ensure that such measures are based on the best scientific evidence available and are designed to maintain and restore stocks at levels capable of producing maximum sustainable yield. As a means to achieve this, the agreement directs attention to the use of reference points intended to constrain harvesting within safe biological limits within which the stocks can produce maximum sustainable yield and the need to implement improved techniques for dealing with risk and uncertainty (ICES 2000).

The yield that can be extracted from a stock depends on the environment and its productivity. The environment dictates an overall upper limit for the stock size and its growth. Environmental management aims to secure a healthy and productive environment, but is restricted to assure that human activities do not impair the health of the ecosystem, mostly by restricting those activities. Fisheries management regulates the fishing effort or catch that is removed from the sea. However, the yield curve is peculiar compared to other production curves in as much as it has an optimum (MSY) at a particular fishing mortality and management goals for fisheries based around this fishing mortality. Therefore, reference points for advisory use in fishery management are based on two main considerations: (1) the maintenance of the reproductive potential of the stock and (2) the maintenance of the growth potential of the stock. While the main considerations remain unchanged over time, the emphasis that ICES has placed on the different components has changed since ICES began to provide fishery management advice for the commercial fish stocks in the Northeast Atlantic (Lassen et al. 2014). The reference points represent long-term averages and do not account for natural variability and uncertainties in assessment or how precisely a policy can be implemented in practical fishery management. ICES introduced buffers that absorb the natural variability and uncertainty in the assessment, i.e., ICES implemented the 'precautionary approach'. This means that the fishing pressure on which the advice is based is less than the actual estimate of the reference points, thereby allowing the system to be resilient in absorbing the variability. The safe biological limits are a key concept in relation to the state of a fish stock and fisheries: a stock is considered to be inside safe biological limits if it, according to our best present knowledge, will be able to sustain its productivity under the present exploitation. The safe biological limits are defined according to reference points which are quantifiable indicators of the state of a stock and the fisheries exploiting it. A reference point is associated with a management objective (e.g., maximizing yield or preserving the reproductive capacity of the stock). Thus, a reference point is a signpost that can be used to evaluate the present situation in relation to objectives and to give guidance for action as to the direction in which we should move.

The reference points introduced above are based on stock considerations, i.e., single species models. However, fish interact – fish eat fish – and therefore, in many systems, an analysis of the dynamic of each species without accounting for this predator–prey interaction or cannibalism – older fish eat the progeny – is not satisfactory, and the reference points for the optimal fishing pressure need to be modified to account for these effects (Gislason 1999).

The reference points presently in use are the spawning stock biomass (SSB) and the fishing mortality (F). The classical Maximum Sustainable Yield concept is an early example of a reference point. The underlying management objective is long-term yield maximization. Determination of this point on the yield per recruit curve is absolutely necessary. Comparing the fishing mortality estimates of a certain year with the values allowing for the maximum sustainable yield, it can be understood whether the actual fishing mortality should be increased or decreased, and by how much, to obtain sustainable exploitation of the stock on the maximum yield level.

The model for stock exploitation on the level of the maximum sustainable yield does not answer two important questions related to the precautionary approach to resources for granting sustainability: (1) the model does not involve the necessity of accounting for the reproduction potential of stocks. The SSB decreases at the increase in fishing mortality, and at the level of maximum yield, it need not grant the reproduction capacity necessary for sustainability; (2) the model represents long-term averages. It does not involve natural variability and the uncertainty related to fishery management. Therefore, it is necessary to add the reference points involving sustainability and the precautionary approach.

Consideration of Reproduction Potential at Stock Assessment The ability of a stock to reproduce itself is associated with the number of fertilized eggs/embryos produced (with the size and composition of the parent population) and the state of the marine environment. ICES uses the Spawning Stock Biomass (SSB) concept for the characterization of the reproductive potential of the stock. It is important to know how small the biomass of the SSB can be in order not to have a negative impact upon the recruitment. ICES applies historical datasets to determine this SSB size. The level of the SSB below which a reduced recruitment has been observed is used as a biological reference point for stock sustainability. This spawning stock reference point was referred to as the Minimum Biologically Acceptable Level (MBAL) (ICES 2000). However, in addition to the size, the age composition of the spawning stock is also extremely important in the estimation of the reproductive potential of the stock. Differences in the quality of eggs and of the spawning time between the first-time spawners and the medium/older age groups allow older spawners to contribute much more (compared with the first-time spawners) to the recruitment. The difference between these age groups varies by species. Besides reducing the absolute value of the SSB, high fishing pressure segregates medium/older age groups from the stock, i.e., the most important specimens at reproduction. These facts substantially complicate the use of the SSB as the definitive measure of the reproductive potential of a stock.

The abundance of fish year classes also very clearly depends on environmental conditions. Environment can have an important impact on the conditions of both a high and a low SSB. In the brackish-water Baltic Sea, with the environment varying in each natural region, accounting for environmental conditions and a sufficient SSB by natural regions and periods of environmental fluctuations is an especially complicated task. It is extremely necessary to have an index to represent the reproduction ability in the assessment of stock conditions; however, understanding how to apply the SSB for this purpose under fluctuating external conditions requires voluminous long-term investigations.

Precautionary Approach As in fisheries management in general, relevant activity in the Baltic Sea involves the risk that management actions will not grant fulfilment of management decisions. This can result from an inadequate perception of the management practice and a wide natural variation of fish populations. Fisheries management involves the following uncertainties:

1. The quality of data on the condition of stocks (catches and discards, age composition, natural and fishing mortality, abundance of the forming year class, etc.) is not as good as that required for a high-quality assessment and catch prognosis.
2. Catch forecasts composed on the basis of low-quality data cannot be of high quality. In addition, the impact of management actions, fishing vessels and other gears is usually mostly unknown.

However, the precautionary approach requires that management take into account the risks which have originated as a result of management decisions.

Sustainable Precautionary Management The most understandable goal of precautionary management is the prevention of a decrease in the population biomass to below the critical size. To achieve this situation, ICES, in its advice, has applied the principle that the level of the SSB should be higher than the M_{BAL} level and the fishing mortality lower than the level that should be avoided (ICES 2000). In the advice, the M_{BAL} level is designated the reference point B_{lim} limiting the spawning stock biomass. To keep the SSB biomass higher than the B_{lim} point, a limiting reference point (F_{lim}) was added relating to fishing mortality. F_{lim} is a supplementary measure to sustain the spawning stock above B_{lim}. This can help to avoid situations in which it is necessary to prevent a rapid reduction of the spawning stock.

For fisheries management to be precautionary, management actions should be taken well before the critical spawning stock sizes are approached and also at the spawning stock level well above B_{lim}. For this, another set of reference points has been introduced which indicates the state in which action must be taken if the probability of passing beyond the limit reference point is to remain low. In the case in which the stock is approaching the limit reference points, ICES has supplemented the limit reference points with 'pa' (precautionary approach) reference points, F_{pa} and B_{pa}. The 'pa' reference points can be related to an anticipatory management strategy which requires specific activity in case the stock falls behind that reference point.

Multispecies Considerations Fishery exploits a number of fish populations. For both biological and technical reasons, fisheries management should involve several stocks. Fishery science and fisheries management have had a certain amount of success in the application of principles of sustainability and precautionarity for the case in which only one population is involved (i.e., when only one system is treated in which no groups with different stock dynamics occur). It is much more complicated to succeed in handling several populations together. The main reason for this is the complexity of the task, both in terms of science and management. Especially problematic is the joint handling of relations between several populations and the environment. It has been possible to develop conceptions and a quantitative framework for reference points, models, the sustainable exploitation of one system (population) and a precautionary approach. The situation is very different if we try to treat ecosystems in an analogous way. The present-day level of knowledge on the complexity and natural variability of ecosystems does not usually enable their usage for prognostic aims. Also, the influence and mechanism of human impact and other impacts causing constant disturbances cannot presently be fully understood. The possibility of solving these problems depends on developments in the corresponding fields of science.

Estimating the Temporal Development of Fish Stocks The status of the fish populations is assessed on a stock-by-stock basis. The foundation for this separation of the fish populations is the unit stock. A unit stock comprises all the individuals of a fish species that are part of the same reproductive process. It is self-contained, with no emigration or immigration of individuals from or to the stock, but the stock may spread over a wider area and may mix with other stocks of the same species in some areas and in some seasons. In an ideal world, a unit stock occupies a well-defined spatial range and is independent of other stocks of the same species, although random dispersal can occur. The impact of fishing on a species cannot be determined without knowledge of this stock structure. However, in the Baltic, there are few if any such isolated unit stocks, but a certain degree of exchange between the populations of the same species exists. During the past 20 years, the significant improvement in DNA analysis techniques has provided a tool that allows for detailed insight into the unity of reproduction and helps to better define these unit stocks.

Having defined the stock, and thereby the subset of the available data relevant for this particular assessment, three types of data (catch, abundance and biology data) are fed into mathematical models that represent the factors causing changes in harvested fish stocks. The models produce estimates of the stock size and the exploitation pressure (fishing mortality). It is important to stress here that the quality of the result does not only depend on models used. A far greater number of mistakes is usually made in the sampling and treatment of biological data, which require high and wide knowledge and practical experience in aquatic biology and ecology. The final result is the establishment of the historic development of the key indicators, i.e., fishing mortality, spawning stock and recruitment. As commercial fisheries are economic activities, the landings are also included in the standard presentation of the status of the stocks.

When possible, stock assessment models include information on ecosystem and environmental effects to improve the interpretation of historical information and the precision of forecasts.

In practice, the assessments are thrown between two evil horns. On the one hand, the assessment aims to deal with the populations of an area in as much detail as can be enabled by knowledge of the stock structure, while on the other, management aims to establish a simple and effective system that is based on fairly large areas. The methods of fish stock assessment for routine management needs are chiefly model-based calculations on existing resources. They allow for projection of the condition of the existing stock some years ahead (usually roughly 2–3). The temporal development of biomass and mortality is traced through data from the commercial fisheries from dedicated research vessel cruises (abundance surveys). Stock development is described through analytical (age-based) or dynamic pool (biomass) models. These models exist in several versions. The analytical model is based on the cohort concept, i.e., it is assumed that the fish inhabiting the sea in a certain year will be the same fish the next year, but a year older. The number is reduced by the number fished and the number removed from the population through the biological processes designated as 'natural mortality'. The biomass is the combined result of the individual growth and the reduced numbers. Each year, a new year class is added to the population based on the number of eggs produced and the environmental and biological conditions that determine survival through the egg, larval and juvenile fish stages. The dynamic model is based on biomass development and growth; mortality and recruitment are modelled together, while the fishery removals are accounted for explicitly. The estimates of fishing mortality and stock size are based on fitting time series of abundance indices and removals into these models. In some cases, the fishing effort is also used as a direct indicator of fishing mortality. There are also models that work with data which include 'holes', i.e. years where no or only a reduced dataset was obtained.

Model versions vary as to how a particular model accounts for the uncertainties that exist in the data and how the observed data relate to population characteristics; for example, it is often assumed that the abundance survey results (CPUE data from research vessels) are proportional to the stock size, although this is not necessarily the case. Another difference is how uncertainty in the catch data (catch-at-age) is handled: whether a full account of the sampling process is included in the assessment model or whether sampling uncertainty is considered to be minimal and can be ignored in the calculations.

Because a key objective of the assessment process for commercial fisheries is to provide a basis for management, it is necessary to establish the link between the expected landings and the regulatory measure, typically the TAC[1] or fishing effort. Therefore, the next step in the assessment process is to provide forecasts for stock development under different fishery scenarios. These scenarios are defined in terms of total yield, but may include an account of how fishing affects the components of the population, e.g., the fishery may operate on the spawning grounds and thereby concentrate its effort on adult fish, or the nursery grounds are fished and the fishery is taking juveniles in particular. Fishing on the feeding grounds may take a mixture of juveniles and adult fish. To establish relevant scenarios, the scientists need to build a model that accounts for how fishing effort generates fishing mortality in the stocks that are a part of the catch mix. Another scenario to be considered is that in which it is relevant whether a changed gear, e.g., a large minimum mesh size, could lead to higher yields.

Scenarios are selected based on long-term considerations, the overall goal being to obtain the long-term Maximum Sustainable Yield (MSY), which includes two key features: that the reproduction in the stock is at maximum capacity and not impaired by a low spawning stock and that the growth potential of the individual fish is used to its fullest extent. Where the fisheries exploit a mixture of stocks and where there are significant interactions between stocks either competing for food or predating on each other, the MSY for a stock depends on the fishing pressure on the other stock. In such cases, the choice among possible scenarios becomes more complicated and the management plan that determines the relevant fishery scenarios must be based on additional socio-economic considerations.

Coastal fish stocks such as vimba bream, roach, perch, pikeperch and pike are assessed by HELCOM (2006, 2008) based on monitoring coastal fish in 15 areas in the Baltic Sea, using multi-mesh gillnets and gillnet series. Fishery statistics covering commercial coastal catches are available, but are often insufficient to understand the removal of the fish from the ecosystems, as there are also, in many cases, significant recreational and other non-commercial catches. The EU DCF programme includes an attempt to estimate these catches through surveys, but at present, the key information comes from the gillnet surveys. The key indicator in the surveys is the Catch per Unit Effort (CPUE) for a number of key species and the species composition in the catches.

[1] TAC: Total Allowable Catch, landing quota for the total fishery of a specific species and area

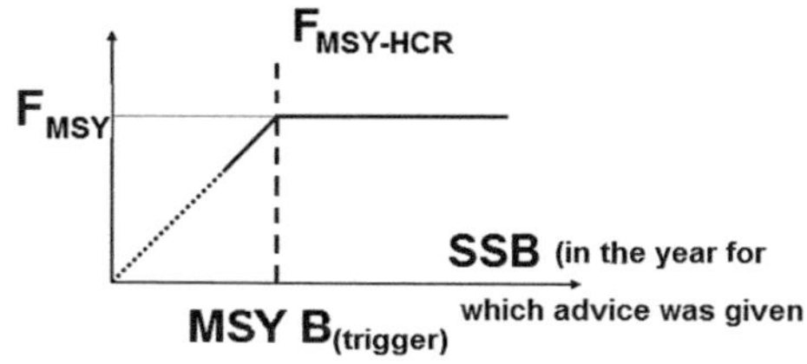

Fig. 5.1 Harvest control rule, relating the exploitation pressure and the size of the spawning stock to the advisable fishing pressure in the coming fishing season

Management Plans and Harvest Control Rules The FAO Code of Conduct for Responsible Fisheries (1995) urges states to establish 'a fishery management plan or other management framework'. In the Baltic Sea and elsewhere in the North Atlantic, this plan has been or is being implemented as agreed–upon Management Plans with imbedded Harvest Control Rules (HCR). These management plans are pre-agreed responses to the variation in stock size and changes in the reproduction capability of the stock. The management plans also include strategies for regulating impacts on non-target species and abiotic conditions. The HCR is the biological core of the management plan that determines what the actual allowed fishery should be. The HCR typically relates the exploitation pressure and the size of the spawning stock to the recommended fishing pressure for the coming fishing season. A detailed description of the HCR behind the ICES advice is given in (ICES 2015b, Fig. 5.1). It is similar to approaches used elsewhere (e.g., Cooke 1999) and considers that the harvest strategy is composed of an operating model and a management procedure. The operating model consists of biological and fishery models which define the stock and fleet dynamics controlled by the strategy. The management procedure consists of the observational activities (e.g., surveys and fishery monitoring) which are used in the stock assessment to provide the indicators used in the HCR. Implementation error (e.g., catch misreporting) is considered here in relation to the observations. A large number (e.g., 1000) of long-term simulations are undertaken in which the stock and fishery dynamics are projected under a range of uncertainties to evaluate the performance of the management procedure. ICES has determined that a harvest strategy is precautionary if the probability of the risk of the SSB falling below B_{lim} and fishing mortality rising above F_{lim} is less than 5%.

ICES most often advises the TAC that will generate the fishing mortality that, in the long-term perspective, is expected to generate MSY, but also helps develop scenarios based on management plans that follow specific sets of socio-economic objectives.

Fisheries affect other parts of the ecosystems besides the target species. Examples of this are trawls with bottom contact; non-commercial bycatches of benthos, fish, marine mammals and sea birds; abiotic effects on the bottom from trawls, etc. These effects are considered as part of the ecosystem approach, but do not influence the reference points.

5.5 Assessment and Management of Ecosystems in the Baltic Sea

Humans affect the Baltic ecosystems significantly, and societies around the Baltic Sea have attempted to control this impact. This is a global approach which was institutionalized by the 1972 Stockholm declaration of the United Nations Conference on the Human Environment. This declaration established that '*A point has been reached in history when we must shape our actions throughout the world with a more prudent care for their environmental consequences. Through ignorance or indifference we can do massive and irreversible harm to the earthly environment on which our life and well-being depend. Conversely, through fuller knowledge and wiser action, we can achieve for ourselves and our posterity a better life in an environment more in keeping with human needs and hopes …* '. However, with the technology that is available today, there is a fundamental conflict between the exploitation of resources and the wish to conserve ecosystems. Regulation of the exploitation is therefore built on political compromises, and as a background to making these compromises, we have built systems which monitor and assess the status of the ecosystems.

During the last 100 years, the state of the biota in the Baltic Sea has basically changed. The organisms that have found their home in the Baltic since the last deglaciation and have adapted to the geological, climatic and other natural factors have become increasingly affected by humans. Anthropogenic influences have unbalanced ecosystems with intemperate exploitation of living resources, pollution of the marine environment, facilitation of invasion of alien species, etc. As the physico-chemical and ecological conditions in the Baltic Sea cannot be identified with the conditions in 'normal' seas, the result of the adaptation of living resources to the natural and human-made environmental conditions in the brackish Baltic Sea is different from the corresponding process in seas of normal salinity. Today, the ecosystems of the Baltic Sea have deviated from their ordinary state, and the ecosystem services and goods of the Baltic Sea are no longer available in the same quality and quantity as they were earlier.

The impacts are transboundary, pollutants are conveyed with water masses, airborne pollution spreads widely, and fish stocks know no boundaries. For these reasons, overall management needs to be coordinated at an international level involving all affected parties, and international organizations have been built to facilitate the necessary cooperation between countries.

Below, we take a look at the monitoring activities in the Baltic Sea, assessment of the status of the Baltic Sea ecosystems, and management of human activity. Based on the assessments, decisions concerning management are made, if required. Management is coordinated internationally through a number of organizations, including the European Commission (Fisheries and Environment), HELCOM (Environment), EU–Russian Commission (Fisheries) and ASCOBANS (Marine Mammals).

Monitoring and assessment of the Baltic Sea ecosystems are coordinated internationally through HELCOM and ICES. These organizations cooperate through a number of joint working groups.

The basis for the assessments is a comprehensive monitoring programme based on the HELCOM Monitoring and Assessment Strategy. The HELCOM programme is designed to take the varying environmental conditions in the Baltic Sea into account and addresses the range of topics that were identified above. The HELCOM programme includes monitoring the status of coastal fish, while the commercial marine fish are covered by national and EU programmes, with ICES as the coordinating organization. The HELCOM programme addresses the Baltic Sea Action Plan, and also addresses requirements set up in the EU Marine Strategy Framework and Water Directives for the EU Member States (all Baltic Sea countries except the Russian Federation).

Monitoring is carried out by national laboratories following standards common for all participating laboratories. The laboratories participate in exchange programmes to assure their analytical capability.

Coordination of the monitoring programmes is organized through ICES and HELCOM. This includes not only the sampling programmes, but also quality assurance programmes and exchange and analysis of the data. The key tool is the use of working groups staffed by scientists working in universities and national laboratories. Based on data and preliminary analyses brought to the table within these groups, the groups review the analyses and develop a consensus assessment. HELCOM-based groups, joint HELCOM/ICES working groups and ICES working groups are involved in this process. HELCOM partly works based on input from ICES, but also makes extensive use of 'lead countries,' a situation in which a country accepts the obligation to provide an initial assessment that is reviewed and revised as appropriate among the HELCOM delegations.

When HELCOM was first established, the status of the Baltic Sea ecosystems was dismal. Elmgren (2001) summarized the situation: 'Populations of top predators, such as, e.g., seal, *Halichoerus grypus*, harbour porpoise, *Phocoena phocoena*, and white-tailed eagle, *Haliaëëtus albicilla*, were found to have heavy body burdens of mercury, the pesticide DDT and polychlorinated biphenyls (PCB). The high mercury content made fish in some coastal areas unfit to eat. Oil spills had become more frequent …, as were toxic emissions from pulp and paper mills and other industries. … The deep basins of the Baltic were deficient in oxygen to an extent unseen during 75 years of study…. Major Baltic fish stocks were beginning to be overexploited'.

In 2010, HELCOM provided a holistic assessment of the general status of the Baltic Sea Ecosystems (HELCOM 2010a). In spite of the improvements in the well-being of the Baltic Sea ecosystems since the early 1970s reported by HELCOM, this assessment also presented rather disappointing conclusions:

1. None of the open basins of the Baltic Sea has an acceptable ecosystem health status at the present time.
2. The Baltic Sea ecosystem is degraded to such an extent that its capacity to deliver ecosystem goods and services to the people living in the nine coastal states is

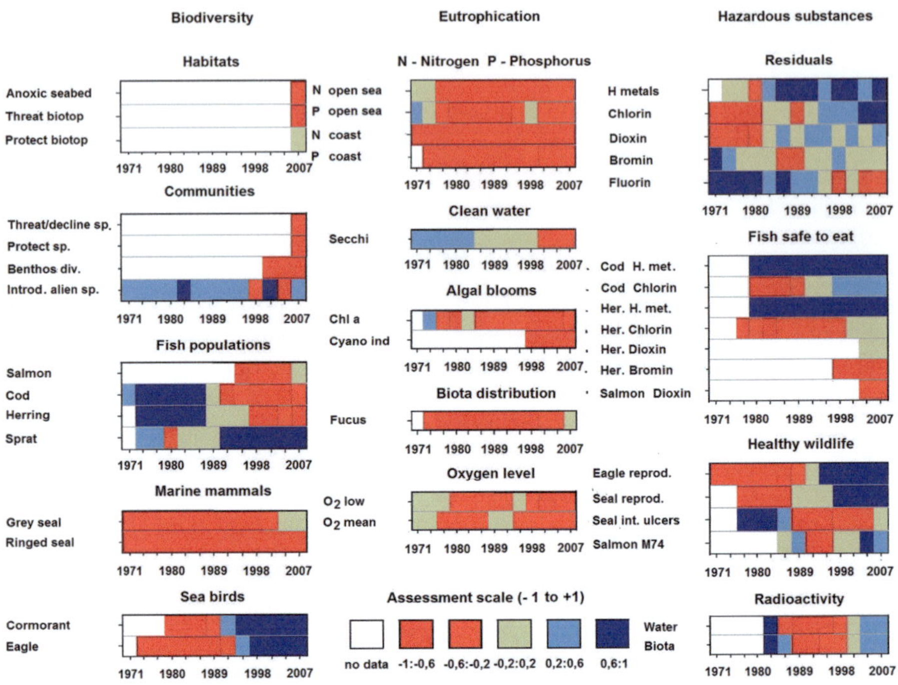

Fig. 5.2 Long-term changes in the state of selected aggregate indicators of the central Baltic Sea. The data are averaged by 3-year periods and the transformed values (the scale from −1 to 1) are grouped into five categories shown by colors (Ojaveer and Eero 2011)

hampered. The resilience of the marine ecosystem has been undermined by the input of contaminants from 85 million people living in the catchment area.

3. Eutrophication, caused by nutrient pollution, is of major concern in most areas of the Baltic Sea.
4. Bothnian Bay and the north-eastern parts of the Kattegat are the only open-sea areas of the Baltic Sea not affected. The only coastal areas not affected by eutrophication are restricted to the Gulf of Bothnia.
5. The entire Baltic Sea area is disturbed by hazardous substances, with the status mainly being classified as moderate. Only at a very few coastal sites and in the western Kattegat is the water still undisturbed by hazardous substances. Key substances of concern are PCBs, heavy metals, DDT/DDE, TBT, dioxins and brominated substances.
6. The biodiversity status was classified as being unfavourable throughout most of the Baltic Sea, since only the Bothnian Sea and some coastal areas in Bothnian Bay were classified as having an acceptable biodiversity status. The results indicate that changes in biodiversity are not restricted to individual species or habitats; the structure of the ecosystem has also been severely disturbed.

Ojaveer and Eero (2011) state (Fig. 5.2) that eutrophication has been and continues to be the most serious problem in the Baltic Sea. The dynamics indicate that rapid improvement cannot be expected in the near future. Since the 1970s, the concentration of toxic compounds has decreased, although in some fish, the content is still above allowed levels. Also, the level of radioactivity has decreased to the pre-Chernobyl level, in both the water and biota. The state of the biological variability varies by the applied indicators, as they have been influenced by both climate and anthropogenic impact. On some indicators of biological variability, data are absent. Also, long-term data on the Baltic Sea organisms are rather rare. Therefore, the reliability of the estimate on biological variability is less than that for eutrophication or toxic pollution.

Reference Points for Environmental Assessment To allow management to determine whether an ecosystem component, e.g., a habitat, is in good health requires comparison between ecosystem components in a normal state and in an impacted state, i.e., a reference against which the actual situation can be compared is necessary. The EU Water Framework Directive indicates four ways of determining such a reference condition: (i) to find an area similar to the one under study but without the pressures; (ii) to hindcast to a time before pressures exerted an influence; (iii) to numerically model an unaffected condition for comparison, and, if none of these are possible, (iv) to use expert judgement. As discussed by Borja et al. (2013), each of these alternatives poses challenges: (i) the unavailability or difficulty of finding unaffected conditions, especially in highly developed areas or within eco-regions; (ii) the question of the baseline conditions for hindcasting (and the basis that some unaffected prior utopia may be unattainable); (iii) the uncertainty or unavailability of numerical models for unbounded marine areas or moving baselines (caused by climate change), and (iv) the perceived reluctance to rely on expert judgement if there is the likelihood of any legal challenge to management mechanisms, such as sanctions on industries likely to impact an area. It is not possible to measure the status of the ecosystems directly. Therefore, the approach is to establish indicators that can be monitored, with each indicator informing on the status of a component of the ecosystem. These indicators are brought together in a holistic assessment of the overall status (HELCOM 2010a). To assess the status, it is necessary to define reference levels for each of these indicators, the concentration by station and season and where biological samples are taken by species and stage, the area of occurrence, or the trend that is desired. Commercial fisheries have long-standing experience with reference levels laid down as the precautionary approach and the MSY target for fishery management. The problem is perhaps more complicated for environmental indicators, but in general, indicators such as ‘no loss of biodiversity whether on the gene, species or population level’, ‘no decline in densities or abundance’, and ‘the area of occurrence for habitats or benthos species’ are applicable. The important requirement is that the reference points or reference trend should be directly measureable by the indicator.

Management of the Environment Management possibilities are limited to bringing inputs under control through the reduction of phosphorus and nitrogen in waste water, regulating the use of pesticides and fertilizers in agriculture, restoring wetlands, and regulating the use of hazardous substances in the industry. These efforts are costly and restrict economic growth. However, the Baltic countries have taken measures coordinated through HELCOM to address these issues, e.g., the use of DDT and PCBs was banned during the 1970s in many Baltic countries. Phosphorus, and to a lesser degree nitrogen, discharges from urban waste water were reduced through the building of urban wastewater treatment plants. Data from HELCOM indicate that the efforts to limit eutrophication and toxic pollution have yielded some results, e.g., threats to the white-tailed sea eagle have now been reduced significantly.

Ecosystems and living resources should be managed in their natural boundaries. To reach the aim of management, one should concentrate on those environmental conditions that influence the structure of the ecosystem and limit its functioning, diversity and natural productivity. The scales of ecosystem treatment may be variable. The boundaries of management are determined by users, managers, scientists and the local society.

Restoration of ecosystems on the scale of the Baltic Sea is an overwhelming challenge. It can be seen that examples claimed to be ambitious concern lakes, which are small compared to the Baltic Sea. An essential precondition is to bring the inputs under control and limit these within sustainable limits. When this has been achieved, there are a number of possible measures that can be introduced and will help the systems to restore their functionalities. These include reclaiming wetlands and the creation of protected marine areas. Complete reclaiming of the wetlands around the Baltic Sea could have a significant impact on the amount of nitrogen discharged into the sea. Also, the possibility of removing dioxins from the Baltic Sea through a directed fishery on herring and sprat has been studied.

The European Commission's policy and strategy for reducing exposure to dioxins has resulted in a range of Community legislation on dioxins in food and animal feed. As mentioned earlier, dioxins cannot be prohibited, as they are not produced intentionally. Production and use of PCBs, on the other hand, has been discontinued in most industrialized countries. In the EU, their use has been prohibited since 1978. Existing PCB-containing equipment was taken out of service by the end of 2010.

5.6 Long-Term Assessments and Forecasts on Ecosystems and Fish Resources

The methods of fish stock assessment for routine management needs are chiefly model-based calculations on existing resources. They allow for estimation of the condition of the existing stock by around 2–3 years ahead. The results of such assessments are applied for solving short-term problems, including the quantification of possible annual catches. However, for correct management of ecosystems and resources, their possible behaviour over longer periods is also very important. Relevant information is necessary for the decisions important to the society and the economy: planning of employment, ordering of fishing gear and vessels, preparations for the marketing of catches, etc. Estimates of ecosystem conditions and the size of fish resources 10–20 years in advance are extremely important for species with a short life cycle and sharp changes in abundance, like sprat and gulf herring. Differently from short-term prognoses, long-term forecasts are of qualitative character, giving information mainly on principal changes. Thus, the field of the application of the long-term forecasts and their aim differ significantly from those of the short-term forecasts. Also, the methods and data required for the composition of short-term and long-term forecasts are different.

Composition of long-term forecasts on ecosystems and fish stocks requires thorough and detailed knowledge about the main environmental factors and the structure and functioning of the corresponding ecosystems, as well as possession of long-term high-quality databases of the regime-forming factors and possible developments in ecosystems and fish stocks during the long-term period in question.

Detailed knowledge of the conditions of the formation of the abundance of year classes of the considered populations, as well as of their diseases, enemies and mortality in various states of the ecosystem, is also extremely important. Special attention should be paid to the character and possible variations in anthropogenic impacts (exploitation, eutrophication and toxic pollution of the environment, invasions of alien species, etc.) during the prediction period.

Due to heavy predation of cod on herring and sprat, increased abundance of cod self-evidently means an increase in the predation pressure on the pelagic species and a drop in their abundance. During the periods of decreasing abundance of cod populations, the pelagic species usually flourish.

Up to the present, the only long-term forecast valid for the Baltic Sea ecosystems and fish stocks was composed in 1989 (Kalejs and Ojaveer 1989) on Baltic cod, sprat, sea and gulf herring up to 2008. The forecast was short, saying that '*the period up to 1992 will generally be unfavorable for the formation of rich cod, autumn herring and Southern Baltic spring herring year classes. However, after some warmer winters, especially at high temperatures in spring and summer, good sprat recruitment and in milder winters average autumn herring year classes will develop. Environment will less limit the abundance of populations better adapted to low salinity and severe winters (e.g. the Northern Baltic herring populations). The period from 1993 until 2008 will favor, probably more than the preceding one, the marine boreal fauna.*'

Correspondence of the forecast to the actual data can be seen in Fig. 5.3. In this figure year class, abundances of the eastern population of Baltic cod, Baltic sprat, herring of the open Baltic, herring of the Gulf of Riga and the Gulf of Bothnia in the periods 1973–1987 (before the composition of the forecast) and 1988–2008 (the forecast period) are presented (data from ICES 2011a). It can be seen that in comparison with the period 1973–1988, in 1989–2008, the abundance of the year classes of Eastern Baltic cod and open Baltic herring significantly decreased, while in sprat, Gulf of Riga and Gulf of Bothnia herring, the abundance of year classes increased. After the strong saline water inflow of 1993, the salinity in the Baltic Sea increased. The ICES Working Group estimated that after the mid-1990s, some cod year classes of comparatively average/good abundance have developed. Consequently, the

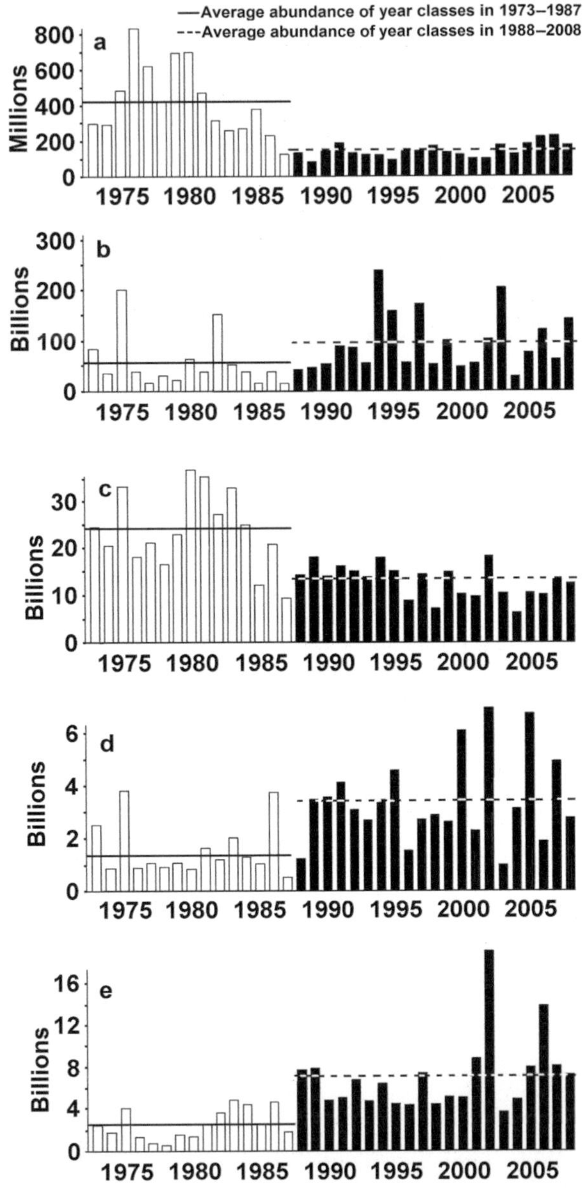

Fig. 5.3 Abundance of the Eastern Baltic cod (**a**), sprat (**b**),sea herring (**c**), Gulf of Riga herring (**d**) and Bothnian Sea herring (**e**) year classes according to ICES evaluations (ICES 2010) in the periods 1973–1987 and 1988–2008. Black bars represent the abundance of the year classes predicted in Kalejs and Ojaveer (1989) Horizontal black line indicates the average abundance of year classes in the period 1973–1987, horizontal broken line in the period 1988–2008 (Ojaveer 2014)

development of the populations/ecosystems considered in the first long-term forecast was predicted actually correctly.

After the occurrence of sharp changes in the ecosystems and fish stocks of the Baltic Sea, the events had been repeatedly analysed (HELCOM 2002; Köster et al. 2003; Alheit et al. 2005; etc). The ecosystem assessments performed in the ICES/HELCOM Working Group on Integrated Assessments for the Baltic Sea covering seven Baltic Sea areas over the last three decades (Bergström et al. 2010; ICES 2009, 2011b, 2013, 2014, etc.) stated that in the Central Baltic Sea, the recruitment and SSB of cod sharply decreased in the mid-1980s and have remained at a low level ever since, while the abundance level of sprat recruitment markedly increased beginning in the early 1990s and the recruitment of sea herring sharply decreased in the second half of the 1980s. In the Gulf of Riga and the Bothnian Sea, the level of gulf herring recruitment has substantially increased since the end of the 1980s–early 1990s. The conclusion summarizes that '… the major period of reorganization in the Baltic was invariably found between 1987 and 1989. In these years distinct strong shifts were detected in all sub-ecosystems and with different types of discontinuity analyses. In several of the systems, abrupt changes were also found during the mid-1990s probably related to the major North Sea water inflow in 1993 following a long stagnation period' (Bergström et al. 2010). Consequently, these evaluations of the changes in the ecosystems and Baltic cod, herring and sprat populations in the period under consideration confirm the changes predicted by Kalejs and Ojaveer (1989).

The methodology of the above forecasts is based on the fact that the fluctuation in the amount of freshwater runoff of the rivers falling into the Gulfs of Riga and Finland has been periodic (Fig. 5.4). This has caused periodicity in the variation of salinity, which is the most important regime-forming factor in the Baltic Sea. The long-term forecasts on the Baltic Sea fish stocks were composed on the basis of the estimated freshwater input into the Gulfs of Riga and Finland.

The regime shifts – large persistent changes in the structure and functions of ecosystems – have gained importance in ecology, because they can substantially affect the state of ecosystems and the services offered by them. Therefore, recogni-

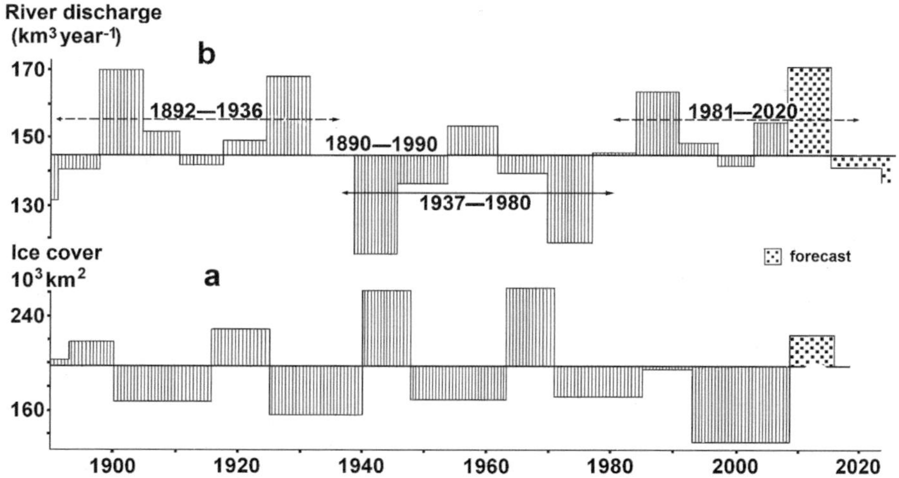

Fig. 5.4 Periodicity of the extent of ice cover (**a**) and river discharge into the Gulfs of Finland and Riga (**b**) during 1890–2009. Foreseeable river discharge and the extent of ice cover are also presented, as they constitute the basis for the prognoses composed by the author's method (Ojaveer 2014)

tion of the arrival of the period of the sequent ecological regime shift is important. Concerning regime shifts, an important aim of science should be forecasting the time of their arrival and their peculiarities. This would enable us to foresee regime shifts and will perhaps help in softening their negative economic and social consequences.

In the periods of increasing freshwater runoff, the salinity in the Bornholm, Gotland and Northern Deeps decreases, reproduction conditions of cod and other organisms of marine background worsen, and their abundance declines. Moderate changes in salinity, which limit the abundance of cod year classes, have usually not substantially influenced the abundance of gulf herring and sprat year classes. Therefore, during such periods, sprat and herring have dominated, as, because of the small abundance of cod, their natural mortality has been low. However, due to the deterioration of their food base (low abundance of valuable food organisms of marine background in times of low salinity), their growth rate has declined.

In the periods of decreasing freshwater runoff and increasing saline water inflows, causing a rise in salinity, cod reproduction has generally been successful and, as a result, its abundance has been high. In such conditions, cod has dominated in the Baltic ecosystems, using rich resources of invertebrates and fish both in the open Baltic and gulfs for feeding. Due to the increased predation mortality caused by cod feeding, the abundance of herring and sprat has been considerably lower. Therefore, their landings have dropped. However, during such periods, the rate of individual growth of pelagic fishes (including herring and sprat) has been good, due to higher salinity/better conditions for their valuable food animals of marine background.

Understanding the potential factors that can evoke regime shifts requires ample knowledge about the impulsions that are capable of altering the system. The periods in the runoff depend on the richness of precipitation; consequently, they are related to climate. It has been suggested that climate is regulated by the factors related to solar/cosmic processes (Svensmark and Friis-Christensen 1997; Svensmark 1998; Ogurtsov et al. 2003; etc.).

The essence of the method is that the state of the Baltic marine environment, ecosystems and living resources can be considered as a part of the global system and their changes can be foreseen based on periodic oscillations in solar/cosmic factors (Schove 1955; Usoskin et al. 2004; etc.). The current long-term solar cycle had the lowest activity values in the 1890s and the maximum number of sunspots in the late 1950s–early 1960s (Usoskin et al. 2004). This maximum coincided with the end of the 1866–1955 Gleissberg cycle and the beginning of the following one (Schove 1955). This was the highest maximum solar activity period during the last millennium. Supposing that the maximum is situated near the middle of the cycle, the length of the current cycle should be about 130 years. According to Usoskin et al. (2004), this is the dominant length of a cycle in the reconstructed series of secular cycles. A cycle of the same length was detected by Hagen and Feistel (2005), who also found a 43-year-long quasi-period nearly equal to the duration of the climate periods in the Northern Baltic suggested by Hela (1966).

We used the periods in the sunspot/climate data as the basic units for the qualitative forecasting of the year classes of fish stocks in the Baltic Sea (Ojaveer and Kalejs 2010, 2012). Periodic changes in the cosmic factors trigger climate alterations. The freshwater input into the Gulfs of Finland and Riga, probably climate-related, hints at a dependence of the basic environmental conditions in the Baltic area on the fluctuation of the Euro-Asiatic continental climate, and probably reflects a link with the Arctic systems where extraterrestrial effects on the marine systems have also been obvious (Yndestad 2004). In the Arctic, historically unprecedented changes started in 1989, temporally not very far from the regime shift in the Baltic Sea (in both areas, the shift resulted in thorough rearrangements in ecosystems and a collapse of cod stocks). They both occurred in the atmospheric regime and cryosphere (Greene and Pershing 2007).

In the Baltic Sea, the second vital environmental factor – the severity of winters – is probably also regulated by climatic conditions. In the Baltic, winter temperatures have not influenced cod resources very much. However, the direction of winds and temperature from January to March exercises a very substantial impact on biological productivity, and the success of reproduction and growth of planktonic organisms, gulf herring and sprat. The spring and autumn spawning herring populations in the Gulfs of Bothnia, Finland and Riga are mainly affected, as they are situated in the zone of the lowest winter temperatures and the longest ice cover duration in the Baltic. Temperature controls the success of herring and sprat reproduction and restricts the main living conditions of the gulf herring and sprat in winter. The area of the thermophilic sprat is especially affected in the NW part of the sea. Cold winters seriously limit the volume of water in which wintering, feeding and reproduction of sprat is possible. Therefore, after severe winters, the abundance of year classes of these stocks has commonly been rather low.

The development of the above method of the long-term qualitative predictions of the year class abundance of fish stocks in the Baltic is complicated by the scarcity of basic information for understanding the mechanism of the variability of solar cycles and other solar/cosmic processes affecting the Earth's climate and other factors determining the conditions in the ecosystems of the Baltic Sea. Therefore, improvement of the method will take time. Clarification of these regularities belongs mainly to the field of specific studies and their progress would allow for advances in the application of these data in the aquatic sciences.

Good coincidence between the abundance changes of the gulf herring, sea herring, cod and sprat year classes in the period 1989–2008 estimated by ICES (ICES 2011a, b) and the prediction by Kalejs and Ojaveer (1989) allows for considering the method worthy of further development for the composition of long-term forecasts.

5.7 Overexploitation of Living Resources

In the overexploitation of fish stocks in the Baltic Sea, two periods should be differentiated: (1) The time when no methods for the analytical assessment of the condition of fish stocks and fishing intensity existed. The causes of fluctuations in the stock size during that time have remained unknown. (2) The time of the composition of analytical assessments and estimates by ICES.

Before the methods were created for the analytical assessments of the fish stock size and its exploitation level, in the Baltic Sea and the estuaries of rivers falling into this sea, an extremely valuable stock of sturgeon *Acipenser sturio* L. was annihilated. Also, the abundance of many populations of salmon and sea trout were reduced to the stage at which they could be estimated as depleted.

Overfishing has seriously concerned some herring populations already before the start of the international collaboration in the ICES Working Groups on the assessment of Baltic pelagic and demersal fish stocks. In the middle and late 1950s, the Gulf of Riga spring herring was depressed due to a high exploitation rate on spawning grounds, and in the late 1960s, the same fate hit the autumn herring in the southern and southwestern Baltic (Rechlin 1971; Ojaveer 1988).

Beginning with the 1970s the Baltic Sea cod and salmon were intensively fished with very high catch rates (ICES 2015a, b). The fishery on the eastern cod stock can illustrate the general trend, with high exploitation rates during the 1980s and the recent drop (Figs. 3.44 and 4.4).

Starting in the second half of the 1970s, the condition (stock size, composition and exploitation rate) of the main fish stocks of the Baltic Sea (herring, sprat, cod) has been assessed by the ICES working groups. This decreased the possibility of overfishing of the most important stocks, but still could not exclude stock depletions. Based on the estimates of these working groups, from the late 1970s to the early 2000s, the TAC was set by the International Baltic Sea Fishery Commission (IBSFC). This commission usually accepted higher allowable catches than those recommended by ICES and did not differentiate them by populations. The principle of the management of fish species by population was not taken into account, although ICES presented its recommendations by population for all regulated fish species.

Both populations of Baltic cod have been under intense exploitation. In the period of good stock condition (1982–1985), no limitation of catches was recommended for the eastern population. Starting in 1989, the catches were limited by the IBSFC, but they soon increased to a very high level. In the period 1980–1985, the fishing mortality in the eastern population had achieved a very high level. However, during the period 1987–1991, it increased to $F = 1.1-1.5$, which is several times higher than F_{max} (=0.25), allowing for the obtainment of the maximum catches. The population fell under a very strong fishing pressure. Beginning in 1985, the SSB and the total abundance of the population, as well as the catches, rapidly decreased. From the late 1970s on, the salinity in the Baltic Sea started to fall; this fact sharply reduced the success of cod spawning. Starting from the second half of the 1980s, no abundant year classes have hatched in the eastern cod population. Due to the substantial decrease in the stock biomass, fishing restrictions were strengthened, up to the fishing ban proposed by ICES in 1993. The catches remained low at the beginning of the second decade of the twenty-first century as well. Due to strict regulations the fishing mortality has decreased to the lowest level in the most recent decades and the population biomass has started to increase (ICES 2011a).

The western cod population is exploited in mixed fisheries mainly by trawls, nets and hooks. The western population (the SSB of which constituted about 5% of the SSB of the eastern population during the time of its highest level) was managed together with the eastern one, although ICES has always stressed the differences between these populations and the importance of their separate management.

No periods of extraordinarily high level of the SSB and catches, comparable to those in the eastern population, have occurred in the western cod stock. In the period 1965–1985, the catches from the western population constituted 45,000–54,000 tonnes, in 1986–1988, they fell to 27,000–29,000 tonnes, and from 1989 to 1992, they made up 17,000–19,000 tonnes. The western population was still more strongly exploited than the eastern one. Its biomass had already sunk to a very low level by the mid-1980s. The abundance of its year classes substantially diminished beginning from 1983 (some year classes of higher abundance formed in the mid-1990s and in the year 2003) and have generally remained at a low level up to the present. The depressive decrease of both the western and eastern cod stocks took place more or less at the same time that a similar phenomenon was occurring in the neighbouring demersal populations (cod in the North Sea, all populations of demersal fishes in the Baltic, except turbot and sole).

The ICES Working Group on the Assessment of the Pelagic Stocks in the Baltic commenced its activity in 1974. In the period 1974–1988, the group developed a system for the assessment of herring and sprat resources by the natural units defined on the basis of earlier studies. The following herring stocks were treated separately: (1) Rügen-type herring of the SW Baltic; (2) Swedish east coast herring; (3) coastal herring of the Southern Baltic; (4) sea herring of the NE Baltic; (5) Gulf of Riga herring; (6) Gulf of Finland herring; (7) Bothnian Sea herring; (8) Bothnian Bay herring. Introduction of this system presumed that the working group members had very high qualifications in the differentiation of populations in the materials. The working group succeeded in coping with the task and presented the results that allowed for reasonable management of the populations.

However, the work of the Working Group on the Assessment of the Pelagic Stocks in the Baltic was reorganized in 1990. Application of the population-based method for the assessment of herring resources in the Central Baltic and the Gulf of Finland was stopped. Separate treatment of herring resources in the Bothnian Sea and Bothnian Bay (and later in the Gulf of Riga) was continued. But the herring groups of the central part of the Baltic Sea and the Gulf of Finland, substantially differing in their abundance dynamics and the average weight-at-age, were summed for the assessment. Several earlier studies – Popiel (1958); Otterlind (1962); etc.— had shown that in this large area, a number of herring populations with different characters exist. These works were ignored.

The first assessments produced according to the new scheme created the impression that in this large area (Central Baltic and the Gulfs of Finland and Riga), herring resources were in good condition and their exploitation could be increased. This was done during several subsequent years (1991–1996). The Working Group predicted unprecedented annual catches from this newly created unit (317,000–463,000 tonnes). The IBSFC added its wish to the catch estimated by the ICES

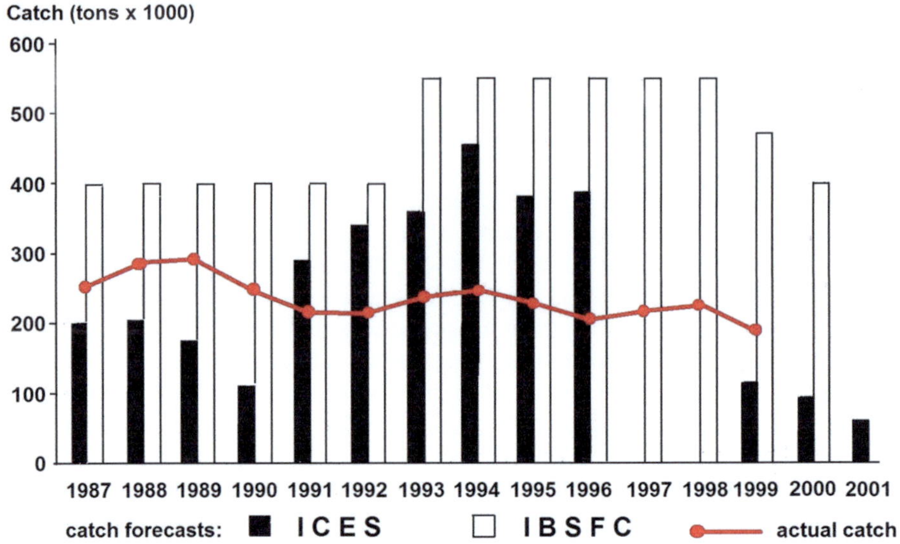

Fig. 5.5 Predicted herring catch for Subdivisions 25–29, 32, IBSFC TAC (for Subdivisions 22–24, and 25–29, 32) and the total catch of herring in Subdivisions 25–29 and 32 in the Baltic Sea. The ICES advice for 1994 was 317–463,000 t and no advice was given for 1997 and 1998 (Ojaveer 2002)

Working Group and approved the herring quota for a number of years (1993–1998, Fig. 5.5) at a level of 550,000 tonnes.

The actual development of the summed biomass in this newly created large unit was very different. During the period 1989–1999, the actual catches dropped from 270,000 to 149,000 tonnes. The populations in this very large area were overexploited and fell to depression for a long time. The cause of the failure in the stock assessment of herring in the Central Baltic and the Gulf of Finland was a failure to take into consideration the basic requirement in the assessment and management of fish stocks – their ecosystem-based treatment.

In salmon fishery, the very high exploitation rate in the offshore fisheries has decreased after introduction of a number of regulatory measures, such as closed areas, changes in the opening time of fisheries and reduced national quotas since 2012, restricting salmon catch in some countries. Furthermore, there are marketing restrictions on large salmon due to their high dioxin levels. Finally, the increased seal populations damage catches and gear. The driftnet ban in 2008 decreased offshore salmon catches in that year to the lowest value recorded since 1972. However, changes in the application of dioxin regulations in 2009, increases in market price for wild caught salmon and reduced opportunities for income in other fisheries resulted in an increase in offshore fishing from 2008 to 2010 (ICES 2015a).

5.8 Marine Spatial Planning and Protected Areas

There are increasing conflicts over the use of marine space, and therefore Maritime Spatial Planning (MSP) has also come to the fore in the Baltic area. The basis of the planning is to agree upon the allocation of the spatial and temporal distribution of human activities in marine areas with the aim of regulating conflicts and achieving ecological, economic and social objectives. MSP includes ecosystem-based, area-based, integrated, adaptive, strategic and participatory characteristics.

MSP is a marine equivalent to terrestrial spatial planning, which has been used to rationally develop, e.g., urban areas, but also to protect environmental and cultural values. However, whereas terrestrial spatial planning has for centuries been an integrated part of national law in many European countries, MSP is a novel, emerging form of legality, having so far been implemented mainly in connection with Marine Protected Areas (MPAs), as well as shipping lanes and traffic separation schemes.

While some parts of MSP are coordinated internationally through a number of initiatives, the main focus is on the coastal areas under national jurisdiction. Therefore, the planning processes are largely nationally based.

An important conservation management tool is the regulation of access to and activities allowed within an area belonging to an MPA. MPAs were introduced with the goal of protecting particular values of ecosystems, both in the open sea and coastal areas. HELCOM has agreed to the MPA regulations that prohibit, restrict or require permission for terrestrial building and construction in the sea, the installing of cables and pipelines, shipping and navigation, fishing, hunting, harvesting, tourism and recreation, military activities, aquaculture/mariculture, the extraction of resources, dredging, dumping, the installation of wind-farms, etc. (HELCOM 2010a).

Spatial protection and management measures have a long history on land, while the idea of conserving marine biodiversity is based on recommendations from the 1982 World Parks Congress in Bali, where it was recommended that the use of protected areas should also be applied to the oceans (Mc Neely Jeffrey and Kenton 1984). Special attention is to be paid to areas of particular ecological importance.

MPAs provide a broad set of tools for protecting ecosystem biodiversity and managing marine resources. They can range from being no-take or no-entry areas to wide, multi-use areas integrating different management practices and incorporating regulatory mechanisms that allow limited use of certain resources like fishing.

HELCOM is working towards the establishment of a coherent network of well-managed MPAs in the Baltic. The creation of protected areas involving the coastal zone, territorial waters and the Exclusive Economic Zone (EEZ) of different countries, and taking into account the dynamics of human activity and ecosystem processes, is a complicated task. The current network of MPAs covers over 12% of the Baltic Sea marine area. It consists of both HELCOM Baltic Sea Protected Areas and EU Natura 2000 sites, providing protection to species and habitats under the EU Habitats and Birds Directives (Fig. 5.6). The areal coverage of the protected areas

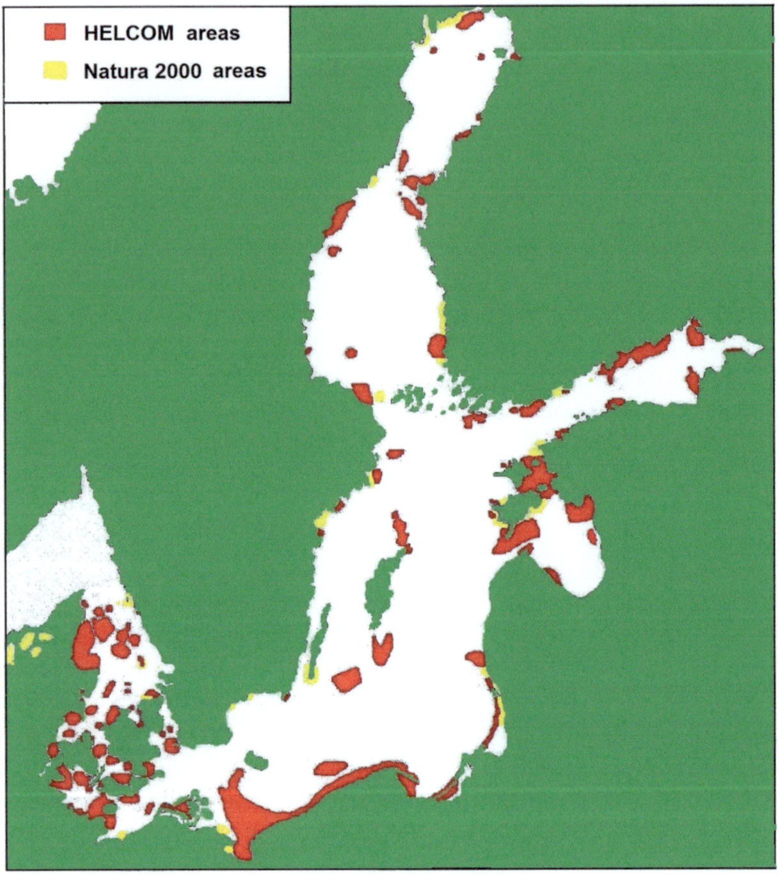

Fig. 5.6 HELCOM and Natura 2000 protected areas in the Baltic Sea (HELCOM 2010b)

has increased from 3.9% of the Baltic marine area in 2004 to 10.3% in 2010. The network is now larger than the target of 10% areal coverage set for regional seas by the UN Convention on Biological Diversity. However, the required 10% coverage has not yet been reached for all regions of the Baltic, in particular, in the Gulf of Bothnia, where few sites have been established.

A major gap in the protection of marine biodiversity is the lack of a sufficient level of management in the system of the network of protected areas. HELCOM considers that the management of the areas under protection in the Baltic Sea requires improvements and that further areas should be added to the system of protected areas (HELCOM 2010b).

References

Alheit J, Möllmann C, Dutz J, Kornilovs G, Loewe P, Mohrholz V, Wasmund N (2005) Synchroneus ecological regime shifts in the Central Baltic and the North Sea in the late 1980s. ICES J Mar Sci 62:1205–1215

Bergström L, Diekmann R, Flinkman J, Gardmark A, Kornilovs G, Lindegren M, Müller-Karulis B, Möllmann C, Plikshs M, Põllumäe A (2010) Integrated ecosystem assessments of seven Baltic areas covering the last three decades. In: Diekmann R, Möllmann C (eds) Coop Res Rep 302

Borja A, Elliott M, Andersen JH, Cardoso AC, Carstensen J, Ferreira J G, Heiskanen A.-S, Marques JC, Neto JM, Teixeira H, Uusitalo L, Uyarra MC and Zampoukas N (2013) Good environmental status of marine ecosystems: what is it and how do we know when we have attained it? Mar Poll Bull 76 (2013): 16–27

Cooke JK (1999) Improvement of fishery-management information through simulation testing of harvest algorithms. ICES J Mar Sci 56:797–810

Draganik B (1996) Polish inshore waters, their properties and role in the coastal area economy. Proc Polish–Swedish Symp Baltic Coastal Fisheries. Sea Fish Inst Gdynia pp 51–70

Elmgren R (2001) Understanding human impact on the Baltic ecosystem: changing views in recent decades. Ambio 30(4):222–231

Garpe K (2011) Ecosystem services provided by the Baltic Sea and Skagerrak. Swedish Environ Prot Agency Rep 5873. Ecosystem services provided by the Baltic Sea and Skagerrak. Econ Mar Inf

Gislason H (1999) Single and multispecies reference points for Baltic fish stocks. ICES J Mar Sci 56(5):571–583

Greene CH, Pershing AJ (2007) Climate drives sea change. Science 315:1084–1085

Hagen E, Feistel R (2005) Climate turning points and regime shifts in the Baltic Sea region: the Baltic winter index (WIBIX) 1659–2002. Boreal Environ Res 10:211–224

Hela I (1966) Fluctuations in the degree of continentality of Northern Europe in 1866–1965. Geophysics 8(4):331–334

HELCOM (2002) Environment of the Baltic Sea area, 1994–1998. Baltic Sea Environ Proc no 82B

HELCOM (2006) Assessment of Coastal Fish in the Baltic Sea. Baltic Sea Environ Proc no. 103 A

HELCOM (2007a) HELCOM Thematic Assessment in 2007. Climate Change in the Baltic Sea Area. Baltic Sea Environ Proc no 111

HELCOM (2007b) Baltic Sea Action Plan, http://www.helcom.fi/baltic-sea-action-plan/action-plan

HELCOM (2008) Guidelines for HELCOM coastal fish monitoring sampling methods

HELCOM (2010a) Ecosystem Health of the Baltic Sea 2003–2007: HELCOM Initial Holistic Assessment. Baltic Sea Environ Proc no. 122

HELCOM (2010b) Hazardous substances in the Baltic Sea. Baltic Sea Environ Proc no 120B

ICES (2000) The Status of Fisheries and Related Environment of Northern Seas. Nord, 2000

ICES (2009) Report of the ICES/HELCOM Working Group on Integrated Assessment of the Baltic Sea (WGIAB), 16–20 Mar 2009, Rostock, Germany. ICES Statutory Meeting, paper CM 2009/BCC:02

ICES (2011a) ICES Advice 2011. Book 8, The Baltic Sea

ICES (2011b) Report of the ICES/HELCOMWorking Group on Integrated assessment of the Baltic Sea (WGIAB), 4–8 Apr., 2011, Mallorca, Spain. ICES Statutory Meeting, paper CM2011/SSGRSP.03

ICES (2013) Report of the ICES Advisory Committee, Book 8, The Baltic Sea

ICES (2014) ICES Advisory Committee 2014, Book 1,

ICES (2015a). Report of the Baltic Fisheries Assessment Working Group (WGBFAS), 14–21 April 2015, ICES HQ, Copenhagen, Denmark. ICES Statutory Meeting, paper CM 2015/ACOM,10

ICES (2015b) ICES Advice 2015. ICES, Copenhagen

Kalejs M, Ojaveer E (1989) Long-term fluctuations in environmental conditions and fish stocks in the Baltic. Rapp P-v Réun Cons int Explor Mer 190:153–158
Köster FW, Möllmann C, Neuenfeldt S, Vinther M, John MA, Tomkiewicz J, Voss R, Hinrichsen HH, Kraus G, Schnack D (2003) Fish stock development in the Central Baltic Sea (1976-2000) in relation to variability in the physical environment. ICES Mar Sci Symp 219:294–306
Lassen H, Kelly C, Sissenwine M (2014) ICES advisory framework 1977 – 2012 from F_{max} to precautionary approach and beyond. ICES J Mar Sci 71:1966–1972
Mc Neely Jeffrey A, Miller Kenton (1984) National Parks, conservation and development: the role of protected areas in sustaining society. Proc of the World Congress on National Parks, Bali, Indonesia, 11–22 October 1982. Smithsonian Institution Press
Ogurtsov MG, Jungner H, Kocharov GE, Lindholm M, Eronen M, Nogovitsyn J (2003) On the link between northern Fennoscandian climate and length of quasi-eleven-year cycle in galactic cosmic ray flux. Sol Phys 218:345–357
Ojaveer EA (1988) Baltiiskije seldi (Baltic herrings). Agropromizdat, Moskva
Ojaveer E (2002) The role of ICES in the assessment and management of resources in the Baltic Sea. ICES Mar Sci Symp 215:582–589
Ojaveer E (2014) Läänemeri. TA Kirjastus, Tallinn
Ojaveer H, Eero M (2011) Methodological challenges in assessing the environmental status of a marine ecosystem: case study of the Baltic Sea. PLoS One 6(4):e19231
Ojaveer E, Kalejs M (2010) Ecology and long-term forecasting of sprat (*Sprattus sprattus balticus*) stock in the Baltic Sea: a review. Rev Fish Biol Fish 20:203–217
Ojaveer E, Kalejs M (2012) Long-term prediction on Baltic fish stocks based on periodicity of solar activity. Rev Fish Biol Fish 22:683–693
Otterlind G (1962) Sillens/strömmingens vandringsvanor vid svenska syd- och ostkusten. Ostkusten 34(1):15–21
Popiel J (1958) Differentiation of the biological groups of herring in the Southern Baltic. Rapp P-v Réun Cons int Explor Mer 143(II):114–121
Rechlin O (1971) Ergebnisse zur Gesamtsterblichkeit beim Herbsthering der Bornholmsee aus den Jahren 1966–1969. Fisch-Forsch, Wissensch Schriftenr 9(1):33–34
Schove DJ (1955) The sunspot cycle, 649 B. C. to A. D. 2000. J Geophys Res 2:127–146
Svensmark H (1998) Influence of cosmic rays on Earth's climate. Phys Rev Lett 81:5027–5030
Svensmark H, Friis-Christensen E (1997) Variation of cosmic ray flux and global cloud coverage – a missing link in solar–climate relationship. J Atm Solar Terr Phys 59:1225–1232
Thoresson G (1996) The Swedish coastal fishery in the Baltic Sea. Bull Sea Fish Inst Gdynia 2(141):67–78
Usoskin IG, Mursula K, Solanki S, Schuessler M, Alanko K (2004) Reconstruction of solar activity for the last millennium using ^{10}Be data. Astron Astrophys 413:745–751
Yndestad H (2004) The lunar nodal cycle influence on the Barents Sea. Dissertation. Norwegian Inst Sci Technol

Chapter 6
International Collaboration in the Assessment and Management of Baltic Ecosystems and Living Resources

Abstract The impact of both natural and human factors on marine ecosystems and biological resources is often transboundary and therefore the corresponding management is based on international agreements and effected by international organizations. Traditionally the international collaboration is based on the scientific knowledge which has been highly appreciated at solving problems related to the Baltic Sea. The issues of regulating human activities at global scale can be addressed by the UN system (United Nations Convention on the Law of the Seas – UNCLOS; International Maritime Organization – IMO, etc.). There are several organizations that establish regional cooperation – Baltic Marine Environment Protection Commission – Helsinki Commission (HELCOM), Baltic Marine Biologists (BMB), Baltic Oceanographers (CBO), fisheries organizations, etc. The most important organization responsible for increasing scientific knowledge in the marine environment and its living resources and to use this knowledge to provide advice to competent authorities is the International Council for the Exploration of the Sea (ICES) which is the first intergovernmental science organization in the world, founded on 22 July 1902.

Marine ecosystems are influenced by a range of natural and human factors with a variety of effects. The impact is often transboundary, and therefore the corresponding management is based on international agreements and effected by international organizations. Traditionally, international collaboration has been based on scientific knowledge, which has been highly appreciated in the solving of problems related to the Baltic Sea. The scale of the activities and their impact are reflected in the institutional setup established to assure international cooperation in regulating human activities. The UN system addresses issues at the global scale, e.g., the following organizations have been founded: the 1982 United Nations Convention on the Law of the Seas (UNCLOS), on the ship traffic (International Maritime Organisation IMO), navigational regulation (IATA) and the International Convention for the Prevention of Pollution from Ships (MARPOL), accepted by the International Maritime Organization (IMO) in 1973. There are Baltic organizations that establish regional cooperation, notably HELCOM. ICES provides coordination of the scientific work, scientific management advice and status reports. Also, there are bilateral cooperation projects, e.g., under the EU–Russian fisheries agreement.

E. Ojaveer, *Ecosystems and Living Resources of the Baltic Sea*,
DOI 10.1007/978-3-319-53010-9_6

6.1 UNCLOS and the Fisheries Agreement

Regulation of human activity in the open sea and management of marine living resources are based on a relevant agreement, the UNFSA (see below). According to UNCLOS, a coastal state has the exclusive right to the resources, e.g., fish harvesting, in its exclusive economic zone (EEZ). This means that fish and other resources in the zone belong to the coastal state as long as it is in the zone. In its EEZ, the coastal state may itself manage the resources by allowing, prohibiting or regulating their exploitation. For fish resources, there are special concerns, because fish are often transboundary, and the same fish may occur in several EEZs during the course of a year, e.g., Western Baltic herring migrating between the Skagerrak, the Kattegat and the Western Baltic Sea (Subdivisions 22–24). Therefore, each state, targeting the same stock, may unilaterally set its own TACs, although several fisheries exploit the same stock and their collective situation depends on the activities of other fleets. For this reason, where the same stock or stocks of associated species occur both within the EEZ and in an area beyond and adjacent to the zone, the coastal state and the states fishing for such stocks in the adjacent area shall seek, either directly or through appropriate organizations, to agree upon the measures necessary for the conservation of these stocks.

When considering the exploitation of living resources, UNCLOS is complemented by the 1995 UN agreement on migratory fish stocks (UNFSA).[1] According to this agreement, the states harvesting the same stock must 'agree on the rights of participants, such as distribution of allowed catch or levels of fishing effort'. UNFSA clearly states that coastal states shall cooperate to ensure effective conservation and management of straddling and highly migratory fish stocks. However, the obligation to cooperate is not an obligation to finalize negotiations with an agreement.

In its Plan of Implementation, the World Summit on Sustainable Development[2] determines that the total (annual) fishing yields from fish stocks must be kept within the MSY. However, the MSY criterion does not in itself deal with national sovereignty as laid down in UNCLOS. Thus, the coastal state alone decides what the MSY is. There is no international court to settle this.[3]

Within the EU, the situation varies depending on whether the regulation of human activity is based on fishery legislation or whether the legal basis is environmental concerns. For issues based on environmental concerns, the EU Member States are required to take conservation or protection measures, if necessary, in

[1] Full name: Agreement for the Implementation of the Provisions of the United Nations Convention on the Law of the Sea of 10 December 1982 relating to the Conservation and Management of Straddling Fish Stocks and Highly Migratory Fish Stocks (dated September 8, 1995). The agreement came to force on December 11, 2001.

[2] Plan of Implementation of the World Summit on Sustainable Development (The Johannesburg agreement) 2002 para 31. See http://www.un.org/esa/sustdev/documents/WSSD_POI_PD/English/WSSD_PlanImpl.pdf.

[3] In the NE Atlantic, ICES is asked to determine MSY and give its opinion as to whether a management agreement based on a certain value is sustainable.

specific areas to fulfil their duties stemming from, e.g., the Habitats and Birds Directives. For fisheries, there is an exclusive competence for the EU formulated in the Common Fisheries Policy of the EU.

This means that in the Baltic Sea, there are only two players when discussing fishery regulation outside the coastal zone: the European Community and the Russian Federation. However, in discussing environmental issues, both the European Community and the individual EU Member States are direct parties to the negotiations. Therefore, management cooperation is regulated through the EU–Russian agreement on fisheries and the EU Common Fisheries Policy, and the cooperation on environmental issues is discussed within HELCOM, in which each of the eight Baltic Sea states is a member in its own right, together with the European Union.

6.2 Scientific Cooperation

The history of Baltic marine science is long and distinguished. Hydrographic, chemical and marine biological studies of the Baltic Sea started in Germany, Sweden and Denmark at the end of the eighteenth century and in the nineteenth century. In German waters, the beginning of the study of macrozoobenthos dates back to the end of the eighteenth century, with more systematic observations being conducted in the second half of the nineteenth century (Zettler et al. 2008). V. Hensen and the Kiel School pioneered quantitative studies of marine plankton, C. Kupffer, of herring embryology, and H. Heincke, of herring intraspecific variations (Kupffer 1877; Wasmund and Siegel 2008; Ojaveer 2014; etc.). Their work inspired the initiation of similar scientific studies in neighbouring areas. In Denmark, the first quantitative samplings of benthic marine organisms were begun by C.G. Johannes Petersen. In Sweden, strong traditions in marine studies were developed by S. Lovén, G. Ekman, and the Stockholm chemistry professor, the prime mover of the Swedish initiative in 1899 to start the International Council for the Exploration of the Sea (ICES), Otto Pettersson, along with others. In the Northern Baltic, the first mention of demersal animals in scientific literature goes back to the year 1769, and the studies of aquatic invertebrates both in the Gulf of Finland and the Gulf of Bothnia were started in the middle and the second half of the nineteenth century by O. Nordquist, K. M. Levander and others (Leppäkoski 2001). K. E. von Baer started his investigations into marine fish overexploitation at Estonian coasts in the early 1850s. After WWI, new independent states came into being at the Baltic Sea, all of which continued the study of that sea. Of them, Finland belonged to the founder states of the ICES. In the 1920s, Estonia, Latvia and Poland were accepted into the ICES under different conditions. In 1939–1940, Estonia, Latvia and Lithuania lost their independence and were incorporated into the Soviet Union. When ICES restarted after WWII, its Baltic activities were hampered by the lack of Soviet, German and Polish (in 1950–1955) participation. By 1955, the Federal Republic of Germany, the Soviet Union and Poland were members of ICES, but the German Democratic Republic was still excluded, becoming an ICES member only in 1974. Marine studies continued around the Baltic, but mostly on a national basis. Normal conditions of marine investigations, including international collaboration, were re-established only step by step, together with political changes, towards the end of the millennium (Ojaveer 2014).

6.2.1 International Council for the Exploration of the Sea (ICES)

Cooperation in marine science between countries developed against a background of applied needs. In the nineteenth century, an important part of the economies of the countries in Scandinavia and around the North Sea relied on fishery. Scientific works by S. Lovén, M. Sars and others had given rise to the understanding that the abundance and distribution of fish in the sea depend on the currents, the availability of fish food (mainly zooplankton), and other conditions related to the marine environment. Studies of marine areas at the coasts of their home countries showed the scientists that the value of the studies would be much higher if larger areas were covered and the studies were consistent. The large-scale investigations by Sweden, Denmark, Norway, Scotland and Germany from the northern waters of Scotland to the southern Baltic in 1893–1894 were successful and had the ability to be continued later on. These studies also supported the idea of organizing permanent international marine investigations, advocated mainly by the chemist, physicist and oceanographer Otto Pettersson, Professor at the Stockholm Technical High School.

The first conference on marine investigations for the preparation of the founding of an international organization dedicated to marine science was arranged in 1899 in Stockholm, with invitations sent out by King Oscar II of Sweden and Norway. The first intergovernmental science organization in the world – the International Council for the Exploration of the Sea – was founded on July 22, 1902. The first members were Denmark, Finland, Germany, the Netherlands, Norway, Sweden, Russia and the United Kingdom. It was agreed that the ICES secretariat would be situated in Copenhagen (Went 1972; Rozwadowski 2002). ICES has been active since its creation, excluding the periods of the world wars. The Council's membership has fluctuated over the years, with nations joining/leaving/rejoining at different times as a result of wars and political decisions. An exchange of letters among the original eight member nations was sufficient to establish the Council in 1902. This type of arrangement continued until the early 1960s, when this informal status became unacceptable in light of the establishment of the United Nations and its subsidiary bodies, as well as other international organizations. Steps were initiated to achieve full international recognition of the Council by the host country Denmark. At a conference convened in Copenhagen on September 7, 1964, a formal Convention was signed, which subsequently came into force on July 22, 1968, following its ratification by the then 17 member nations. Presently, ICES has 20 Member Countries surrounding the Baltic Sea, North Sea and North Atlantic Ocean. The Member Countries of ICES are Belgium, Canada, Denmark, Estonia, Finland, France,

Germany, Iceland, Ireland, Latvia, Lithuania, the Netherlands, Norway, Poland, Portugal, the Russian Federation, Spain, Sweden, the United Kingdom and the United States of America. In addition to the full members, a number of institutions from Australia, Peru, Chile, New Zealand and the South African Republic are affiliated with ICES. Formal observer status has also been granted to the international organizations Worldwide Fund of Nature and Birdlife International. Every Member Country is represented in the Council by two delegates.

The main objective of ICES is to increase scientific knowledge of the marine environment and its living resources and to use this knowledge to provide advice to competent authorities. ICES Science and Advice considers both how human activity affects marine ecosystems and how ecosystems affect human activity. The function of ICES is now based on a convention signed in 1964 and ratified in 1968. According to the convention, ICES shall

(a) promote and encourage research and investigations for the study of the sea, particularly those related to the living resources thereof;
(b) draw up programmes required for this purpose and organize, in agreement with the Contracting Parties, such research and investigation as may appear necessary;
(c) publish or otherwise disseminate the results of research and investigations carried out under its auspices or encourage the publication thereof.

The principal decision and policy-making body of ICES is the Council. The Council consists of a President and two Delegates from each of the 20 member nations. Delegates elect the President, First Vice-President, and five additional Vice-Presidents (all for 3-year terms) to comprise the Bureau or Executive Committee. The Bureau is responsible for carrying out the Council's decisions, preparing and convening Council meetings, formulating Council budgets, appointing a Secretariat staff, and performing other tasks as assigned by the Council. A Finance Committee consisting of five Delegates provides oversight of the Council's fiscal matters. Delegates are also responsible for appointing a General Secretary, who serves as the Council's chief executive officer and is charged with managing the Council's Secretariat facilities and staff, finances, meetings, reports, publications, and communications.

ICES comprises a network of more than 4000 scientists from almost 300 institutions, including a number of institutions from non-member countries (ICES 2014). This network is activated through more than 100 working groups supplemented by workshops, and in this way, ICES ensures that the best available science is accessible for decision-makers to make informed choices on the sustainable use of the marine environment and ecosystems.

The ecosystem approach is being operationalized through the Marine Strategy Framework Directive, which aims to achieve good environmental status in all European marine waters by 2020. To achieve this objective, ICES prioritizes, organizes, delivers and disseminates research needed to fill gaps in marine knowledge related to issues of ecological, political, societal and economic importance at the pan-Atlantic and global levels. ICES conducts integrated ecosystem assessments in collaboration with the North Pacific Marine Science Organization (PICES), the institutions of the Mediterranean Science Commission (CIESM) and the groups active in Arctic science, using biological data collected by the research programmes and landing records of national research institutes to assess the state of the main commercial stocks. The main ICES deliverables are scientific publications, scientific information and management advice requested by Member Countries, and also international organizations and commissions such as the Oslo–Paris Commission (OSPAR), the Helsinki Commission – Baltic Marine Environment Protection Commission (HELCOM), the North East Atlantic Fisheries Commission (NEAFC), the North Atlantic Salmon Conservation Organization (NASCO) and the European Commission (EC). Importantly, these products are unbiased, non-political in nature and based on the best available science (ICES 2014).

ICES is the prime source of scientific advice for governments and international regulatory bodies that manage the North Atlantic Ocean and adjacent seas. The Advisory Committee (ACOM) oversees the process of preparing scientific advice on fisheries and marine ecosystem issues. The Scientific Committee (SCICOM) oversees a broader research programme, which is a longer term investment into the capability to give scientific advice in the future.

ICES coordinates the monitoring and advisory work that is the foundation for the management advice and status reports. This includes a number of abundance surveys that cover the living resources in the Baltic Sea and the Northeast Atlantic Ocean. It also includes sampling of the fisheries, most of which is done, in the Baltic Sea, under the EU Data Framework Programme. However, the standards used in the sampling programmes had already been laid down with the initialization of the ICES statistical programme at the beginning of the twentieth century. The ICES advice plays an important role in the management setup for the Baltic Sea. This advice is produced through cooperation between the countries that monitor and research the Baltic Sea. However, the ICES advice is based on a wider scientific and advisory community through a review system that involves researchers from all 20 ICES Member Countries and the joint formulation of the advice by the ICES Advisory Committee.

ICES produces a variety of publications reflecting its commitment to excellence and the diversity of ICES concerns, like *ICES Journal of Marine Science* (*Journal du Conseil*), *ICES Advice, ICES Cooperative Research Report, Techniques in Marine Environmental Sciences, ICES Identification Leaflets for Plankton, ICES Identification Leaflets for Diseases and Parasites of Fish and Shellfish,ICES Fisheries Statistics,ICES Marine Science Symposia,* etc.

6.2.2 Baltic Marine Biologists (BMB)

After WWII, international cooperation between the democratic Baltic Sea states and the Soviet Union was very poor, especially in the period of the Cold War. However, the necessity for common studies and the exchange of data and materials was extremely important for understanding the causes of the deterioration of the quality of the Baltic Sea environment since the middle of the twentieth century. Therefore, under the leadership of well-known marine biologists, a strong effort for improving collaboration was made. The first symposium of the Baltic marine biologists was arranged in Rostock in 1968. There, the Committee of the Baltic Marine Biologists was founded with the function of aggregating the marine biologists working in the Baltic area and intensifying contacts between marine biologists of the Western and Eastern blocs. The first Chairman of the committee elected was Ernst Albert Arndt from Rostock and the first Secretary was Bernt I. Dybern from Lysekil. It was decided that the committee would be the leading body of the organization of the Baltic Marine Biologists unifying the marine biologists of the Baltic Sea. BMB closely collaborated with HELCOM. The main tasks of the committee was keeping contacts between the Baltic marine biologists, arranging symposia, common investigations, collaboration with other organizations, etc. (Dybern 2004). The Committee of the BMB consists of three representatives from each Baltic Sea country (Denmark, Estonia, Finland, Germany, Latvia, Lithuania, Poland, Russia and Sweden), co-opted members and honorary members. The main function of the committee is leading the organization. The basis of the activity of the committee is the work of its national sections.

The prevailing political situation after the foundation of BMB made it difficult for scientists to meet and exchange knowledge. Later, the role of BMB changed, but the goals of the organization remain the same: to promote research in the Baltic Sea area and to facilitate contacts between marine biologists working in the region. BMB arranges inter-calibrations in which different methods for sampling, analysis, etc., are compared, making it possible to interpret research results from different countries. Another important field of work is the holding of scientific symposia. Sixteen symposia have so far been arranged by BMB, presenting more than 1400 papers. Membership in BMB is free of charge and open to all individuals who are involved in scientific marine biological research, management and education in the Baltic Sea area.

6.2.3 Conference of Baltic Oceanographers (CBO) and Baltic Marine Geologists (BMG)

The situation after WWII also seriously obstructed the work of the Baltic Sea hydrologists. Possibilities for collaboration were discussed by the Hydrography Committee at the ICES Annual Scientific Conference in 1956. A subcommittee (I. Hela, J. Krey, B. Kullenberg and V. A. Lednjov) was created, and in March 1957, a meeting of about 30 Baltic oceanographers was arranged in Helsinki. Interest towards oceanographic investigations in the Baltic Sea was high. The meeting was treated as the First Conference of Baltic Oceanographers (CBO). A number of recommendations for investigations in the Baltic Sea were accepted. CBO has closely collaborated with ICES (having the status of consultants to the Hydrography Committee of ICES), BMB and HELCOM. Baltic oceanographers have arranged/participated in a number of study programmes, including International Baltic Year 1969/1970; Baltic Open Sea Experiment, BOSEX '77, Patchiness Experiment, PEX-86.

Baltic geologists participated in the international collaboration in a joint staff with the Baltic Oceanographers. Later on, they formed their own organization – Baltic Marine Geologists (BMG). In the 1990s, BMB, CBO and BMG joined together. The first common Baltic Sea Science Congress (BSSC) was held in Rönne (Bornholm) in 1996, followed by conferences in Warnemünde (1999), Stockholm (2001), Helsinki (2003), Sopot (2005), Rostock (2007), Tallinn (2009) and St. Petersburg (2011). Today, cooperation is carried out in many contexts, both as EU-funded projects under the framework programmes and as coordinated national programmes involving organizations such as ICES and HELCOM (Ojaveer 2014).

6.3 Collaboration in the Management of Ecosystems

6.3.1 Baltic Marine Environment Protection Commission (Helsinki Commission)

With developments in industry, agriculture, transportation, peoples' living standards, etc., the pressure to the environment, including that of the Baltic Sea, has increased. In the western areas of the Baltic Sea, social groups and institutions dealing with the increased anthropogenic impact on the environment appeared, and the first environmental programmes (e.g., air pollution and wastewater management programmes) were started in the 1960s. In the eastern part, especially in the Soviet Union, the quality of the environment was much less appreciated, and information on the activities dangerous to the environment was frequently concealed for economic or political reasons. Successful work for fighting environmental pollution started much later there than in the west. However, the quality of the environment is important in all communities. The informally-begun collaboration between the western and eastern marine scientists (which resulted in foundation of the organisations *Baltic Marine Biologists* and *Baltic Oceanographers*) were the most important events accelerating origination of the international organisation for the protection of the Baltic Sea.

In 1974, representatives of Denmark, Finland, the German Democratic Republic, the German Federative Republic, Poland, Sweden and the Soviet Union signed the Convention on the Protection of the Marine Environment of the Baltic Sea in Helsinki. The Baltic Convention was not the first convention in those years, but it was unusual at that time as collaboration between the East and West. The initial convention primarily concerned land-based pollution, the elimination of pollution by hazardous substances and pollution from ships. Its focus was mostly the open sea marine environment. The Convention formally entered into force in May 1980. The Convention was the first in the world to take into consideration the whole range of the marine environment and its protection. The steering agency of the Convention is the Baltic Marine Environment Protection Commission – Helsinki Commission (in short, HELCOM). HELCOM is an intergovernmental organization. For the most part, Contracting Parties choose experts from environmental departments as their representatives in the Commission. Since 1980, ministerial meetings have also been arranged every fourth year.

The second and third periodic assessments of the ecosystems in the Baltic Sea (for the periods 1984–1988 and 1989–1993, correspondingly) indicated that the nutrient concentration was high, and that in most parts of the sea, they were still increasing. It was learned that the nutrient runoff from arable land may have been a significant source of eutrophication.

In the latter half of the 1980s, HELCOM adopted various recommendations aimed at limiting nutrient pollution from municipal wastewater treatment plants, agriculture and industry. Total nitrogen and phosphorus loads entering the Baltic Sea amounted to 891,000 and 51,000 tonnes, respectively. In September 1988, Ministers of the Environment of the Baltic Sea countries decided that by the year 1995, nutrient discharges should be reduced by 50% from the late 1980 levels. The 1988 target for phosphorus was achieved by almost all countries, but most countries did not reach the target for nitrogen. From 1990 to 1995, the total phosphorus load decreased by 41% and the total nitrogen load by 20%. During the decade from 1990 to 2000, the direct point source P went down by 68% and the N load by 60% (HELCOM 2009).

In 1992, the Convention was revised to embrace new environmental principles and the changed geopolitical situation. Also, in 1992, the Baltic Sea Joint Comprehensive Environmental Action Plan was established. According to the new convention, its members are Denmark, Estonia, the European Union, Finland, Germany, Latvia, Lithuania, Poland, Russia and Sweden. It came into force in 2000. The new convention has a somewhat stricter and wider scope than its precursor. It also concerns internal water bodies, which were excluded in the first convention.

The ecosystem approach to the management of human activity was adopted by the Contracting Parties of HELCOM in 2003 as the basic principle for HELCOM's decisions towards attaining a good ecological status for the Baltic Sea. Eutrophication continued to be a major environmental problem in the early 2000s. Principles of adaptive management were adopted in November 2007 (HELCOM 2007B).

In 2007, HELCOM estimated that for good environmental status to be achieved, the maximum allowable annual nutrient pollution inputs into the Baltic Sea had to be 21,000 tonnes of phosphorus and about 600,000 tonnes of nitrogen. Annual reductions of some 15,000 tonnes of phosphorus and 135,000 tonnes of nitrogen were required to achieve the plan's crucial "clear water" objective.

The HELCOM Nutrient Reduction Scheme was a regional approach to sharing the burden of nutrient reductions to achieve the goal of a Baltic Sea unaffected by eutrophication agreed to by the Baltic Sea countries.

The Scheme was introduced and first agreed upon in 2007, in the HELCOM Baltic Sea Action Plan. At that time, the countries agreed upon provisional nutrient reduction targets and decided that the figures would be revised using a harmonised approach and the most updated data available, as well as enhanced modelling. The revision process started in year 2008 and was completed in 2013.

HELCOM's eight main Working Groups implement policies and strategies and propose issues for discussion at the meetings of the Heads of Delegations, during which decisions are made.

The five permanent groups address different aspects of HELCOM's work:

- Group on the Implementation of the Ecosystem Approach (Gear);
- Maritime Working Group (Maritime);
- Working Group on Reduction of Pressures the Baltic Sea Catchment Area (Pressure);
- Response Working Group (Response);
- Working Group on the State of the Environment and Nature Conservation (State and Conservation).

The following time-limited groups complement the work of the permanent groups:

- Group on Sustainable Agricultural Practices (Agri);
- Group on Ecosystem-based Sustainable Fisheries (Fish);
- Joint HELCOM-VASAB Maritime Spatial Planning Working Group (HELCOM-VASAB MSP WG).

In addition, HELCOM has created a number of expert groups, e.g.,

AIS EWG – Expert Working Group for Mutual Exchange and Deliveries of AIS data;

EWG OWR – Expert Working Group on Oiled Wildlife Response;

EWG SHORE – Expert Working Group on Response on the Shore, etc.

HELCOM's vision for the future is a healthy Baltic Sea environment with diverse biological components functioning in balance, resulting in a good ecological status and supporting a wide range of sustainable economic and social activities. HELCOM's main roles (http//ww.helcom.fi) involve acting as the following:

1. a coordinating body, ascertaining a multilateral response in case of major maritime incidents;
2. an environmental policymaker for the Baltic Sea area, developing common environmental objectives and actions;
3. an environmental focal point providing information about the state of and trends in the marine environment, the efficiency of measures to protect it and common initiatives and positions, which can form the basis for decision-making in other international forums;
4. a body for developing, according to the specific needs of the Baltic Sea, recommendations of its own and recommendations supplementary to measures imposed by other international organizations;

5. a supervisory body dedicated to ensuring that HELCOM environmental standards are fully implemented by all parties throughout the Baltic Sea and its catchment area.

HELCOM is partly a forum for environmental management and partly a coordinating body for monitoring the environmental status of the Baltic Sea. Work under HELCOM includes implementing joint recommendations to curb pollution originating from land and marine sources, ensuring safer maritime traffic, and protecting biodiversity, for example, by setting up a network of protected areas in the Baltic Sea.

HELCOM's recommendations and other decisions are made unanimously – each member state commanding one vote – and are implemented through national legislations and other national initiatives. The EU Commission will, in cases which fall under community competence within the EU, cast the number of votes corresponding to the number of EU Member States, and these states will not vote.

HELCOM publishes diverse reports, brochures and other material about the Baltic Sea and its environment. The elaboration of internationally harmonized assessments of the Baltic Sea has been central to HELCOM's work since the beginning of the 1980s. The main objective of the HELCOM assessment product is to provide policy-relevant information for targeted users at national and Baltic-wide levels, as well as to provide input to panEuropean and global fora (EU, UNEP, IMO). This is necessary in order to make sound decisions for restoring the Baltic Sea ecosystem, for achieving and maintaining good ecological status and for supporting the implementation of the HELCOM objectives and actions (HELCOM 2009).

In HELCOM activity, chief attention has been turned to the eutrophication of the Baltic Sea, dangerous compounds, transport, and the environmental conditions limiting the abundance and quality of fish resources. HELCOM has presented more than 200 recommendations for improving the conditions of the Baltic Sea environment and living resources. As a result of these efforts, the amount of waste and organic pollutants/nutrients, as well as halogen compounds (dioxins, furans), has substantially decreased, and control over decreasing environmental pollution in industry has become more severe.

European policies on water quality are aimed at the mitigation of eutrophication. In the Baltic Sea area, corresponding directives concern all countries, with the exception of the Russian Federation: Urban Waste Water Treatment, Nitrates Directive, etc.

The Baltic Sea Environment protection Commission fulfills its commitments in close cooperation with a number of international and governmental bodies: the International Commision for the Exploration of the Sea (ICES), the Oslo and Paris Commission (OSPARCOM), the World Meteorology Organization (WMO), the United Nations Environmental Programme (UNEP), the International Maritime Organization (IMO), WWF International, the Intergovernmental Oceanographic Commission (IOC), etc.

6.3.2 ASCOBANS

Small cetaceans are considered separately, although, of course, marine mammals are part of the ecosystems in the Baltic Sea. There is an intergovernmental organization based on the 'Agreement on the Conservation of Small Cetaceans in the Baltic, North East Atlantic, Irish and North Seas' (ASCOBANS). This is a regional agreement on the protection of small cetaceans, which was originally concluded as the 'Agreement on the **Conservation** of Small Cetaceans of the Baltic and North Seas' under the auspices of the UNEP Convention on Migratory Species, or Bonn Convention, in September 1991, and came into force in March 1994. In February 2008, an extension of the agreement area came into force, which changed it to its current name.

ASCOBANS covers all species of toothed whales (Odontoceti) in the Agreement Area, with the exception of the sperm whale (*Physeter macrocephalus*). In the Baltic Sea, the main emphasis is on harbour porpoise (*Phocoena phocoena*).

By 2015, not all Baltic Sea states were party to the Agreement. The European Union has signed the Agreement, but never ratified it. Estonia, Latvia and the Russian Federation have not joined the Agreement, while the other Baltic Sea states have.

Monitoring and assessment of harbour porpoise is coordinated through HELCOM and ICES. The secretariat is based in Bonn, Germany.

6.4 Fishery Organizations

International management of the fish resources in the Baltic Sea was established with the creation of the International Baltic Sea Fisheries Commission (IBSFC) in 1974. This organization struggled to get the fisheries under control and within sustainable limits, but disagreements among the IBSFC member states over allocation issues and what would constitute appropriate TACs, combined with periods of major unreported fisheries, resulted in the general overexploitation of fish resources throughout the lifetime of the IBSFC. When in 2004, Poland, Lithuania, Latvia and Estonia joined the EU, the IBSFC became dysfunctional, but the assessment machine has continued unchanged and provides advice for the EU Common Fisheries Policy and the Russian Federation. Cooperation between the EU and the Russian Federation on fishery management is organized through the EU–Russian agreement signed in 2009.

EU Common Fisheries Policy The Common Fisheries Policy was created to manage the fish stock for the European Union as a whole. Article 38 of the 1957 Treaty of Rome, which created the European Communities (now the European Union), stated that there should be a common policy for fisheries.

The Common Fisheries Policy (CFP) is the fishery policy for the entire European Union (EU). It sets quotas designating the amounts of each fish the Member States are allowed to catch, and also encourages the fishing industry though various market interventions. The Treaty of Lisbon formally enshrined fishery conservation policy as one of the handful of 'exclusive competences' reserved for the EU, to be decided by Qualified Majority Voting. However, general fishery policy remains a 'shared competence' of the EU and its Member States, and decisions are made primarily by the Council of the EU. Within the Common Fisheries Policy, the more general agreements, however, are co-decisions that also involve the European Parliament.

The CFP has four components:

- Regulation of production, quality, grading, packaging and labelling;
- Encouraging producers' organizations to protect fishermen from sudden market changes;
- Setting minimum fish prices and financing the buying up of unsold fish;
- Setting rules for trade with non-EU countries.

For the purpose of this book, it is only the regulation of production that is of concern.

The 2014 reform includes an element of subsidiarity – the principle that political decisions should be handled at the lowest, least centralized competent level. This was introduced in the form of regionalization of management. The Baltic Sea states had foreseen this development and created BALTFISH as a forum for discussions among the states involved.

International Baltic Sea Fishery Commission The International Baltic Sea Fishery Commission (IBSFC) was a regional fisheries management organization established by the 1973 Gdansk Convention on Fishing and Conservation of the Living Resources in the Baltic Sea and the Belts. IBSFC was responsible for management of the shared Baltic fishery resources for the period 1974–2004, and was established pursuant to the Convention on Fishing and Conservation of the Living Resources in the Baltic Sea and the Belts (the Gdansk Convention) signed in 1973 by the Governments of the Baltic Sea states. In addition to the Baltic Sea, the jurisdiction of the IBSFC included the waters of the Belts and the Øresund. The secretariat was based in Warsaw, Poland.

The Baltic Sea was the first region in the world to be covered by common regional goals for sustainable development. Accounting for the political situation in the region and the possibilities for the management of the highly exploited Baltic fish stocks, the constitution of the IBSFC was extremely necessary. Based on the ICES estimates on the condition of the fish stocks, the Commission defined the catch quota by fish species in the Baltic Sea for the following year and distributed it to the countries. Contracting Parties agreed to promote close cooperation '… with a view to preserving and increasing the living resources of the Baltic Sea and the Belts and obtaining the optimum yield, and, in particular to expanding and coordinating studies towards these ends ...'. Activities of the IBSFC were determined annually on the basis of its recommendations and decisions made during the Annual Sessions (IBSFC 1975). The IBSFC, established as the first international organization for the Baltic Sea, also acted as the Sector Task Force. The IBSFC's Extraordinary Session in 1998 adopted a set of biological, economic and social core indicators ('Baltic 21' Indicators of the Sector Fisheries) to highlight trends in biological systems and the economics of fishing communities around the Baltic Sea. The IBSFC saw ecological sustainability as a basic premise for the economic and social future of European fisheries, and fishing the Maximum Sustainable Yield was a target for achieving sustainability.

The activities of the IBSFC provided valuable experience in managing the shared Baltic fishery resources over almost three decades. Regarding ecosystem-based management, UNEP considered the IBSFC as a pioneer in applying the ecosystem approach to fishery management. IBSFC began cooperating with HELCOM in 1992 and was part of the first meeting between an international fishery organization and an environment commission. The implementation of 'Agenda 21 for the Baltic Sea Region' was an important driving force in further developing the management objectives and strategies for Baltic salmon, cod, sprat and herring, as well as for other measures aimed at sustainable use and recovery of the shared fishery resources of the Baltic. Integration of environmental protection requirements and ecosystem considerations into fishery management was considered an important precondition for keeping the resources in a state where the social and economic objectives could be obtained sustainably with high likelihood.

With the enlargement of the EU in 2004, Russia became the only Party that was outside the EU. The IBSFC ceased to exist on January 1, 2007. Decisions on the TAC and their allocation are now made during bilateral negotiations between the EU and Russia, formalized in a fishery agreement signed in 2009.

EU–Russian Federation Fisheries Agreement 2009 The Joint Baltic Sea Fisheries Committee (EU–Russia) is a forum for fishery discussions between the EU and Russia. The annual quota discussions are handled through this arrangement. The formal members are the EU, represented by the Commission, and the Russian Federation. The fisheries in the Baltic Sea are dominated by EU fisheries, which land about 95% of the total landing.

Scientific, Technical and Economic Committee on Fisheries (STECF) The Commission's own Scientific, Technical and Economic Committee on Fisheries (STECF) has been in place since 1979. It is the formal source of all advice given to the Commission and has its legal basis in the EU Common Fisheries Policy. It consists of national experts, who examine scientific advice (e.g., from ICES or commissioned studies) and issue an opinion. Furthermore, STECF presents an annual report on the economic status of the EU fisheries, a role it maintains up to the present day.

The Baltic Sea Fisheries Forum (BALTFISH) Fishery is among the numerous human activities that impact the Baltic Sea, and therefore should be considered in the light of other initiatives to improve the Baltic Sea's ecological status. HELCOM therefore discusses fishery issues, although management competence falls outside its remits. The Baltic Sea Fisheries Forum (BALTFISH) is a regional body organized by HELCOM in 2009, providing a platform for discussions on important fishery issues in the Baltic Sea. BALTFISH is based on the regionalization of the

EU Common Fisheries Policy. Its main objective is to promote cooperation among fishery administrations and other key stakeholders in developing sustainable fisheries in the Baltic Sea region.

Non-governmental Organizations (NGOs) There are numerous NGOs that have taken an interest in the Baltic Sea; some of these are represented as observers at the HELCOM meetings. These organizations are generally considered to be contributing positively to improvement of the ecological health of the Baltic Sea ecosystems. The Nordic Council of Ministers established a programme in 2006 that supports cooperation between NGOs and the governments of the Nordic Countries, but also of Belarus, the Russian Federation and Poland. The aim of the programme is to assure that NGOs can influence the activities in the Baltic Sea.

References

Dybern BI (2004) The Baltic marine biologists. The history of non-governmental Baltic organization. BMB Publication no.17, 2004

HELCOM (2007) HELCOM Baltic sea action plan. Baltic sea environmental proceedings No 11B

HELCOM (2009) Eutrophication in the Baltic Sea. Baltic sea environmental proceedings No 115 B, p 148

IBSFC (International Baltic Sea Fishery Commission) (1975) Proceedings of the first Session, Warsaw, October 28–31, 1974. Warsaw, Poland

ICES (2014) ICES Annual Report, 2014

Kupffer C (1877) Über Laichen und Entwicklung des Herings in der westlichen Ostsee Jahresber der Commn zur Wissenschaftl Unters dtsch Meere in Kiel für die Jahre1874, 1875, 1876, 24–35

Leppäkoski E (2001) From INSECTA, PISCATORIBUS ... NOXIA to ecological models – history of brackish – water biology in Finland. Proc Estonian Acad Sci Biol Ecol (2001) 50(2):66–84

Ojaveer E (2014) Läänemeri. TA Kirjastus Tallinn

Rozwadowski HM (2002) The sea knows no boundaries. A century of marine science under ICES. University of Washington Press, Seattle/London

Wasmund N, Siegel H (2008) Phytoplankton. In: Feistel R, Nausch G, Wasmund N (eds) State and evolution of the Baltic Sea, 1952–2005. Wiley, Hoboken, pp 441–481

Went AEJ (1972) Seventy years agrowing. A history of the international council for the exploration of the sea 1902–1972. Rapp P-v Reun Cons int Explor Mer 165:249

Zettler ML, Schiedek D, Glockzin M (2008) Zoobenthos. In: Feistel R, Nausch G, Wasmund R (eds) State and evolution of the Baltic Sea, 1952–2005. Wiley, Hoboken, pp 517–541

Index

E. Ojaveer, *Ecosystems and Living Resources of the Baltic Sea*,
DOI 10.1007/978-3-319-53010-9

I

J

K

L

M

MIX
Papier aus verantwortungsvollen Quellen
Paper from responsible sources
FSC® C105338

FSC
www.fsc.org

If you have any concerns about our products,
you can contact us on
ProductSafety@springernature.com

In case Publisher is established outside the EU,
the EU authorized representative is:
Springer Nature Customer Service Center GmbH
Europaplatz 3, 69115 Heidelberg, Germany

Printed by Libri Plureos GmbH
in Hamburg, Germany